Frost Resistance of Concrete

Frost Resistance of Concrete

Proceedings of the International RILEM Workshop on

Resistance of Concrete to Freezing and Thawing
With or Without De-icing Chemicals

University of Essen, September 22–23, 1997

EDITED BY

Max Josef Setzer

IBPM Institute of Building Physics
and Materials Science
University of Essen
Essen (Ruhr)
Germany

Rainer Auberg

IBPM Institute of Building Physics
and Materials Science
University of Essen
Essen (Ruhr)
Germany

CRC Press
Taylor & Francis Group
Boca Raton London New York

CRC Press is an imprint of the
Taylor & Francis Group, an **informa** business
A TAYLOR & FRANCIS BOOK

CRC Press
Taylor & Francis Group
6000 Broken Sound Parkway NW, Suite 300
Boca Raton, FL 33487-2742

First issued in paperback 2019

© 1997 RILEM
CRC Press is an imprint of Taylor & Francis Group, an Informa business

No claim to original U.S. Government works

ISBN-13: 978-0-419-22900-1 (hbk)
ISBN-13: 978-0-367-86374-6 (pbk)

A catalogue record for this book is available from the British Library

Publisher's Note This book has been prepared from camera-ready copy provided by the individual contributors in order to make the book available for the workshop.

Visit the Taylor & Francis Web site at
http://www.taylorandfrancis.com

and the CRC Press Web site at
http://www.crcpress.com

CONTENTS

Part II   Chemical Parameters

Part III   Physical Parameters and Testing

PREFACE

In 1990 RILEM TC 117 FDC „Freeze Thaw and Deicing Salt Resistance" had its inaugural session here in Essen. Since then we have had meetings in Great Britain/Brighton, Sweden/Lund, France/Lyon, Switzerland/Dübendorf, Canada/Québéc, Norway/Trondheim, Japan/Sapporo and Finland/Espoo.

The committee consists of: **Chairman**: Max J. Setzer, Germany; **Secretary**: Rainer Auberg, Germany; **Members**: Dirch Bager, Denmark; Gjöran Fagerlund, Sweden; Volker Hartmann, Germany; Stefan Jacobsen, Norway; Don Janssen, USA; Heikki Kukko, Finland; Jaques Marchand, Canada; Takashi Miura, Japan; Per-Eric Petersson, Sweden; Michel Pigeon, Canada; Terje F. Rônning, Norway; Eric Sellevold, Norway; Eberhard Siebel, Germany; Jochen Stark, Germany; Werner Studer, Switzerland; **Corresponding Members**: Christian Clergue, France; J.R. Clifton, USA; Corinne Dubois, France; Geoffrey Frohnsdorf, USA; Y. Guerpillon, France; J. Prost, France; A. Reymond, France; Kenneth A. Snyder, USA; Rupert Springenschmid, Germany.

During the tenure of this committee remarkable research efforts have been given and round robin testing has been done by the committee members covering all aspects of basic research in frost action, application and test procedures. It has been a great honour for me to chair this committee. I want to thank all the members for their constructive co-operation, excellent discussions and personal support. The work has been a fruitful exchange of ideas, experience and experimental results.

The workshop here in Essen should be a culminating effort which has its basis in the work of this RILEM TC 117 FDC. The resonance of consensus researchers indicates that the topic of this work is still of high interest both for people in practice confronted with durability problems and for researchers in the basic and applied field. The proceedings of this workshop should be a basis and starting point for the future research work in freeze thaw and deicing salt resistance. There are still unsolved problems in the basic understanding of the dynamic process of freezing and thawing and deicing salt attack, in the chemical and physical processes involved and in measuring the internal damage due to these attacks. Therefore, RILEM General Council decided to start a new technical committee: RILEM TC IDC „Internal Damage of Concrete due to Frost Action".

I am honored to be the host of this international workshop. I thank all the authors for their contributions. Many thanks to the sponsors of the workshop: the Deutsche Forschungsgemeinschaft, the Ministerium für Wissenschaft und Forschung des Landes Nordrhein-Westfalen, the Readymix AG, the E. Schwenk Zementwerke KG, the Bundesverband Deutsche Beton- und Fertigteilindustrie e.V., the Katzenberger GmbH & Co. KG, the Dyckerhoff Zement-GmbH and the University GH Essen. And I thank my secretaries of RILEM TC 117 FDC, Dr. Hartmann (until 1992) and Mr. Auberg (since 1992) for their work and for their dedication.

I hope that both the workshop and this book are a source of insight for the participants and the readers of these proceedings.

Max J. Setzer, June 9, 1997

PART I

Material Parameters and Concrete Design

The influence of material parameters on freeze-thaw resistance with and without deicing salt

D.J. JANSSEN
Department of Civil Engineering,
University of Washington, Seattle, WA, USA

Abstract

The production of concrete that is resistant to freezing and thawing requires that careful attention be paid to the material parameters used to specify the concrete. The specifications must consider the aggregates, the cementitious binder, and the entrained air-void system. Each of these components is examined with respect to its primary contribution to the frost resistance of concrete, and secondary contributions of some of the components are also described.

Keywords: D-cracking, entrained air, freezable moisture, Philleo factor, spacing factor, specific surface.

1 Introduction

Bryant Mather once stated that "Concrete will be immune to the effects of freezing and thawing ... if it is made using sound aggregate, has a proper air-void system, and has matured so as to have developed a compressive strength of about 4,000 psi" (28 Mpa) [1]. While simple in appearance, this statement can be rather complicated in its actual execution. The purpose of this paper is to examine this statement with respect to the material parameters of a concrete mixture. This examination will include the influence of the aggregates, the influence of the cementitious binder, and the influence of the air-void system.

2 Aggregates

Aggregates are generally used in a continuous distribution of sizes from the coarsest to the finest. The influence of aggregates on frost resistance, however, can generally be evaluated in terms of the influence of coarse aggregates (particles larger than approximately 5 mm) and the influence of fine aggregates (particles smaller than approximately 5 mm).

Frost Resistance of Concrete, edited by M.J. Setzer and R. Auberg. Published in 1997 by E & FN Spon, 2–6 Boundary Row, London SE1 8HN, UK. ISBN: 0 419 22900 0.

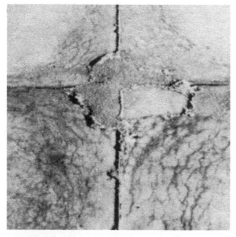

Fig. 1a. D-cracking at a joint in a concrete pavement.

Fig. 1b. D-cracking in a concrete curb.

2.1 Coarse aggregate

2.1.1 D-cracking

The primary influence of coarse aggregate in frost resistance is with respect to frost damage in the aggregate pieces themselves. This is generally termed D-cracking, and can occur even if the concrete has an adequate cementitious binder and entrained air-void system [2]. Many types of coarse aggregate have been identified as D-cracking susceptible, while other sources of the same type of rock have not been found to be susceptible. The pore structure of the coarse aggregate is thought to be the primary contributing factor to the D-cracking susceptibility of the aggregate.

D-cracking is characterized by cracks through both the coarse aggregate and the mortar portion of the concrete. Away from the cracks, both the mortar and the coarse aggregate are strong and show no signs of deterioration. The development of D-cracking requires considerable moisture and repeated cycles of freezing and thawing. As a result, D-cracking usually appears close to joints, Figure 1a, and cracks, edges, and corners, Figure 1b where moisture is available from more than one direction. D-cracking generally appears as a series of cracks approximately parallel to the primary moisture source.

Most coarse aggregates identified as being susceptible to D-cracking are sedimentary rocks, though many sedimentary rocks have not been found to be D-cracking susceptible. Igneous rocks are generally not considered to be susceptible to D-cracking, unless the rocks are quite weathered. Such weathered rocks would probably be undesirable for concrete production due to their low strength and likelihood to break down from handling. Most metamorphic rocks have not shown D-cracking susceptibility, however some partially metamorphosed sedimentary rocks have been identified as D-cracking susceptible [3]. The maximum aggregate size is also important in the development of D-cracking in concrete made with a D-cracking susceptible aggregate. Numerous studies [3, 4, 5, 6] have shown that the D-cracking susceptibility of a given aggregate is reduced by reducing the nominal maximum size of the aggregate. Normally, reducing the maximum aggregate size to below 19 mm is sufficient to significantly reduce the D-cracking potential of an aggregate, but in extreme cases the top size may need to be reduced to 12 mm. Unfortunately, reducing the nominal maximum aggregate size can have less desirable side effects including increased paste demand to maintain workability at a given strength level.

D-cracking can require a number of years to fully develop, and a considerable amount of concrete from a susceptible source could be placed before a problem was identified. This, combined with natural variability of aggregate sources, leads to the need for identification of D-cracking susceptible aggregates prior to their use in concrete exposed to moisture and repeated cycles of freezing and thawing.

2.1.2 Surface appearance
Coarse aggregate can also influence the surface appearance of concrete exposed to freezing and thawing by increasing the potential of thin surface mortar coatings to scale off. While this problem has not often attracted significant attention in the past, the development of scaling tests with high levels of repeatability [7] should allow detailed study of aggregates that are more susceptible to scaling.

2.2 Fine aggregate

The primary influence of fine aggregate is to assist in the retention of entrained air in concrete [8]. While this may not seem like a significant contribution to the frost resistance of a concrete mixture, the various activities that make up concrete construction including transporting, placing, consolidating, and finishing could result in a loss of entrained air. The retention of entrained air voids is especially important with respect to scaling at the concrete surface.

3 Cementitious binder

Mather [1] simplified the consideration of the cementitious binder by stating that the concrete should achieve a compressive strength of 28 MPa prior to exposure to repeated cycles of freezing and thawing. The primary purpose of this requirement is to reduce the amount of freezable water likely to be in the concrete at the time of frost exposure. An implication of this requirement is that adequate curing conditions be maintained prior to frost exposure. Some high-strength concrete mixtures may achieve this minimum strength in less than 24 hours, while lower strength concrete may require weeks or months of curing. This leads to questions concerning whether higher strength is better, and whether there may be advantages to the use of alternate cementitious binders to achieve the minimum strength requirement.

Considerable work has been done in examining the effects of water-cement ratio (w/c) and supplementary cementitious materials on pore sizes and pore size distribution, as measured by mercury porisimetry. While the issues of whether these measurements were made on cement paste or concrete along with physics of converting mercury porisimetry measurements into a quantifiable amount of freezable water in concrete is beyond the scope of this paper, a simplified look at freezable moisture can be of some use in examining the role of the cementitious binder on frost resistance.

Powers and Brownyard [9] determined that the water in concrete that would freeze at -20°C would also evaporate out of a saturated concrete specimen if the surrounding relative humidity were reduced to approximately 85 %. Figure 2 shows typical results of measurements of equilibrium moisture contents in concrete at various relative humidities [10]. Above about 50 % relative humidity the measurements are close enough to linear to permit the interpolation of the moisture content at 85 % relative humidity. Figure 2 shows the difference in moisture contents between 85 % relative and 100 % relative humidity (saturation) as the "freezable moisture".

Similar measurements of freezable moisture from a number of concretes were conducted on a number of concretes with w/c's of 0.40, 0.45, and 0.52 [10]. Figure 3 shows the average freezable moisture for these concretes, normalized as a percentage of the paste volume in the concrete. This figure clearly shows that decreasing w/c decreases the freezable moisture in the paste portion of the concrete.

Limited tests were also conducted on mixtures containing supplementary cementitious materials [10]. Figure 4 shows the effect of various supplementary cementitious materials on the volume fraction of freezable moisture in 0.45 water-cementitious ratio (w/(c+p)) concretes. Replacing cement with supplementary cementitious material appears to reduce the volume fraction of freezable moisture at the replacement levels studied.

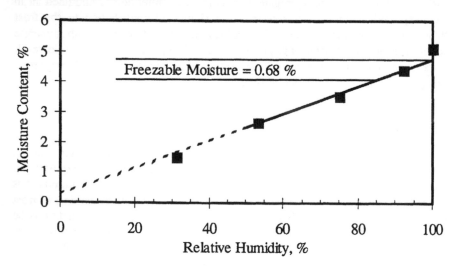

Fig. 2. Equilibrium moisture content at various relative humidities.

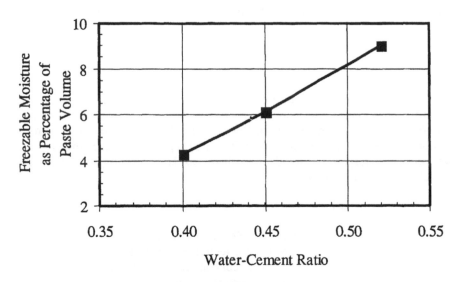

Fig. 3. Volume fraction of freezable moisture in concrete paste.

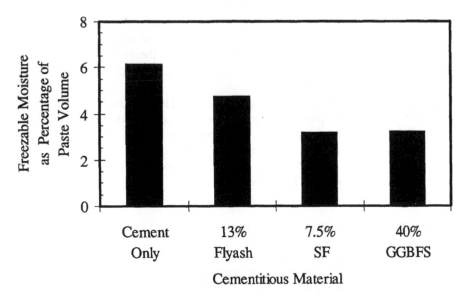

Fig. 4. Effect of supplementary cementitious materials on freezable moisture.

Though the volume fraction of freezable moisture in the cementitious binder portion of the concrete paste is reduced by decreasing the w/c and also by judicious replacement of a portion of the cement with supplementary cementitious materials, it must be kept in mind that any mixture changes which increase the volume fraction of the cementitous binder portion of a concrete mixture may reduce the resistance of the concrete to frost damage. This is at least in part due to the probabilistic nature of frost damage (as well as most other durability-related issues) and is discussed further in section 4.2.

4 Air-void system

Mather [1] states that if the cementitious binder portion of the concrete contains sufficient freezable moisture to require protection in the event of frost exposure, it is the air-void system that must provide that protection. Theories on the frost damage mechanism(s) have been published for over 50 years [11], and while they differ on the actual mechanism responsible for damage there has long been agreement on the beneficial role of entrained air [12].

4.1 Air-void parameters

The air-void system is most often specified in terms of a minimum volumetric percentage of air in the concrete, with some recognition that factors such as smaller coarse aggregate that lead to higher paste contents require more entrained air [13]. Volumetric air content is the parameter related to the air-void system that is easiest to measure at the time of concrete placing, with typical required air contents in the range of 5-6 %. What is more important, however, is the spacing and size of the air voids.

The spacing of the air voids is generally referred to as the spacing factor ($\overline{L}$). A typically-used maximum value for $\overline{L}$ is 0.20 mm. Because of simplifications used in the calculation of $\overline{L}$ (see ASTM C 457-90 for equations), the void size may also be specified. This is done by specifying the minimum specific surface (α). A typical minimum value is 25 mm^2/mm^3.

Another air-void parameter is the Philleo factor ($\overline{P}$) [14]. Though not used much to date, this parameter is perhaps a better model to use for the air-void system because it is probabilistic in nature. The parameter identifies the maximum distance from an air void for a given percentage of the concrete paste. Thus for a given percentage of the paste, say 90 %, there is a distance from an air void that includes that percentage of the paste. For a concrete with a $\overline{L}$ of 0.2 mm and a α of 25 mm^2/mm^3, the $\overline{P}_{90}$ would be about 0.04 mm.

4.2 Probabilistic nature of air-void protection

A few simple calculations of volumes indicate that very little void space is necessary to accommodate the 9 % expansion of the freezable moisture in the paste portion of the concrete described in Figures 3 and 4. Additional calculations of distances and volumes quickly suggest that less than 1 % of entrained air is required to provide a $\overline{L}$ of 0.2 mm and a α of 25 mm^2/mm^3 if the voids are uniformly-sized and evenly spaced (as assumed in the ASTM C 457-90 equations). But the air voids are of different sizes and not uniformly spaced. In some cases, air voids may be clustered at aggregate surfaces, and the placing, consolidating, and finishing operations may change the distribution of the air voids at the concrete surface.

The paste may not be uniform either. In addition to normal variations in the concrete itself, placing, consolidating, finishing, and curing may result in variations in the amount of freezable moisture in the concrete. When climatic exposure including the minimum temperature and number of freeze-thaw cycles, the drying and wetting prior to freezing, and the use of deicing salts are included into the consideration of requirements for frost-resistant concrete, a certain degree of conservatism with regards to entrained-air requirements seems reasonable. This realization of potential variability should also be kept in mind when interpreting the behavior of concrete exposed to freezing and thawing, especially freezing and thawing in the field. Acceptable performance should be repeatable under the most severe expected conditions.

5 Secondary material effects

Before closing this overview discussion of the primary material parameters affecting the resistance of concrete to freezing and thawing, possible secondary effects should be mentioned. There is no guarantee that concrete will be affected by only one durability problem at a time, and combined effects could accelerate the deterioration of concrete.

One secondary effect that should be considered is the alkali-silica reactivity (ASR) of the fine aggregate. Evidence of ASR products have been found in entrained air voids in concrete. This could reduce the effectiveness of the air voids in providing protection from freezing and thawing [15].

Delayed ettringite formation (DEF) could also result in the partial filling of entrained air voids. DEF, sulfate reaction, ASR, or any durability-related issue that caused an increase in the amount of freezable water held in concrete pores could have the secondary effect of reducing the frost resistance of concrete. Evidence has even been found indicating that the exposure of certain limestone aggregates to deicing salts could change their pore structure enough to cause them to become D-cracking susceptible

6 Summary and conclusions

Concrete that is resistant to repeated cycles of freezing and thawing has been made, and will continue to be made as long as proper attention is paid to the influence of the material parameters on the performance of the concrete. Continued research, and the associated advances in our knowledge of the aggregates, cementitious binder, and air voids, will improve our ability to produce frost resistant concrete. This will allow the production of concrete with a greater probability of frost resistance, in environmental exposure conditions that are more severe, and at a lower relative cost than we are able to today.

7 References

1.   Mather, B. (1990) How to make concrete that will be immune to the effects of freezing and thawing. *Paul Klieger Symposium on Performance of Concrete*, ACI SP-122. pp. 1-18.

2.   Schwartz, D.R. (1987) *D-Cracking of Concrete Pavements.* National Cooperative Highway Research Program, Synthesis of Highway Practice No. 134, Washington, D.C.

3.   Stark, D. (1976) *Characteristics and Utilization of Coarse Aggregates Associated with D-Cracking.*   Research and Development Bulletin RD047.01P, Portland Cement Association, Skokie, Illinois.

4.   Stark, D. and Klieger, P. (1973) Effects of maximum size of coarse aggregate on D-cracking in concrete pavements. *Highway Research Record* Vol. 441. pp. 33-43.

5.   Klieger, P., Monfore, G., Stark, D., and Teske, W. (1974) *D-Cracking of Concrete Pavements in Ohio.* Final Report, Ohio-DOT-11-74

6.   _____ (1990) *Influence of Design Characteristics on Concrete Durability.* NCHRP 84-2, Missouri Highway and Transportation Department, Jefferson City, Missouri.

7.   Setzer, M.J., Fagerlund, G. and Janssen, D.J. (1996) CDF test - test method for the freeze-thaw resistance of concrete - tests with sodium chloride solution (CDF).   developed by RILEM TC 117-FDC: Freeze-Thaw and Deicing Resistance of Concrete, *Materials and Structures*, Vol. 29. pp. 523-528.

8.   Walker, S and Bloem, D.L. (1946) Studies of concrete containing entrained air. *Journal, American Concrete Institute*, Vol. 42. pp. 629-639.

9.   Powers, T.C. and Brownyard, T.L. (1947) Studies of the physical properties of hardened cement paste, part 8. *Journal of the American Concrete Institute*, Volume 18, Number 8. pp. 933-969.

10.   Janssen, D.J. and Snyder, M.B. (1994) *Resistance of Concrete to Freezing and Thawing.* SHRP-C-391, Strategic Highway Research Program, National Research Council, Washington, D.C.

11.   Powers, T.C. (1945) A working hypothesis for further studies in frost resistance of concrete. *ACI Journal, Proceedings*, 41, No. 4., pp. 245-272.

12.   Powers, T. C. (1949) The air requirement of frost-resistant concrete. *Proceedings, Highway Research Board*, Vol. 29. pp. 184-202.

13.   Kosmatka, S.H. and Panarese, W.C. 1988, *Design and Control of Concrete Mixtures*, Thirteenth Edition, Portland Cement Association, Skokie, Illinois.

14.   Philleo, R. E. (1955) *A Method For Analyzing Void Distribution In Air-Entrained Concrete.*   Portland Cement Association, Research and Development Division, Skokie, Illinois.

15.   Jensen, A.D., Chatterji, S, Christensen, P. and Thaulow, N. (1983) Studies of alkali-silica reaction - part II: effect of air-entrainment on expansion. *Cement and Concrete Research*, Vol. 14. pp. 311-314.

16.   Dubberke, W., and Marks, V.J. (1985) The effect of deicing salt on aggregate durability. *Transportation Research Record*, No. 1031. pp. 27-34.

Effects of fly ash on microstructure and deicer salt scaling resistance of concrete

J. MARCHAND, Y. MALTAIS,
Y. MACHABÉE, C. TALBOT and M. PIGEON
Centre de Recherche Interuniversitaire sur le Béton,
Université Laval, Québec, Canada

Abstract

Over the past decades, numerous laboratory studies have clearly indicated that the use of supplementary cementing materials, and particularly fly ash, can significantly reduce the scaling durability of properly air-entrained concrete. Despite the great deal of research done on the topic, the reasons behind the detrimental influence of fly ash remain unclear. In order to bring more information on the subject, a comprehensive investigation of the influence of fly ash on the deicer salt scaling resistance of concrete has been recently carried out in our laboratory. This report summarizes the main results of the project. The effects of fly ash (originating from various sources) on the pore structure of concrete, cement hydration and ice formation are presented. The influence of fly ash on concrete microstructure and deicer salt scaling resistance is also discussed.
Keywords: Fly ash, microstructure, deicer salt scaling, ice formation, durability.

1 Introduction

Given the increasing importance of the problem, the deicer salt scaling deterioration of concrete has been the subject of a great deal of applied research in recent years [1, 2]. Most of the published data available on the influence of supplementary cementing materials indicate that fly ash, whatever its origin and chemical composition, can significantly reduce the scaling resistance of properly air-entrained concrete [1, 2]. Despite the great deal of research done on the topic, the reasons behind the detrimental influence of fly ash remain unfortunately unclear.

The influence of various supplementary cementing materials on the scaling resistance of concrete has recently been the subject of a 4-year research project in our laboratory. As part of this project, the effects of various sources of fly ash on the microstructure and deicer salt scaling resistance of concrete was investigated in the laboratory using standardized procedures and by field-trial tests. This report summarizes the main results of the laboratory-portion of the project.

Frost Resistance of Concrete, edited by M.J. Setzer and R. Auberg. Published in 1997 by E & FN Spon, 2–6 Boundary Row, London SE1 8HN, UK. ISBN: 0 419 22900 0.

2 Test program

The main objective of the experimental work was to understand why fly ash concrete mixtures are, in appearance at least, more susceptible to deicer salt scaling. The test program was designed to investigate two potential effects of fly ash on the concrete pore structure and its deicer salt scaling resistance:

• the presence of fly ash has a dilution effect on the cement paste microstructure. This explanation rests on the hypothesis that fly ash particles have not reacted at the time of testing (at 14 or 28 days), and are only acting as "inert" grains which increases the effective water/cement ratio of the mixture;

• fly ash affects the pore solution chemistry and modifies the ice formation process.

The test program included thermal analyses on paste mixtures performed at various times in order to investigate the evolution of cement and fly ash hydration process. Low-temperature calorimetry and mercury intrusion porosimetry tests were also performed on mortar mixtures to investigate the effect of fly ash on the material pore structure. Finally, ASTM C 672 deicer salt scaling resistance test were conducted on sawed concrete surfaces. The complete mixture program is given in Table 1.

Table 1. Mixture program

	Mixture	W/B ratio	Paste mixture	Mortar mixture	Concrete mixture
1	T10-0.40	0.40	√	√	√
2	T10-0.50	0.50	√	√	√
3	T10-0.66	0.66			√
4	T10-20%FA1-0.40	0.40	√	√	√
5	T10-20%FA2-0.40	0.40	√	√	√
6	T10-20%FA3-0.40	0.40	√	√	√
7	T10-40%FA2-0.40	0.40		√	√

Note: T10, CSA type 10 cement; FA1 to FA3, fly ash 1 to 3

3 Materials, sample preparation and experimental procedures

3.1 Materials

Mixtures were produced with an ordinary CSA type 10 portland cements and three different North-American ASTM fly ashes (Two Class F and one Class C ash). The chemical and physical analyses of cement and fly ashes are given in Table 2. Despite their high calcium content, fly ash 1 and 2 comply with the ASTM requirements for a Class F ash. Granitic fine and coarse aggregates from a local source were used in the preparation of the mortar and concrete mixtures. The absorptivity for the fine aggregate was 0.7%, and that of the coarse aggregate 0.8%. The maximum size of the coarse aggregate was 5 mm for the mortar mixtures and 20 mm for the concrete mixtures.

Table 2. Chemical and physical analysis of cement and fly ash

Oxides (%)	Cement Type 10	Fly ash FA1	FA2	FA3	Physical properties of fly ash	FA1	FA2	FA3
SiO_2	20.43	46.25	52.68	32.45	Specific surface	2720	3270	3850
Al_2O_3	5.13	20.64	23.51	19.49	(cm^2/g)			
Fe_2O_3	2.66	5.60	3.78	5.89	Density	2.63	2.09	2.65
CaO	63.19	18.10	12.52	28.19	Pass. 40 μm	84.7	83.9	90.7
					(%)			
MgO	2.22	4.50	1.18	4.58	BET (m^2/kg)	0.62	1.2	1.5
SO_3	3.53	1.75	0.27	2.90				
Na_2O	0.17	0.51	3.62	1.90	Pozzolanicity*			
K_2O	0.85	0.30	0.51	0.30	7 days	0.80	0.76	0.94
TiO_2	0.22	1.02	0.85	2.21	28 days	0.94	0.89	0.98
MnO	0.00	0.22	0.11	0.03	91 days	0.99	1.08	1.03
P_2O_5	0.09	0.31	0.11	1.00				
Cr_2O_3	0.00	0.01	0.01	0.02	Physical properties of cement			
SrO	0.00	0.56	0.12	0.46	Blaine (m^2/kg)	356		
					% passing 45	94		
LOI	2.26	0.25	0.79	0.41	μm sieve			

* Measured according to the French standard (NF EN 196-1)

3.2 Sample preparation

All mixture were prepared at a water/binder ratio of 0.4. The neat paste and mortar mixtures were prepared in an ASTM standard Hobart mixer, and the concrete mixtures were produced in a counter-current pan type mixer. The cementitious materials, the water, and the water reducer were first introduced into the mixer, and mixed until a homogeneous paste was obtained. When required (i.e. for the mortar and concrete mixtures), the fine and the coarse aggregates were then added along with the air-entraining admixture (for the concrete mixtures only). Table 3 gives the exact composition of all concrete mixtures.

Table 3. Concrete mixture characteristics

Mixture	Cement (kg/m^3)	Water (kg/m^3)	Fly ash (kg/m^3)	Sand (kg/m^3)	Stone (kg/m^3)
T10-0-40	408	163	—	732	1098
T10-0-50	358	179	—	720	1081
T10-0-66	262	173	—	775	1163
T10-20%FA1-0.40	301	151	75	727	1090
T10-20%FA2-0.40	305	153	76	735	1104
T10-20%FA3-0.40	317	158	79	736	1104
T10-40%FA2-0.40	233	155	155	735	1117

AEA dosage: 20 ml/kg of cement for OPC mixtures, 25 to 45 ml/kg of binder for fly ash mixtures

The neat paste mixtures were cast in plastic molds (diameter = 95 mm, height = 200 mm). The molds were sealed and rotated during the first 24 hours in order to prevent segregation. The mortar and concrete mixtures were cast in molds and finishing operations were performed. Demolding took place after 24 hours. All specimens were cured in lime saturated water until testing.

3.3   Thermogravimetric analyses

Thermogravimetric analyses were performed using a Perkin-Elmer thermobalance. Tests were carried out on paste samples after 1, 14 and 90 days of curing. All experiments were performed on crushed samples that had been vacuum-dried for 24 hours. Generally, 40-mg samples were heated, in a nitrogen atmosphere, at a rate of 10° C/min.

3.4   Evaporable water content and mercury intrusion porosimetry

For each mixture, two mortar specimens were cured for 14 days, and then dried at 110° C until equilibrium. The evaporable water content (We) was obtained by weighing each specimen before and after drying. All samples were tested after 14 days of curing.

The pore size distribution of the mortar mixtures after 14 days of curing was measured by mercury intrusion porosimetry. The instrument was capable of a minimum intruding pressure of 2.6 kPa (20 mmHg) and a maximum of 212 MPa. In order to minimize drying-induced pore structure alterations, water-saturated specimens were immersed in reagent-grade anhydrous isopropanol for a three week period prior to testing [3]. At the end of the three week period, specimens were vacuum dried for 48 hours, and tested.

3.5   Low-temperature calorimetry

After 14 days of curing, mortar specimens (diameter = 12 mm, height = 70 mm) were cored from the central portion of larger cylinders previously cast. All specimens were placed in a desiccator kept at 20° C and 50% relative humidity for 21 days.

At the end of the drying period, all specimens were vacuum-saturated with a saturated lime solution. Ice formation measurements were conducted at the Aalborg Portland Cement laboratory using a Setaram Low Temperature Microcalorimeter. Experiments were performed in the 0° C to -60° C range. Previous studies have indicated that only negligible amounts of ice are formed below -60° C [4]. Heat flow data were collected during both cooling (at a rate of 3.3° C/h) and heating (at a rate of 4.1° C/h). At the end of the low-temperature calorimetry run, the water content of each specimen was determined by weight loss measurements.

3.6   Deicer salt scaling

All deicer salt scaling tests were performed in accordance with ASTM C 672. For each mixture, three sawed surface samples (cut from 150 x 300 mm cylinders) were tested. All samples for the scaling tests were cured for 14 days in lime saturated water and then allowed to dry 14 days at 23° C and 50% R.H. At the end of the drying treatment, the test surface of each specimen was covered with fresh water for 7 days. At the end of the resaturation period, water on top of each specimen was flushed and the test surface was covered with a 3% NaCl solution (28 g/l). All specimens were subjected to 50 daily cycles of freezing and thawing (Tmin = -18° C, Tmax = 18° C).

4   Test Results

4.1   Thermal analyses

Results of the thermal analyses are given in Table 4. The portlandite ($Ca(OH)_2$) contents of the tested samples were derived from the thermogravimetric curve by measuring the weight loss between 450° C and 550° C which corresponds to the portlandite dehydroxylation. As could be expected, the portlandite contents of the various fly ash pastes are consistently lower than that of the companion OPC samples.

This reduction of the $Ca(OH)_2$ content is mostly related to the dilution effect. The replacement of cement by fly ash contributes to reduce the total amount of material that can produce portlandite.

Table 4. Portlandite content (% of total mass) derived from the thermal analyses

Mixture	Time	Portlandite content (% of total mass)	Portlandite consumption by the pozzolanic reaction (% of portlandite content)	Wn(t) (%)
T10-0.40	1d	14.6	-	15.6
	14d	15.5	-	18.0
	90d	16.9	-	19.1
T10-0.50	1d	14.8	-	16.6
	14d	16.9	-	20.0
	90d	18.2	-	21.2
T10-20%FA1-0.40	1d	10.9	0.78	17.6
	14d	11.4	1.00	20.6
	90d	11.4	2.12	22.8
T10-20%FA2-0.40	1d	10.7	0.98	17.6
	14d	11.7	0.70	19.1
	90d	11.5	2.02	21.6
T10-20%FA3-0.40	1d	10.1	1.58	18.1
	14d	10.5	1.90	22.8
	90d	11.9	1.62	24.3

Table 4 also gives the portlandite consumption by the pozzolanic reaction. For all fly ash mixtures, the amount of portlandite consumed by the pozzolanic reaction can be estimated by subtracting the portlandite content measured in the blended paste from the total portlandite content of the reference mixture and by considering the dilution effect of the fly ash. For instance, at 1 day, the portlandite consumed by the pozzolanic reaction for the mixture T10-20%FH2-0.40 would be (14.6 x 0.8) - 10.7% = 0.98%.

The slow consumption of portlandite is in contradiction with the usual ideas of significant pozzolanic reaction. The values of Table 4 rather indicate a low consumption of $Ca(OH)_2$ which can be attributed to a very slow reaction of the fly ashes. These results are in apparent contradiction with those of some previous investigations [5]. They are, however, in good agreement with the data reported by Diamond et al. [6] who found evidence of minimal pozzolanic reaction of various fly ashes during the period of major strength development (i.e. for the first six months of hydration).

The low alkali contents of the cement ($Na_2O = 0.17\%$, $K_2O = 0.85\%$), compared to the usual content of other CSA type 10 cements, may have contributed to retard the pozzolanic reaction. This hypothesis is supported by the results reported by Diamond et al. [6]. The importance of alkali ions in the reaction of supplementary cementitious materials has been recently emphasized by many authors [6, 7].

Table 4 also provides the non-evaporable water contents (Wn) calculated from the thermal analysis data. The chemically bound water content of the various mixtures was calculated according to equation 1:

$$W_n(\%) = \frac{\left(W_{105°C} - W_{1000°C}\right)}{W_{1000°C}} \times 100 \tag{1}$$

where $W_{105°C}$ represents the mass of the sample at the beginning of the thermogravimetric test and $W_{1000°C}$ the final mass of the sample at the end of the test. These results indicate that the non-evaporable water content produced in the blended pastes, despite the dilution effect, are greater than those of the reference mixture. Such results are in good agreement with those reported by other authors [6]. According to the results of Table 4, this increase in the non-evaporable water content cannot be attributed to the reaction of the fly ashes. The greater amounts of chemically bound water measured for the fly ash pastes tend to indicate that the presence of fly ash has contributed to accelerate the cement hydration.

4.2 Total porosity and mercury intrusion porosimetry data
The results of the evaporable water content measurement and mercury intrusion pore size distributions of mortar mixtures are given in Table 5. The evaporable water contents are expressed in terms of mass of water per unit mass of dry concrete (g/g_{dry}). Each result represents the average value obtained for two specimens.

The use of fly ash as partial replacement for portland cement, even after only 14 days of curing, appears to have little influence on the evaporable water content. At a replacement level of 20%, the use of fly ash generally decreases the value obtained, but it increases slightly this value at 40%. On the other hand, mercury intrusion test results show that the addition of supplementary cementing materials has contributed to refine the mortar pore structure. This effect is unexpected considering the slow pozzolanic reaction rates measured for these materials. It should be emphasized that the mercury intrusion tests were performed on samples that had been cured for only 14 days.

Table 5. Evaporable water content and pore size distribution data

| Mixture | Evaporable water cont. (g/g_{dry}) | Hg intrusion porosimetry data | | | | |
		Total (%)	Class 1 (%)	Class 2 (%)	Class 3 (%)	Class 4 (%)
T10-0.40	0.0590	8.5	0.43	3.65	3.53	0.91
T10-0.50	0.0629	10.0	0.46	4.20	4.32	1.00
T10-0.66	0.0761	12.0	0.95	5.49	4.87	0.67
T10-20%FA1-0.40	0.0547	8.8	0.4	3.80	3.86	0.70
T10-20%FA2-0.40	0.0535	8.4	0.49	2.5	4.3	1.12
T10-20%FA3-0.40	0.0583	8.2	0.38	3.6	3.57	0.63
T10-40%FA2-0.40	0.0643	9.4	0.29	1.33	6.75	1.02

Class 1: > 0.9 μm   Class 2: 0.9 μm < > 0.06 μm
Class 3: 0.06 μm < > 0.009 μm   Class 4: 0.009 μm < > 0.006 μm

4.3 Ice formation measurements
The calculated amounts of ice formed at -10° C, -20° C and -35° C are summarized in Table 6. All ice formation results are expressed in mg/g of saturated surface-dry concrete basis. Results in Table 6 indicate that the addition of fly ash has no clear effect on ice formation. In some cases, fly ash tends to slightly decrease the amount of ice. In other cases, a small increase of the freezable water content is noted. Overall, the use of fly ash and the replacement ratio (20% or 40%) do not seem to have any markedly detrimental effect on ice formation.

Table 6. Low-temperature calorimetry test results

Mixture	Cumulative amount of ice formed (mg/g ssd)		
	-10° C	-20° C	-35° C
T10-0.40	3	6	9
T10-0.50	10	14	18
T10-0.66	19	23	26
T10-20%FA1-0.40	4	6	12
T10-20%FA2-0.40	2	3	11
T10-20%FA3-0.40	5	8	12
T10-40%FA2-0.40	7	11	17

4.4 Deicer salt scaling test results
The results of the scaling tests (on sawed surfaces) are given in Table 7. Test results indicate that the deicer scaling resistance, measured on sawed surfaces, is excellent for all OPC concretes (even for the 0.66 water/cement ratio). Results in Table 7 also indicated that the scaling residues appears to be more important for the fly ash mixtures.

Table 7. Deicer salt scaling test results

Mixture	5 cycles (kg/m^2)	25 cycles (kg/m^2)	50 cycles (kg/m^2)
T10-0.40	0.03	0.06	0.09
T10-0.50	0.01	0.06	0.09
T10-0.66	0.04	0.08	0.13
T10-20%FA1-0.40	0.03	0.12	0.48
T10-20%FA2-0.40	0.02	0.14	0.45
T10-20%FA3-0.40	0.04	0.23	0.63
T10-40%FA2-0.40	0.04	0.25	0.74

Results of Table 7 clearly emphasize the detrimental effect of all fly ash on the deicer salt scaling resistance of sawed samples. The use of sawed surface (in comparison with troweled surfaces usually used in scaling test) allows us to disregard the influence that fly ash can have on the formation of the surface microstructure of concrete [8]. According to the results of Table 7, the negative influence of fly ash on salt scaling resistance is not solely related to the surface microstructure of concrete. The use of fly ash appears to have an intrinsic effect on the deterioration mechanisms.

The effect of fly ash content on scaling is also show in Figure 1. Test results seems to indicate that the scaling process is almost linear. The detrimental effect of fly ash on deicer salt scaling seems to increase with the fly ash content for mixtures made with fly ash 2. The value obtained for the mixture containing 20% of fly ash 2 is a little less than 0.50 kg/m^2 after 50 cycles, but that obtained for the mixture produced with 40% of this fly ash approaches 0.75 kg/m^2 after 50 cycles. These results are in good agreement with those of Machabée [9].

5 Discussion

Although widely studied in the past decades, the effects of fly ash on the concrete microstructure and physical properties are still imperfectly understood. More particularly, the relationship between the microstructure of fly ash concretes and their deicer salt scaling resistance has been the subject of very little investigations.

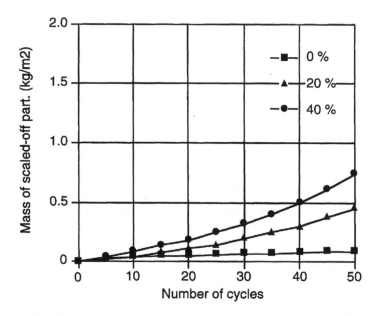

Fig. 1. Influence of the fly ash 2 content on the scaling of sawed surfaces

According to Xu et al. [10], fly ash can have five different effects on the concrete microstructures: (1) replacing cement by fly ash dilutes the cement concentration of the mixture which reduces the heat of hydration at early ages, and subsequently the hydration rate; (2) the addition of fly ash increases the water/cement ratio of the mixture which not only affects the initial porosity of the system but also influences the degree of ion supersaturation in the hydration products and the diffusion of ions (mainly calcium ions) to nucleation sites; (3) on the long term, the dissolution of fly ash releases silicate and aluminate ions into solution which may hinder the hydration of residual cement grains; (4) small fly ash particles can act as nucleation sites for hydration products and improve the hydrated cement paste microstructure; (5) the presence of fly ash particles increases the number of interfacial transition zones (fly ash/paste, paste/aggregates, fly ash/aggregates) which tend to be more porous (at least at early ages).

Considering the evaporable water content measurements, the mercury intrusion porosimetry and ice formation tests, it is clear that the use of fly ash at a replacement level of 20% did not very significantly influence the pore system of concrete as determined after 14 days of curing. At 40%, a small increase in evaporable water content, total porosity and amount of ice formed at -20°C was observed. This shows that during the first 14 days, fly ash does not simply act as an inert filler which would simply dilute the system (i.e. increase the water/cement ratio), but that they have an influence on the amount of hydration products that are formed. This phenomenon apparently occurs despite very little pozzolanic activity, which thus indicates that, as previously mentioned, clinker hydrates more rapidly in the presence of fly ash (this is confirmed by the results of the measurements of the non-evaporable water content).

The positive influence of fly ash during the first 14 days is particularly well confirmed by the results for the 0.66 reference mixture. If fly ash simply acted as inert fillers, a 0.40 water/binder ratio fly ash mixture containing 40% of Class F ash would be similar to a 0.66 portland cement concrete. The results of the evaporable water content (0.0643 g/gdry versus 0.0761 g/gdry), the total Hg porosity (9.4% versus

12.0%), and the ice formation (0.011 g/gssd at -20° C versus 0.023 g/gssd) show that this is not the case.

Deicer salt scaling test results clearly indicate that the use of fly ash reduces the durability of sawed surfaces. The results of this investigation indicate that, at the same water/cement ratio (i.e. considering that fly ash act as inert fillers), the reference mixtures have a much better scaling resistance than the mixtures containing fly ash. In fact, the amount of residues from the sawed surface of the 0.66 mixture is very low. The simple assumption that the water/cement ratio increase (due to the dilution effect of fly ash) is the basic cause of the lower durability of fly ash concretes is thus insufficient to explain the results obtained, and there has to be another reason for the negative influence of fly ash.

This reason could be a progressive water saturation of the micro air-voids network during the 7 days saturation period prior to the deicer salt scaling test. In fact, on the basis of Fagerlund's work [11], the deicer salt scaling resistance of concrete seems to be related the degree of water saturation of the porous network. In a recent paper, Fagerlund [12] has indicated that, for a cement paste with a water/cement ratio of 0.40, the capillaries (up to 30 μm) were saturated in only a few days. The water saturation of the small air-voids increases the freezable water content of concrete and the effective spacing factor.

Maltais et al. [15] have recently analyzed (with a special image analysis technique) the air-void characteristics of OPC and fly ash concretes. They have observed the presence of numerous micro air-voids (in the 10 to 30 μm range) in the fly ash mixtures (as compared to OPC mixtures). In this context, the detrimental effect of fly ash on scaling resistance could be related to the increased number of saturated small air-voids. The larger number of small saturated air-voids may increase the amount of freezable water content and the scaling of fly ash concretes.

6 Conclusion

- Test results indicate that all fly ashes had a slow reaction rate;
- Test results also indicated that fly ash tends to refine porosity (i.e. to promote the formation of smaller pores), even after only 14 days;
- Data indicate no very significant influence of fly ash on the freezable water content (at -20°C), and on the total porosity. However, fly ash was found to significantly increase the non-evaporable water content. This confirms that fly ash does not simply act as "inert filler" but rather influence the early hydration of cement;
- Finally, considering the dense pore structure of the fly ash pastes and concretes tested, it must be concluded that fly ash does not detrimentally affect the concrete microstructure. Then, the specific mechanism causing scaling deterioration of fly ash concrete (on sawed surfaces) is most probably not related to the effect of the supplementary cementing material on the microstructure of concrete.

7 References

1. Marchand, J., Pleau, R., Gagné, R. (1995) Deterioration of concrete due to freezing and thawing, Materials Science of Concrete, Volume IV, American Ceramic Society, pp. 283-354.
2. Marchand, J., Sellevold, E.J., Pigeon, M. (1994) The deicer salt scaling deterioration of concrete — An overview, ACI Special Publication SP-145, pp. 1-46.

3. Feldman, R.F., Beaudoin, J.J. (1991) Pretreatment of hardened hydrated cement pastes for mercury intruding measurements, Cement and Concrete Research, Vol. 21, N° 2-3, pp. 297-308.
4. Sellevold, E.J., Bager, D.H. (1985) Some implications of calorimetric ice formation results for frost resistance testing of cement products, Technical Report 86/80, Technical University of Denmark, Building Materials Laboratory, 28 p.
5. Patel, H.H., Pratt, P.L., Parrott, L.J. (1989) Porosity in the microstructure of blended cements containing fly ash, Materials Research Society Symposium Proceedings, Vol. 137, Edited by L.R. Roberts and J.P. Skalny, pp. 381-390.
6. Diamond, S., Sheng, Q., Olek, J. (1989) Evidence of minimal pozzolanic reaction in a fly ash cement during the period of major strength development, Materials Research Society Symposium Proceedings, Vol. 137, Edited by L.R. Roberts and J.P. Skalny, pp. 437- 446.
7. Fraay, A.L.A., Bijen, J.M., de Haan, Y.M. (1989) The reaction of fly ash in concrete — A critical examination, Cement and Concrete Research, Vol. 19, N° 2, pp. 235-246.
8. Pigeon, M., Talbot, C., Marchand, J., Hornain, H. (1996) Surface microstructure and scaling resistance of concrete, Cement Concrete Research, Vol. 26, N° 10, pp. 1555-1566.
9. Machabée. Y. (1997) Study of the influence of various parameters on the deicer salt scaling resistance of concrete, M.Sc. thesis, Université Laval, Québec, Canada, 209 p., (in French).
10. Xu, A., Sarkar, S.L., Nilsson, L.O. (1993) Effect of fly ash on the microstructure of cement mortar, Materials and Structures, Vol. 26, pp. 414-424.
11. Fagerlund, G. (1995) Moisture uptake and service life of concrete exposed to frost, International Conference on Concrete under severe conditions: Environment and Loading, Edited by K. Sakai, N. Banthia and O.E. Gjørv, pp. 221-232.
12. Fagerlund, G. (1996) Predicting the service life of concrete exposed to frost action through a modelling of the water absorption process in the air-pore system, The Modelling of Microstructure and its Potential for Studying Transport Properties and Durability, Edited by H. Jennings et al., pp. 503-537.
13. Maltais, Y., Machabée, Y., Marchand, J. (1997) Influence of supplementary cementing materials on the air-void characteristics of concrete, Scientific report GCS-08, Université Laval, Québec, Canada (in preparation).

Laboratory and field studies of salt scaling in fly ash concrete

M.D.A. THOMAS
Department of Civil Engineering,
University of Toronto, Toronto, Ontario, Canada

Abstract
This paper presents findings from a review of laboratory and field studies related to the salt scaling resistance of concrete containing fly ash. Laboratory studies using accelerated scaling tests, such as ASTM C 672, invariably show that the partial replacement of Portland cement with fly ash reduces the scaling resistance of concrete, particularly at high levels of replacement and in lean concretes (low cementitious materials content). This apparent drawback has been a barrier to the wider use of fly ash concrete in highway structures in northern climates. However, the poor performance of fly ash concrete in laboratory tests is not corroborated by field experience. A field survey including more than 20 sites (mainly highway structures) where fly ash concrete had been exposed to freezing and thawing in the presence of deicing salts was recently completed by the author. Observations from these field studies indicated no problem due to salt scaling of the concrete even after more than 20 winters in some cases. Fly ash concrete samples taken during a number of field placements showed poor performance in accelerated laboratory tests in contrast to excellent field performance. It is concluded that the ASTM salt scaling test is not appropriate for evaluating the scaling resistance of fly ash concrete.
Keywords: Concrete, de-icing salts, field studies, fly ash, laboratory studies, scaling.

1 Introduction

It is well established that fly ash concrete will be resistant to cyclic freezing and thawing provided the aggregate is frost-resistant, sufficient strength is attained prior to first freezing and an adequate air-void system is present. However, a number of

Frost Resistance of Concrete, edited by M.J. Setzer and R. Auberg. Published in 1997 by E & FN Spon, 2–6 Boundary Row, London SE1 8HN, UK. ISBN: 0 419 22900 0.

laboratory studies have indicated increased mass loss in fly ash concrete (compared to Portland cement concrete) when subjected to cycles of freezing and thawing in the presence of de-icing chemicals such as sodium chloride and calcium chloride (1-5). The mass loss occurs gradually as small pieces of cement paste or mortar spall away from the concrete surface, eventually leaving coarse aggregate particles exposed. This phenomenon is commonly referred to as scaling (or 'salt scaling'). The lower scaling resistance of fly ash concrete is more pronounced in lean concretes (low cementitious material content) or concretes with high levels of cement replaced with fly ash.

Concrete flatwork exposed to frequent de-icer salt applications and freeze-thaw cycles is particularly susceptible to scaling, this includes sidewalks, curb and gutter sections, exposed bridge decks and pavements, and to a lesser extent structures adjacent to highways such as median barriers and bridge substructures which may be regularly splashed with salts. Due to the high vulnerability of highway structures to scaling and concerns over the scaling performance of fly ash concrete, some transport agencies have placed limits on the maximum level of fly ash for use in highway structures (5).

Despite the poor performance of fly ash concrete in accelerated laboratory scaling tests, there have been few (if any) reported instances of poor scaling resistance in the field that have been directly attributable to the use of fly ash. However, the field performance of fly ash concrete exposed to freezing and thawing in the presence of de-icing salts has not been adequately documented.

This paper presents a brief review of the effect of fly ash on the scaling resistance of concrete in laboratory tests and provides a summary of an ongoing field survey of fly ash concrete exposed to aggressive scaling conditions.

2 Laboratory Studies

2.1 Test Methods

There are various accelerated laboratory tests for evaluating the scaling resistance of concrete exposed to de-icing salts. The most commonly used test in North America is ASTM C 672 which essentially involves ponding the 'finished' surface of a concrete slab with a 4% solution of calcium chloride and cycling the temperature from -18°C to 23°C on a daily basis. Concrete performance is based on a visual rating of the surface condition usually after 50 freeze-thaw cycles. Common variations on this test procedure include the substitution of *NaCl* for the specified *CaCl* and the use of mass loss as a method of assessment in preference to the visual rating. A commonly used acceptance criteria is a maximum mass loss of 0.8 kg/m^2 after 50 cycles (5). There is no universally accepted relationship between mass loss and visual rating, however in the author's experience, specimens which exhibit mass losses of the order of 0.8 kg/m^2 generally have visual ratings in the range 3 to 4.

The C 672 test method generally has poor reproducibility even when the scaled mass loss is used in place of the subjective visual rating (6) and this is probably due to the sensitivity of the test result to surface preparation techniques and specimen freezing rates. The ASTM method requires the final finishing to be done after bleeding has stopped and interpretation of this instruction is a potential source of operator variability. The only specified requirement of the freezing equipment is that it be capable of lowering the temperature of the specimens to -18°C within 18 hours; the actual rate of cooling is not controlled. The size and power rating of the freezer are likely sources of inter-laboratory variation, whereas changes in the number of specimens within the freezer during test will affect within-laboratory repeatability.

2.2 Effect of Fly Ash

Table 1 presents a brief summary of published data (1-5, 7-13) from scaling tests carried out on fly ash concrete. There is considerable conflict in the findings from different researchers due in part to the variation in materials used and differences in curing regimes, but also due to the poor reproducibility of the test methods.

Both Whiting (3) and Johnston (4) showed that the scaling performance was related to the water-cement ratio, W/C, rather than the water-cementitious ratio, W/CM, of the concrete mix. This implies that fly ash does not contribute to scaling resistance. Indeed, Johnston (4) suggested that specifications should limit the W/C ($\leq$ 0.45) of concrete exposed to de-icing salts with no account being taken of the fly ash in the mix. A similar philosophy was put forward by Whiting (14); he proposed that the minimum Portland cement content of concrete should be 500 lb/yd^3 ($\sim$ 300 kg/m^3) to "ensure good durability" regardless of the presence of fly ash. These conclusions are not sustained by the results of other studies which have shown satisfactory performance at levels of fly ash up to 30% (1, 7, 9) or even higher (10, 11).

A few general trends can be discerned from the published data; these are:

- Scaling increases as the W/CM increases
- Scaling mass loss generally increases with fly ash content, especially at high levels of replacement (i.e. $\geq$ 40 to 50%)
- Class C fly ash performs better than Class F ash
- Results from concrete containing fly ash tend to be more variable
- The use of curing compounds (membranes) reduces scaling; this is particularly noticeable for fly ash concrete

The balance of evidence would suggest that fly ash concrete is likely to provide satisfactory scaling performance (i.e. mass loss < 0.8 kg/m^2 and visual rating $\leq$ 2-3) provided the water-cementitious material, W/CM, does not exceed 0.45 and the level of fly ash does not exceed 20 to 30%. This, of course, assumes an adequate air-void system is present in the concrete.

The higher variability in test data for fly ash concrete may be partly due to the variability encountered between fly ash from different sources. However, the performance of concrete containing a single source of fly ash tends to be more variable and more sensitive to specimen preparation and curing conditions than similar grade Portland cement concrete. Klieger and Gebler (1) ascribed some of the variability in fly ash performance to the organic matter content and its effect on air-entraining requirement and air-void stability. They found that the loss of air increased with air-entraining dosage which in turn increased with the organic content of the fly ash. Generally, Class C fly ashes retained more air and had lower $\overline{L}$ values than Class F ash and this may be attributed to the lower organic content of Class C fly ash (1). Air loss

Table 1 Summary of Published Data from Laboratory Scaling Tests on Fly Ash Concrete

Ref	Test Details	Results and Conclusions
Gebler & Klieger, 1986 (1)	10 FA's (Class F and C) at 25% replacement. CM = 307 and $0.40 \leq$ W/CM ≥ 0.45. ASTM C 672 (300 cycles). Different curing regimes, including membrane.	VR = 1-2 for PC and 2-3 for FA concrete (300 cycles). No difference between FA and PC concrete when membrane curing applied.
Whiting, 1989 (3)	6 FA's (Class F and C) at 25 and 50% replacement. $250 \leq$ CM ≥ 335 and $0.41 \leq$ W/CM ≥ 0.62. ASTM C 672 (150 cycles). 1 or 7 days moist curing.	VR = 1-2 for PC concrete, 2-4 for 25% FA (dependent on CM, type of ash and curing), and 4-5 for 50% FA (150 cycles). Relationship between VR and W/C. Suggests FA limited to 25%.
Johnston, 1987; 1994 (2, 4)	2 FA's (Class C) at 20, 35 and 50% replacement. $340 \leq$ CM ≥ 440 and $0.36 \leq$ W/CM ≥ 0.50. ASTM C 672 with 3% *NaCl*.	ML < 0.01 for PC concrete and ML > 0.80 for all FA concrete (top surfaces - 50 cycles). Relationship between ML and W/C. Suggests W/C $\leq$ 0.45 (fly ash not included) for good scaling resistance.
Langan et al. 1990 (8)	7 FA's (Class F and C) at 50% replacement. CM = 300 and W/CM = 0.47. ASTM C 672.	VR = 0 for PC concrete (50 cycles) and 5 for FA concrete (5 or 10 cycles!).
Bilodeau et al. 1991 (9)	1 FA (Class F) at 20 and 30% replacement. $290 \leq$ CM ≥ 460 and $0.35 \leq$ W/CM ≥ 0.55. ASTM C 672. Different curing regimes, including membrane.	No trend in VR with FA content. ML generally increased with higher FA content. ML < 0.80 for most FA concrete (50 cycles). Membrane curing improved resistance of all concrete, but especially FA concrete.
Naik and Singh, 1995 (10)	1 FA (Class C) at 15, 30, 40, 50 and 70% replacement. No other mix details.	VR = 0 for concrete with 0 to 40% FA, and 2-3 for 50% FA, and 5 for 70% FA (50 cycles)
Bilodeau et al. 1994 (12)	8 FA's (Class F and C) at 58% replacement. CM ~ 363 and W/CM = 0.33. ASTM C 672 with 3% *NaCl*.	VR = 5 and ML = 2.30-9.34 for all concretes (50 cycles).

FA = fly ash (kg/m^3); PC = Portland cement (kg/m^3); CM = cementitious material (kg/m^3); VR = visual rating; ML = mass loss (kg/m^2)

may be compounded at the surface of concrete. A further factor that may influence the performance of the fly ash concrete is the time to finishing. Fly ash generally delays the set and cessation of bleeding in concrete. However, reduced bleeding rates (and total bleed water) are also encountered with fly ash and the point at which bleeding stops is more difficult to detect visually. Thus the actual time to finishing may become a more important operator-dependent variable in fly ash concrete. If bleeding continues beyond the final finishing operation, this will have a profound adverse effect on the scaling resistance.

3 Field Studies

A number of fly ash concrete field placements in Ontario, Minnesota, Wisconsin and Michigan were visited by the author between 1996 and 1997. Structures included pavements, sidewalks, curb and gutter sections, median barriers and exposed bridge decks, and ranged in age from 1 to 23 years. The fly ash content of the concretes ranged from 13 to 70% and examples of both Class F and C ash were included. All of the structures had been exposed for at least one severe winter (1995/6) and most for longer periods. Generally, concrete with poor scaling resistance will show some level of noticeable deterioration (often severe) after just one winter's exposure.

It is beyond the scope of this paper to discuss these field placements in detail. A summary of the structures visited and the condition of the exposed surface is presented in Figure 1. In most cases, the concretes examined were in excellent condition. Poor performance was only observed in the field concretes with 50 to 60% Class F fly ash or 70% Class C fly ash. Severe scaling of concrete with 50% Class F fly ash was observed in certain paved sections of a parking lot ('h' in Fig. 1); this was due to poor practice with additional water being sprayed on the surface to assist hand-finishing. Machine placed and finished sections of the same mix are still in excellent condition. Surface loss was also observed in concretes with high levels of Class F (60%) or Class C (70%) fly ash in trial pavements at a thermal generating station ('f' in Fig. 1), but were confined to areas subjected to heavy wheel loads. Adjacent areas were in good condition despite severe freeze/thaw and de-icing salt exposure.

4 Correlation between Laboratory and Field Performance

An essential requirement of any accelerated laboratory test method used for evaluating construction materials is that the test result give a reliable indication of the field performance of the material. Litvan et al. (15) compared the field performance of 10 different types of paving slab after 5 winters at four different exposure sites with laboratory scaling test data using a test essentially similar to the current ASTM C 672. They concluded that:

> "The conventional laboratory test, under conditions prevailing in the present program, yielded predictions which show little, if any, agreement with actual field performance."

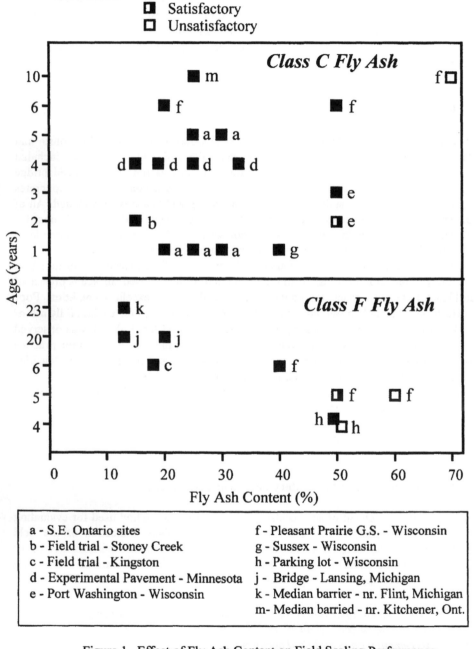

Figure 1   Effect of Fly Ash Content on Field Scaling Performance

There has been little attempt to compare ASTM C 671 test data with field performance since this time. However, in five of the field placements shown in Figure 1, small scale slabs were cast in the field for laboratory scaling tests. This permits a direct comparison of the field and laboratory performance. Full details of the concrete mixes used, methods of construction, field exposure conditions, and laboratory test data have been presented elsewhere (5, 7, 11, 13, 16). In four of the five cases, control placements with plain Portland cement concrete are also available for comparison.

Brief details of the placements, mixes used and the results from laboratory scaling tests are presented in Table 2. In all cases, the fly ash concrete showed inferior performance in laboratory tests compared with the control concrete. Some of the fly ash concretes failed to meet the acceptance limit of a maximum 0.80 kg/m^2 mass loss after 50 cycles (5).

The poor performance of the fly ash concrete in the accelerated laboratory tests is not consistent with the field performance of the same concrete mixes. All of the fly ash concretes have given satisfactory performance in the field and are generally in good to excellent condition. Furthermore there is little consistent difference in the surface appearance of the fly ash and control concretes at each location.

5 Discussion

Whilst it is widely recognized that the incorporation of fly ash in concrete generally improves its durability, there is a perception that salt scaling resistance is reduced. There is conflicting evidence from laboratory tests regarding the scaling resistance of fly ash concrete. This is partially due to inadequacies in the test method itself. However, many studies demonstrate satisfactory performance for concrete with up to 30% fly ash provided the W/CM is kept below about 0.45 and the concrete has an adequate air-void system. This is certainly corroborated by field evidence which clearly demonstrates that properly proportioned fly ash concrete can be resistant to salt scaling in severe exposure conditions even when the replacement level is significantly above 30%.

There is clearly a lack of consistency between the performance of fly ash concrete under accelerated laboratory and field conditions. A number of field placements containing fly ash are continuing to provide good service in severe conditions despite the poor performance of samples of the same concrete taken during construction and tested in the laboratory. Previous workers have reported that laboratory scaling tests are inadequate for predicting the field performance of Portland cement concrete (15). However, it would seem that the test is unduly harsh for fly ash concrete as it frequently indicates inferior performance of fly ash concrete compared with Portland cement concrete even though the same concretes show similar performance under field conditions. A similar trend has been found with slag concrete (17).

The poor resistance of fly ash concrete in laboratory scaling tests has led to a perception that fly ash may not be suitable for use in highway structures and has led at least one transport agency to limit the fly ash level to a maximum of 10% by mass of total cementitious material (5). Such limits are clearly not supported by field performance data. The 'optimum' replacement level for maximizing concrete

Table 2 Results from Laboratory Tests on Concrete used in Field Trials

Project	Mix details	Laboratory Scaling Data
Median barrier, nr. Kitchener, Ont.	Control: PC = 355, W/C = 0.41	ML = 0.306 (avg. for two loads)
constructed 1986 (16)	25% Fly Ash: PC = 267; FA (C) = 88; W/CM = 0.41	ML = 1.714 (avg. for two loads)
Sidewalk slabs, Kingston, Ont.	Control: PC = 415, W/C = 0.39	ML = 0.41 (standard cure)
constructed Fall 1991 (5)	18% Fly Ash: PC = 351; FA (F) = 77; W/CM = 0.37	ML = 1.31 (standard cure)
Paving slabs, Stoney Creek, Ont.	Control: PC = 355; W/C = 0.42	ML = 0.13 to 0.14
constructed June 1994 (13)	15% Fly Ash: PC = 302; FA (F) = 53; W/CM = 0.42	ML = 0.36 to 1.24
Curb & gutter, Univ. Calgary, Alb.	Control: PC = 330; W/C 0.37	VR = 1
constructed 1984 (7)	15% Fly Ash: PC = 297; FA (C) = 50; W/CM = 0.35	VR = 2
	30% Fly Ash: PC = 264; FA (C) = 99; W/CM = 0.33	VR = 2
Pavement, Wisconsin,	40% Fly Ash: PC = 218; FA (F) = 146; W/CM ≅ 0.36	VR = 2+; ML = 0.608
constructed Sept. 1990 (11)	20% Fly Ash: PC = 287; FA (C) = 66; W/CM = 0.34-0.40	VR = 1-2; ML = 0.511
	50% Fly Ash: PC = 177; FA (C) = 177; W/CM = 0.34-0.37	VR = 4; ML = 2.296

FA = fly ash content (kg/m^3); PC = Portland cement content (kg/m^3); VR = visual rating; ML = mass loss (kg/m^2)

performance in terms of chemical resistance (e.g. chloride ingress, alkali-silica reaction, sulphate attack) is generally well above 20% and the imposition of maximum replacement limits below such a level prevents the full technical benefits of using fly ash from being realized. Furthermore, low replacement limits do not provide sufficient financial incentive for many concrete suppliers to add an additional silo for fly ash and, as such, have effectively become a barrier to the wider use of fly ash.

Further work is required to develop an accelerated test method to reliably predict the scaling resistance of concrete under field conditions. It is important that such a test does not unduly penalize materials that have a history of satisfactory field performance.

References

1. Gebler, S.H. and Klieger, P. "Effect of fly ash on the durability of air-entrained concrete." Proceedings of the 2nd International Conference on Fly Ash, Silica Fume, Slag, and Natural Pozzolans in Concrete (Ed. V.M. Malhotra), ACI SP-91, Vol. 1, American Concrete Institute, Detroit, 1986, pp. 483-519.

2. Johnston, C. "Effects of microsilica and Class C fly ash on resistance of concrete to rapid freezing and thawing and scaling in the presence of deicing agents." In Concrete Durability (Ed. J.M. Scanlon), ACI SP-100, Vol. 2, 1987, pp. 1183-1204.

3. Whiting, D. "Deicer scaling resistance of lean concretes containing fly ash." Proceedings of the 3rd International Conference on Fly Ash, Silica Fume, Slag, and Natural Pozzolans in Concrete (Ed. V.M. Malhotra), ACI SP-114, Vol. 1, American Concrete Institute, Detroit, 1989, pp. 349-372.

4. Johnston, C.D. "W/CM code requirements inappropriate for resistance to deicer salt scaling." Proceedings of the 3rd International Conference on the Durability of Concrete (Ed. V.M. Malhotra), ACI SP-145, American Concrete Institute, Detroit, 1994, pp. 85-105.

5. Afrani, I. and Rogers, C. "The effects of different cementing materials and curing on concrete scaling." Cement, Concrete, and Aggregates, Vol. 16 (2), 1994, pp. 132-139.

6. Marchand, J., Sellevold, E.J. and Pigeon, M. "The deicer salt scaling deterioration of concrete - An overview." Proceedings of the 3rd International Conference on the Durability of Concrete (Ed. V.M. Malhotra), ACI SP-145, American Concrete Institute, Detroit, 1994, pp. 1-46.

7. Gifford, P.M., Langan, B.W., Day, R.L., Joshi, R.C. and Ward, M.A. "A study of fly ash concrete in curb and gutter construction under various laboratory and field curing regimes." Canadian Journal of Civil Engineering, Vol. 14, 1987, pp. 614-620.

8. Langan, B.W., Joshi, R.C. and Ward, M.A. "Strength and durability of concretes containing 50% Portland cement replacement by fly ash and other materials." Canadian Journal of Civil Engineering, Vol. 17, 1990, pp. 19-27.

9. Bilodeau, A., Carette, C.G., Malhotra, V.M. and Langley, W.S. "Influence of curing and drying on salt scaling resistance of fly ash concrete." Proceedings of the 2nd International Conference on the Durability of Concrete (Ed. V.M. Malhotra), ACI SP-126, Vol. 1, 1991, pp. 201-228.

10. Naik, T.R. and Singh, S.S. "Use of high-calcium fly ash in cement-based construction materials." Proceedings of the 5th International Conference on Fly Ash, Silica Fume, Slag, and Natural Pozzolans in Concrete, Supplementary Papers, 1995, pp. 1-44.

11. Naik, T.R., Ramme, B.W. and Tews, J.H. "Pavement construction with high-volume Class C and Class F fly ash concrete." ACI Materials Journal, Vol. 92 (2), 1995, pp. 200-210.

12. Bilodeau, A., Sivasundaram, V., Painter, K.E. and Malhotra, V.M. "Durability of concrete incorporating high volumes of fly ash from sources in the U.S." ACI Materials Journal, Vol. 91 (1), 1994, pp. 3-12.

13. Boyd, A.J. "Salt scaling resistance of concrete containing slag and fly ash." MASc Thesis, University of Toronto, Ontario, 1995.

14. Whiting, D. "Strength and durability of residential concretes containing fly ash." Portland Cement Association, Research and Development Bulletin RD099T, Skokie, Illinois, 1989, p. 40.

15. Litvan, G.G., MacInnis, C. and Grattan-Bellew, P.E. "Cooperative test program for precast concrete paving elements." In Durability of Building Materials (Ed. P.J. Sereda and G.G. Litvan), ASTM STP 691, American Society for Testing and Materials, Philadelphia, 1980, pp. 560-573.

16. Chojnacki, B. and Northwood, R.P. "Fly ash concrete - laboratory and field trials in Ontario." MTO Report EM-86, Engineering Materials Office, Ministry of Transportation, Downsview, Ontario, 1988.

17. Hooton, R.D. Private communication, 1997.

Influence of the type of cement on the freeze-thaw resistance of the mortar phase of concrete

C. GIRODET, M. CHABANNET, J.L. BOSC and J. PERA
Unité de Recherche Génie Civil – Matériaux,
Institut National des Sciences Appliquées de Lyon, France

Abstract
A lot of literature deals with the influence of the type of cement and mineral admixtures on the freeze-thaw resistance of concrete. However, the analyses are contradictory. The aim of the present work is to classify the influence of cement characteristics on the behavior or mortars subjected to a thermo-mechanical loading.
Two parameters are investigated : the chemical analysis and the specific surface area of cement and the type of pozzolan (silica fume or metakaolin). The mortars are cast with a cement : sand ratio of 3.375 and present the same workability.
The damage of prismatic samples (40 x 40 x 160 mm^3) subjected both to three-point flexure and freeze-thaw cycles is evaluated. The rate of loading is 20 % of the 28-day tensile strength. A durability factor is therefore defined, allowing the rating of the different parameters.
The best results are obtained with a cement presenting those characteristics : low specific surface area, low C_3A content and high C_2S amount. Pozzolanic admixtures and low water to cement ratios increase the durability of mortars subjected to flexural loading. The investigation of microstructure is interesting to understand the evolution of their mechanical properties.

Keywords : Cement, damage, durability, freeze-thaw, mortar, pozzolan, tensile strength

1 Introduction
The relative influence of the type of cement and mineral admixtures on the freeze-thaw durability of concrete is not yet clearly established. It is very difficult to choose a

Frost Resistance of Concrete, edited by M.J. Setzer and R. Auberg. Published in 1997
by E & FN Spon, 2–6 Boundary Row, London SE1 8HN, UK. ISBN: 0 419 22900 0.

cement on the basis of its Bogue potential composition or specific surface area [1 to 4].

Due to the deviations observed in strength, the role of pozzolans may be contradictory with that of air-entraining agents [5 to 8].

The mortar phase of concrete is particularly sensitive to freezing effects due to its high porosity and the interfaces which are developed inside [3], [4], [8], [9], [11] Consequently, the present work tries to rate the influence of some components of the binders used in specific mortars subjected to freeze-thaw cycles in presence of a mechanical strength which allows an acceleration of the degradation [10]. The parameters investigated are : the chemical composition of cement and its Blaine specific surface area, the nature of the pozzolanic admixture, and the water to cement ratio.

2 Cementitions and pozzolanic materials

2.1. Cements

As shown in Table 1, five ordinary portland cements were used. Their Bogue potential composition and Blaine specific surface area were different.

Table 1. Composition and fineness of the cements

Admixture	Bogue potential composition (%)				Blaine surface area (m^2/kg)
	C_3S	C_2S	C_3A	C_4AF	
1	67.7	5.7	8.0	9.9	375
2	67.0	10.0	5.6	8.8	307
3	66.3	11.2	3.3	12.0	363
4	63.6	14.7	0.9	14.6	286
5	57.6	20.2	2.0	13.8	443

The C_3A content of cements 4 and 5 was very low.

2.2. Pozzolans

A non-condensed silica fume (SF) and two metakaolins (MK1 and MK2) were used as pozzolanic materials. Their chemical composition and BET specific surface area are reported in Table 2. MK1 is slightly more rich in metakaolinite than MK2 (higher Al_2O_3 content) but its specific surface area is lower.

Table 2. Characteristics of pozzolans

Cement	Chemical composition (W_t %)								BET surface area (m^2/g)
	SiO_2	Al_2O_3	Fe_2O_3	CaO	K_2O	TiO_2	Na_2O	LOI	
SF	96.8	0.1	0.1	0.2	0.5	0.0	0.0	2.0	23.4
MK1	51.5	44.8	0.6	0.0	0.1	2.1	0.3	0.7	11.9
MK2	54.3	39.3	1.5	0.0	1.3	1.5	0.0	2.1	18.7

3. Specific mortars

The mortar phase of concrete cannot be represented by a standard mortar which is too rich in cement. Therefore, a specific mortar prepared with a binder to sand ratio of 0.296 instead of 0.333 was utilized in the tests in order to better represent the behavior of concrete and specially the role played by matrix-aggregate interfaces [8]. Three types of binder were investigated :
- 100 % ordinary portland cement (OPC),
- 80 % OPC + 20 % SF,
- 80 % OPC + 20 % MK (MK1 or MK2).

The water to binder ratio of mortars was maintained constant for plain OPC and adjusted for each mortar containing blended cement in order to obtain the same workability, measured by the flow-table test.

Prismatic samples (40 x 40 x 160 mm^3) were cast according to the French standard NFP 15-403. The mortar bars were demoulded after 24 hrs and immersed in lime-saturated water at 20°C for 27 days.

4. Test procedure and expression of test results

4.1. Test procedure

Freeze-thaw testing was carried out in presence of mechanical loading in order to accelerate the mortar damage [8], [12]. Freeze-thaw cycles in air started at the age of 28 days. The specimens were covered by a water-saturated sheet of paper and sealed in a plastic foil. In the climate chamber, the temperature varied from - 25°C to + 20°C . Thus, the temperature at the core of the sample reached - 22°C after 120 min of freezing ; this temperature was maintained for 30 min. 75 min were necessary to thaw the sample and reach + 18°C, which was maintained for 15 min. Therefore, the total length of the cycle was 4 hrs and 6 cycles were done per day. The total amount of cycles reached 40 for each sample.

During the free-thaw testing, a three-point flexural stress was simultaneously applied. The equipment was described in a previous paper [12] and a load (Q) equal to 20 % of the average flexural strength (Q_R) of each mortar was applied. Furthermore, a LVDT sensor recorded the deflection at the mid-span of the sample.

In order to separate the effect of freezing and that of mechanical loading, six apparatus were used : three of them were placed in the climate chamber, and three others were stored at 20°C and 90 % R.H., so as to get the creep of the sample under flexural loading.

Before and after the freeze-thaw cycles, the porosity was assessed by mercury intrusion porosimetry and the microstructure identified by scanning electron microscopy (S.E.M.).

4.2. Expression of test results

4.2.1. Damage factor

A damage factor (D) was defined as follows :

$$D = \frac{\Delta f / f_0}{1 + \Delta f / f_0}$$

where : f_0 = initial deflection obtained as the load was applied,
         Δf = increase in deflection due to freezing.

4.2.2. Durability factor

A durability factor (DF) derived from ASTM C 666 was defined :

$$DF = \frac{N}{M} . P$$

where : M = total amount of cycles (M = 40),
        P = relative dynamic modulus (E_N/E_0) after N freeze-thaw cycles,
        here, P = 1 - D,
        N = amount of cycles leading to a specified value of P (N < M), or total
        amount of cycles (N = M).
   Standard ASTM C 666 fixes a specified value :P = 0.60, for concrete subjected to
freeze-thaw cycling. In our case, as we worked on mortars, we specified P = 0.20, this
value allowed the rating of the relative durability of tested mortars [10].

5. Results and discussion

5.1. Plain OPC mortars
The main properties of plain OPC mortars are shown in Table 3.

Table 3. Properties of plain OPC mortars

OPC	W/C	Spread (mm)	Air content (%)	Flexural strength (MPa)	Compressive strength (MPa)
1	0.54	140	4.2	9.0	65.4
2	0.54	170	4.2	8.5	58.5
3	0.54	175	5.3	7.9	51.0
4	0.54	155	6.8	7.2	45.5
5	0.54	145	4.7	7.7	64.8

5.1.1. Damage and durability factor

The evolution of the resistance loss with freezing cycles , P = 1 - D, is presented in Fig.
1.
   For each cement, the durability factor (DF) obtained for (1 - D) = 0.20 is given in
Table 4.
   The definition of the ratio DF_i/DF_1, which is the relative durability of cement i versus
cement 1, allows the rating of the different cements. The best durability is obtained for
cement 4, which presents the lowest Blaine surface area.

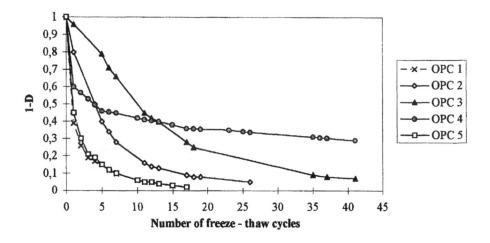

Fig.1. Influence of the type of cement on the mortar damage

Table 4. Durability factor obtained for (1 - D) = 0.20

Cement	Blaine specific area (m²/kg)	N	DF	DF$_i$/DF$_1$
1	375	3	0.015	1.0
2	307	11	0.055	3.7
3	363	22	0.110	7.3
4	286	40	0.290	19.3
5	443	4	0.020	1.3

5.1.2. Correlation between freezing resistance and cement characteristics.

As shown from Tables 1 and 4, except for cement 5, the higher the C_2S and C_4AF contents of cement, the higher the durability of mortar. The durability factor decreases as the C_3S and C_3A contents increase.

The hydration of C_3S and C_2S leads to the same hydrates (portlandite and C-S-H), but C_3S generates three times more portlandite than C_2S, in a shorter period of time. In presence of gypsum, C_3A produces ettringite which contains 32 molecules of water. If the transformation of ettringite into monosulphate is not achieved during cement hydration, the water contained in this crystal risks to freeze during the cycles and damage the microstructure.

However, the Bogue potential composition is not sufficient to determine the freezing resistance of cement. Cement 5, whose composition is close to that of cement 4, is less durable. In this case, its higher specific surface area seems to play a negative role in the freezing resistance.

5.1.3. Evolution of mortar porosity

As shown in table 5, the porosity changes after freeze-thaw cycles, but not always in the same way.

Table 5. Porous distribution of mortars

Cement	State	Air content (%)	Total porosity (%)	Micropores < 0.1 µm (%)	Mesopores (%)	Macropores > 0.6 µm (%)
2	Initial	4.2	11.1	81.1	14.1	4.9
	Final		8.7	71.1	12.2	16.7
4	Initial	6.8	16.9	63.7	25.2	11.1
	Final		17.6	58.9	27.7	13.4
5	Initial	4.7	14.0	65.7	28.9	5.4
	Final		13.3	58.6	14.5	26.9

The porosity of OPC 4-based mortar is high and regularly distributed between micro, meso and macropores. This higher porosity explains lower strengths, as seen from Table 3. The quantity of entrapped air is higher and this cement is coarser than OPC 2 and 5. After freezing cycles, the porous distribution slightly shifts towards meso and macropores and the total porosity increases.

For OPC 2-based mortar, the initial porosity is lower and micropores are abundant. After freezing cycles, the total porosity decreases but the porous distribution shifts towards macropores, which means the formation of micro-cracks during those cycles.

For OPC 5-based mortar, the initital porous distribution is close to that of OPC 4 with a lower total porosity, certainly due to its higher fineness (better initial reactivity). After freezing cycles, mesoporosity drops and macroporosity increases. Micro-cracks were observed by S.E.M.

5.2. Blended OPC mortars

In order to improve its durability, OPC 1 was blended either with silica fume (SF) or metakaolin (MK1 or MK2). Each blended cement contained 80 % OPC 1 and 20 % mineral additive. Such amount of substitution leads to the better strengths of metakaolin-blended cement [13].

The substitution of OPC by pozzolans implied to cast mortars with variable water to binder ratios (W/B) to get about the same workability (a spread of 130 mm in the flow-table test). In addition, the presence of non condensed silica fume needed the use of a superplasticizer (1 % of the weight of cement).

Table 6 gathers the initial properties of blended OPC mortars. In order to study the influence of water content, two W/B were used for MK2 containing cement. OPC 3 was also blended with 20 % of MK2.

Table 6. Properties of blended OPC mortars

Binder (B)	W/B	Spread (mm)	Air content (%)	Flexural strength (MPa)	Compressive strength (MPa)
100 % OPC 1	0.540	140	4.2	9.0	65.4
80 % OPC 1 + 20 % MK2	0.580	125	4.3	7.7	58.2
80 % OPC 1 + 20 % MK1	0.625	125	4.8	7.8	56.4
80 % OPC 1 + 20 % SF	0.625	125	5.3	7.2	58.0
80 % OPC 1 + 20 % MK2	0.625	150	-	7.8	54.0
80 % OPC 3 + 20 % MK2	0.625	155	-	8.7	56.5

The 28-day compressive strength of blended OPC 1 mortars is about 12 % less than that of the control one.

5.2.1. Damage and durability factor

The evolution of (1-D) versus the number of freeeze-thaw cycles is given in Fig.2.

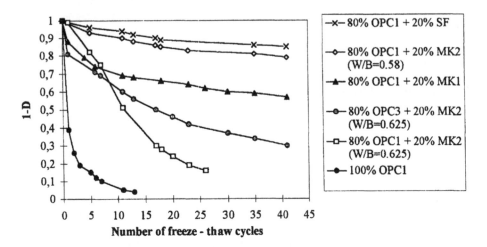

Fig. 2. Evolution of (1-D) for blended OPC mortars.

The durability factor (DF) obtained for (1-D) = 0.20 is given in Table 7.

The presence of pozzolans enhances the durability of OPC 1. SF plays a particularly beneficial role and the blended cement (80 % OPC 1 + 20 % SF) presents a DF factor (0.86) three times higher than that of the best OPC, which was 0.29 for OPC 4.

Table 7. Durability factor DF of blended OPC mortars (1-D = 0.20)

Binder (B)	W/B	N	DF	DF/DF$_{OPC1}$	Lime consumption at 28 days (%)
100 % OPC 1	0.540	3	0.015	1	-
80 % OPC 1 + 20 % SF	0.625	40	0.860	57	56
80 % OPC 1 + 20 % MK2	0.580	40	0.790	53	52
80 % OPC 1 + 20 % MK1	0.625	40	0.580	39	55
80 % OPC 3 + 20 % MK2	0.625	40	0.300	20	58
80 % OPC 1 + 20 % MK2	0.625	23	0.115	8	52

5.2.2. Correlation between freezing resistance and pozzolan characteristics

As seen from Tables 2 and 7, for a given workability, the higher the pozzolan BET surface area, the higher the freezing resistance of blended OPC 1 mortars. From Table 7, it is also clearly established that there is no direct relationship between the lime consumption of the pozzolan and the freezing resistance of the mortar. This resistance also depends of the type of cement : when OPC 3 is used instead of OPC 1 in the same conditions (20 % MK2 - W/B = 0.625), the durability factor is 2.5 times higher.

The water to binder ratio is a very important factor in the freezing resistance of mortars : as W/B increases of 7 %, DF decreases of about 85 % (80 % OPC 1 + 20 % MK2 - W/B = 0.58 or 0.625).

5.2.3. Evolution of blended OPC mortar porosity

The characteristics of the porous distribution before (initial) and after the freeze-thaw cycles (final) are presented in Table 8.

From Table 8, two trends are pointed out :
- for the more durable mortars prepared with blended OPC 1 (SF, MK2-W/B = 0.58, MK1), the total porosity decreases after the freeze-thaw cycles and the porous distribution is slightly shifted from micro to meso and macropores.
- for the less durable mortars, the total porosity increases and micropores are slightly transformed into mesopores.

The use of metakaolin leads to a lower quantity of mesopores and increases the amount of micropores, providing a better compactness to mortars which enhances their freezing behavior.

The decrease of the total porosity in the more durable mortars may be explained by a secondary pozzolanic reaction which occurs during the freeze-thaw cycles.

The use of silica fume leads to a different porous distribution, more regular and less affected by the frost action. The presence of the superplasticizer certainly plays a role in such microstructure.

Table 8. Porous distribution of OPC blended cements

Binder	State	W/B	Total porosity (%)	Micropores < 0.1 µm (%)	Mesopores (%)	Macropores > 0.6 µm (%)
100 % OPC 1	Initial	0.540	12.9	76.6	16.4	6.9
80 % OPC 1 + 20 % SF	Initial	0.625	14.4	63.9	22.2	13.9
	Final		13.1	66.9	17.2	15.9
80 % OPC 1 + 20 % MK2	Initial	0.580	14.9	86.7	6.0	7.3
	Final		12.7	62.1	24.8	13.1
80 % OPC 1 + 20 % MK1	Initial	0.625	17.4	90.0	5.9	4.1
	Final		15.3	83.2	9.5	7.3
80 % OPC 3 + 20 % MK2	Initial	0.625	15.2	83.8	9.1	7.1
	Final		16.9	80.5	14.3	5.2
80 % OPC 1 + 20 % MK2	Initial	0.625	15.7	88.2	7.0	4.8
	Final		18.9	80.4	11.0	8.6

6. Conclusion

The original procedure developed in the present research (freeze-thaw cycles in presence of mechanical loading) allows the measurement of specific mortars damage. The durability factor DF, which is easily calculated from the damage value, is very useful to rapidly rate the relative importance of the different components of the mortar phase of concrete subjected to freezing.

The results obtained point out an increasing influence of the following parameters :

i) for plain cements, the best behavior is obtained for a low Blaine specific surface area, a low C_3A content and a high C_2S content (the durability factor may be multiplied by 19).

ii) for blended cements containing 20 % of pozzolan, the durability factor of the less durable plain cement can be multiplied by 57, as the average 28-day compressive strength is only reduced by 12 %. The higher the BET surface area of the pozzolan, the higher the freezing resistance of mortar. The use of pozzolans needs to be very careful with the water to binder ratio : a increase of W/B by 7 % leads to a decrease of 85 % in the durability factor.

The good performances observed with blended cements are not linked to the pozzolanic activity of the addition, but to the porous distribution. Other microstructural investigations (morphology and type of hydrates, interfaces) should be developed to better point out the main parameters affecting the freezing resistance of mortars.

7. References

1. Gagne, R., Pigeon, M. et Aitcin, P.C. (1990). Durabilité au gel des bétons de hautes performances mécaniques. *Materials and Structures*, Vol. 23, pp. 103-109.
2. Miller, E.W. (1993). Blended cements - Applications and implications. *Cement and Concrete Composites*, Vol. 15, pp. 237-245.
3. Pigeon, M. (1984). *Microstructure et résistance au gel des ciments et bétons.* Thesis, Paris, (1989), 340 p. (in French).
4. Tenoutasse, N. et Marion, A.M. (1989). Influence of industrial by-products on the porosity of hydrated Portland cement. In. *Proceedings of the 3rd International Conference on Fly Ash, Silica Fume, Slag and Natural Pozzola, in Concrete*, ACI SP-114 (ed. V.M. Malhotra), Vol. 1., pp. 33-40.
5. Hooton, R.D. (1993). Influence of silica fume replacement of cement on physical properties and resistance to sulfate attack, freezing and thawing, and alkali-silica reactivity. *ACI Materials Journal*, Vol. 90, n° 2, pp. 143-151.
6. Ollivier, J.P., Carles-Gibergues, A. et Hanna, B. (1988). Activité pouzzolanique et action de remplissage d'une fumée de silice dans les matrices et bétons de haute résistance. *Cement and Concrete Research*, Vol. 18, pp. 438-448.
7. Larbi, J.A. et Bijen, J.M. (1990). Orientation of calcium hydroxide at the Portland cement paste-aggregate interface in mortars in the presence of silica fume. *Cement and Concrete Research*, Vol. 20, pp. 461-470.
8. Chabannet, M. (1994). *Gel interne de matrices cimentaires sous sollicitation mécanique - Intérêt du métakaolin.* Thesis, Lyon, 314 p. (in French).
9. Alexander, M.G. (1993). Two experimental techniques for studying the effects of the interfacial zone between cement paste and rock. *Cement and Concrete Research, 1993, Vol. 23, pp. 567-575.*
10. Girodet, C. (1996). *Endommagement des mortiers sous sollicitations thermo-mécaniques-Influence des caractéristiques des matériaux sur la résistance au gel interne des bétons.* Thesis, Lyon, 299 p. (in French).
11. Okada, E., Hisaka, M., Kazama, Y. et Hattori, K. (1981). *Freeze-thaw resistance of superplasticizer concretes.* American Concrete Institute Special Publication 68, pp. 215-231.
12. Bosc, J.L., Chabannet, M. et Péra, J. (1996). Comportement au gel interne de bétons sous contraintes mécaniques. *Materials and structures.* Vol. 29, août-sept., pp. 395-400.
13. Ambroise, J., Maximilien, S. et Péra, J. (1994). Properties of metakaolin blended cements. *Journal of Avanced Cement Based Materials*, Vol. 1, pp. 161-168.

Frost resisting and waterproof fine-grained slag ash concrete for roofs of residential structures

S.I. PAVLENKO, A.A. PERMYAKOV and V.K. APHANASIEV
Siberian State Academy of Mining and Metallurgy (SSAMM),
Novokuznetsk, Russia

Abstract

The frost resistance and waterproofness of the concrete developed are 500 cycles and 8 to 12 W which is considerably higher than the requirements of the Building Code 2.03.01-84 for roof structures (300 cycles and 6 W, respectively). The concrete does not require waterproof and frost resisting coatings. The fine-grained concrete developed consisted of the following materials: M 500 and 600 portland cement (520 - 640 kg/m^3), slag sand (925 - 1085 kg/m^3), hydra-removed ash (185 - 220 kg/m^3), water (215 - 250 kg/m^3) and 136-157 M hydrophobic admixture (0.1 percent by weight of cement). The 1-year investigation on physic-mechanical properties of the concrete demonstrated its advantage over the ordinary concrete. The prism-to-cube strength ratio for the concrete developed and ordinary concrete were 0.8 to 0.85 and 0.70 to 0.75, respectively. At 1 year, the shrinkage strains and creep for the fine-grained and ordinary concrete were 0.10 to 0.15 mm/m, 0.22 to 0.25 mm/m and 0.14 to 0.24 mm/m , 0.27 to 0.34 mm/m, respectively. The modulus of elasticity of the concrete investigated was 1.1 to 1.3 higher than that of the ordinary concrete. The concrete is used at the Tom-Usinsky large-panel house-building plant.
Keywords: Fine-grained slag-ash concrete, fly ash, frost resisting, roofs of residential, slag sand, waterproof.

1 Introduction

Until recently, the Tom-Usinsky large-panel house-building plant has produced precuts reinforced concrete roof structures from ordinary heavy concrete covered with waterproof and frost resisting coatings. This complicated the technology of the production and considerably increased its cost. The SSAMM under a contract to the above plant has developed the composition and technology of a fine-grained slag ash concrete based on waste products from the Tom-Usinskaya Power Plant (T-UPP).

According to the requirements of the Building Code 2.03.01-84 [1], frost resistance and waterproofness of a concrete used for reinforced concrete roof structures should be no less than F 300 and W 6, respectively. The objective of this work was to 1) develop concrete

Frost Resistance of Concrete, edited by M.J. Setzer and R. Auberg. Published in 1997
by E & FN Spon, 2–6 Boundary Row, London SE1 8HN, UK. ISBN: 0 419 22900 0.

with a higher frost resistance and waterproofness, 2) exclude the use of a protective coating for the roof structures and 3) use waste products from thermal power plants (slag and ash) as replacement for natural aggregates. The concrete developed was patented [4]. Mixture proportions and technology of the frostresisting fine-grained slag ash concrete were successfully used at the Tom-Usinsky reinforced concrete plant.

2 Experimental program

2.1 Materials

The granulometric composition, physical properties and chemical analysis of ash and slag sand used in this study are given in Tables 1, 2 and 3.

Table 1. Grading of aggregates

Sieve size	Cumulative percentage retained		
(mm)	Fly ash	Hydro-removed ash	Slag sand
5	-	-	-
2.5	0.1	0.1	18.25
1.25	0.3	0.4	38.50
0.63	3.7	5.2	58.25
0.315	13.7	18.9	73.75
0.14	49.7	52.3	87.50
>0.14	100.0	100.0	100.0

Table 2. Physical properties of ash and slag sand from Tom-Usinskaya power plant

	Fly ash	Hydro-removed ash	Slag sand
Bulk density (kg/m^3)	900	870	1400
True density (kg/m^3)	2000	2000	2200
Fineness (cm^2/g)	2900	3500	-
Residue on sieve No 008	25	22	-
Fineness modulus	-	-	2.9
Compressive strength ($MPa)$	-	-	1

Table 3. Chemical analysis of ash and slag sand from Tom-Usinskaya power plant

Oxides (%)	Fly ash	Hydro-removed ash	Slag sand
SiO_2	51.93	57.55	56.32
Al_2O_3	25.57	25.15	22.18
Fe_2O_3	8.27	6.34	8.55
CaO	6.83	5.98	6.33
MgO	2.62	2.03	2.02
MnO	0.20	0.20	0.25
(Na_2O+K_2O)	2.29	2.71	3.00
TiO_2	0.50	0.69	0.68
P_2O_5	0.12	0.56	0.30
SO_3	0.18	0.27	1.01
Loss on Ignition	4.63	4.02	0.00

2.2 Slag sand

Slag sand from the Tom-Usinskaya power plant, 0 to 5 mm in size, was used as a main aggregate. It is an amorphous alumina-silicate glass [3]. It has low water demand, high frost resistance and does not contain unburnt particles.

Besides, its granulometric composition (Table 1) corresponds to the highest compactness of particles in concrete [4]. All these properties of the slag sand contribute to the production of a waterproof and frost resisting concrete.

2.3 Fly Ash and hydro-removed ash from the Tom-Usinskaya power plant

The properties of the two ashes are similar (Tables 1, 2, 3). In this work, the comparison was made between them and the possibility of use of the Hydro-removed ash was studied. Both ashes are acidic and do not have any binding properties.

They were used as an active mineral admixture to cement (10 to 15% by weight of cement). The use of ash as a partial replacement of cement increased density, waterproofness and resistance to salt of a concrete.

2.4 Admixtures

Technical grade lignosulphonate (TGL) and 136-157 M hydrophobic liquid proportioned as 1 : 2 were used as an admixture. The lignosulphonate is a waste product of a pulp and paper industry having a hydrophilic effect, which is commonly used as a plasticizing and air-entraining admixture. The hydrophobic liquid is a combination of polymethyl-hydrosiloxanes having a formula:

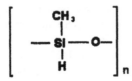

where: n = 15 to 20.

This complex admixture reacts with the products of the cement hydration making the fine slag aggregate (amorphous alumina-silicate glass) more active. Besides, it is a very effective plasticizer. Its use in a slag ash concrete increased its frost resistance by 2 to 3 times (850-1000 cycles). The frost resistance of the concrete was 300 to 500 cycles when the lignosulphonate and the hydrophobic liquid were used separately [4].

3 Mixture proportioning

Frost resistance of a concrete depends on the volume and the nature of a porosity of a cement clinker [5-7]. It is a capillary porosity that determines the frost resistance of a concrete. Water does not freeze in gel pores at low temperatures and does not penetrate into air pores during saturation.

A considerable increase in frost resistance of a fine-grained concrete may be attained through changing the nature of the porosity by additional introduction of conventionally-closed pores which do not absorb water. These reserve pores do not permit of microcracks in concrete clinker and destruction of the concrete during the formation of ice in capillaries.

With the introduction of the plasticizing admixtures, the frost resistance increases due to the low water-to-cement ratio and additional air-entrainment in the concrete mixture. The use of the air-entraining admixtures is effective provided the volume of the entrained air and the size of the air pores in the concrete are controlled.

Testing for the frost resistance of the fine-grained slag ash concrete was carried out on 10-cm cube specimens according to the State Standard 10060-76 [8]. The specimens were examined, weighed and placed in water baths at 20° for 4 days. After saturation with water, the reference specimens were tested for compressive strength. The other specimens were frozen in a refrigerating chamber at -50°C and then thawed in the water baths.

Freezing and thawing of the specimens took 24 h. At the end of a definite number of cycles, the specimens were weighed and tested for compressive strength. The testing was terminated when the strength of the specimens was reduced by more than 15 % of the initial strength.

Water penetration tests were carried out on 150-mm-thick cylinders according to State Standard 12730.5-78 [9]. The initial pressure of water was 0.1 MPa. Every eight hours, the pressure was increased by 0.1 MPa until leakage of water was registered at the upper end of the specimen.

To determine the optimum mixture proportions of the concrete, 32 mixtures were considered with the following quantities of the materials: 250 to 800 kg/m^3 cement, 150 to 250 kg/m^3 ash for M 500 concrete and 200 to 700 kg/m^3 cement, 150 to 300 kg/m^3 ash for M 600 concrete; the amount of the admixtures ranged from 0.1 to 0.5% by weight of cement. Using the method of experimental planning [10, 11], the optimum mixture proportions of the frost resisting concrete were determined. The proportioning of the concrete mixtures is summarized in Table 4.

As can be observed from the data given in Table 4, the use of the high grade cements and the complex admixture (lignosulphonate + hydrophobic liquid) significantly increased the frost resistance (2.5 to 3 times) and waterproofness (1.5 to 2 times) of the concrete developed. This makes it applicable to roof structures of buildings without any protective coatings.

4 Physico-mechanical and deformation properties of concrete

4.1 Cube and prism strengths

Testing for compressive strength was performed in accordance with State Standard 10180-78 [12]. 10-cm cubes and prisms, 10 x 10 x 40 cm in size, were cast. The data on the frost resistance and waterproofness from the mixtures having the highest compressive strength of the concrete are given in Table 5.

It is evident from the Table 5, that the compressive strength of the concrete developed increased by 20% over a 1-year period, whereas of the ordinary sand concrete, it increased by 10 to 12% in accordance with the Building Code 2.03.01-84 [1]. The prism strength of the concrete developed was also considerably higher. The R prism-to R cube ratios were 0.76 to 0.84 and 0.70 to 0.75 for the concrete developed and the ordinary sand concrete. It

Tai le 4: Optimum Mixture Proportions of Frost Resisting Concrete

Concrete Strength Grade	Quantities (kg/m³)				Admixtures (%)		Workability		Water-to-Cement, Water-to-Binder Ratios		Frost-resistance (cycle)	Water-proofness, (W)
	Portland-cement	Ash	Slag Sand	Water	TGL	TGL+ 133-157M	Slump (10⁻² m)	Hardness (sec)	W/C	W/C+Ash		
Based on M 500 Portland Cement (50 MPa)												
M 300 (30 MPa)	490	200	1000	210	0.3	-	6 - 7	-	0.43	0.30	420	8
	450	210	1050	190	0.3	-	2 - 3	-	0.42	0.29	480	9
	410	222	1108	160	0.3	-	-	30	0.39	0.25	440	6
M 500 (50 MPa)	640	185	925	250	-	0.1	4 - 6	-	0.39	0.30	1250	12
	590	195	980	235	-	0.1	1 - 3	-	0.39	0.30	1160	12
	540	210	1035	215	-	0.1	-	20-40	0.40	0.29	975	10
Based on M 600 Portland Cement (60 MPa)												
M 500 (50 MPa)	580	195	975	250	-	0.1	4 - 6	-	0.43	0.32	1320	12
	520	210	1035	235	-	0.1	1 - 3	-	0.45	0.32	1270	11
	480	220	1085	215	-	0.1	-	20-40	0.45	0.31	1150	10

indicates that the slag ash concrete sustains greater loads in thin-walled high-rise structures than the ordinary concrete with the same concrete strength grade.

Table 5. Summary of compressive and prism strengths of a fine-grained slag ash concrete

Mixture No		1	2	3
Quan-tities, (kg/m^3)	M 500 portland cement	450	640	-
	M 600 portland cement	-	-	580
	Hydroremoved ash	210	185	195
	Slag sand	1050	925	975
	Water	190	250	250
	Admix-tures	TGL	0.3	-	-
	TGL+136-157M Liquid (%)	-	0.1	0.1
Slump (cm)		2-3	4-6	4-6
Age (days)		28/360	28/360	28/360
Cube strength (MPa) 28/360		31/38	66/70	58/70
Prism strength (MPa) 28/360		24/29	46/57	49/59
R_{prism}-to-R_{cube} Ratio, 28/360		0.77/0.76	0.82/0.81	0.84/0.84

4.2 Shrinkage and creep strains and modulus of elasticity Tests for drying shrinkage and creep were performed on prisms, 10x10x40 cm in size, using the NIIZhB methods [13, 14]. The drying shrinkage was measured by the clock indicator devices with 0.01 mm scale attached to the specimens. Spring devices with four guide threaded rods were used for the creep tests. The load applied to the prisms was half of the prism strength of the concrete investigated. The tests were performed at room temperature at 50% humidity. The drying shrinkage and creep strains of the concrete developed at 360 days were compared with those of the ordinary heavy concrete based on coarse aggregate, sand concrete and ash-sand concretes. The data are given in Table 6.

Table 6: Drying shrinkage, creep and initial modulus of elasticity at 360 days

Mixture No and concrete type	Compressive strength (MPa)	Average density (kg/m^3)	Drying shrinkage (mm/m)	Creep, (mm/m)	Modulus of elasticity $(E \cdot 10^{-3})$
1	38	1900	0.15	0.25	287
2	70	2000	0.13	0.23	370
3	70	2000	0.10	0.22	361
Ordinary heavy concrete	62	2400	0.32	0.27	296
Fine-grained concrete	50	2400	0.46	0.36	137
Ash-sand concrete	32	1600	0.42	0.46	230

As it is evident from the data of Table 6, this concrete presented better deformation properties than ordinary and sand concretes.

The initial modulus of elasticity was also determined on prisms using the methods developed by NIIZhB [13, 14]. The data are given in Table 6. As can be seen, the concrete produced satisfactory performance.

CONCLUSIONS

The concrete in this study developed has high frost resistance (up to 1300 cycles) and waterproofness (W=10 - 12). It produced better physico-mechanical and deformation properties than ordinary and sand concretes.

The concrete mixture can be used in the conditions of alternate freezing and thawing requiring no protective coating. It can be used for roof structures, in the construction of roads, tunnels, canals etc.

REFERENCES

1. Building Code 2.03.01-84 (1985) *Concrete and Reinforced Concrete Structures,* Gosstroy USSR, Moscow.
2. Pavlenko, S. I. (1989) Concrete on the Basis of the Steam Electric Station Ash and Crushed Slag, *Third CANMET/ACI International Conference on Fly Ash, Silica Fume, Slag and Natural Pozzolans in Concrete, Supplementary Papers,* Norway, pp. 738-755.
3. Pavlenko, S. I. and Rekhtin, I. V (1991) Fine-Grained Slag Sand and Fly Ash Concrete with Higher Frost Resistance and Waterproofness, *ACI International Conference on Evaluation and Rehabilitation of Concrete Structures and Innovations in Design,* December 2-6, 1991, Hong Kong, ACI, Detroit, Michigan, USA, Vol., pp. 559-576.
4. Patent No 2008293 (1994) *Slag-ash-concrete Mixtures,* Rospatent, 28.02.
5. Powers, T. C. and Helmuth, R. A. (1953) *Theory of Volume Changes in Hardened Portland Cement Paste During Freezing,* Highw. Res. Board Proc. Annu. Meet., p. 285.
6. Litwan, G.G. (1969) *Freezing of Water in Hydrated Cement Paste,* RILEM International Symposium on the Durability of Concrete, B 153-B 160.
7. Zotkin, A. G. (1983) *Providing the Frost Resistance of Concrete,* Irkutsk, pp.4-10 and 20-21.
8. State Standard 10060-76 (1976) *Concretes, The Methods for Determining Frost Resistance,* Moscow, Gosstroy USSR, Izdatelstvo Standartov.
9. State Standard 12730.5-78 (1978) *Concretes, Methods for Determining of Water Permeability,* Moscow, Gosstroy USSR, Izdatelstvo Standartov.
10. Protodyakonov, M. M and Tedder, R. I.(1970) *Methods of Rational Experiment Planning,* Nauka Publishing House, Moscow, p.70.
11. Voznesensky, V. A. (1974) *Statistical Methods of Experiment Planning in Technico-economical Investigations,* Izdatelstvo Statistici, p.192.
12. State Standard 10180-78 (SMEA 3978-83) (1985) *Concretes, Methods and Regulations of Strength Controle,* Gosstroy USSR, Izdatelstvo Standartov, Moscow.
13. NIIZhB (1975) *Recommendations for Study of Shrinkage and Creep of Concrete,* MP-1-75, Moscow.
14. NIIZhB (1975) *Recommendations for Determining Strength and Structural Characteristics of Concretes under Short-and Long-Term Loading,* P-10-71, Moscow.

Water resistant low water consumption plaster binder

K.K. ABDRAKHMANOVA
Academy of Civil Engineering, Almaty, Kazakstan

Abstract
Study of theoretical basis of negative temperature effect onto structure formation and solidification processes of various concrete's based on various binders allows to formulate a working hypothesis that reads the following. High water-resistant plaster binder hydration rate accompanied with chemical binding of large amount of mixing water and with significant exothermic effect and with accelerated formation of primary plaster stone crystal structure, pre-determines quick strength growth rate of the binders and concrete's made on the basis of binders at negative temperatures and thus provides high structure early-age freeze strength. Provided reduction of concrete mixture water consumption and provided formation of fine-porous structure one may enable low primary moisture and solidification of concrete at negative temperatures.

Presently, one the most effective and technologically acceptable for monolithic construction ways to reduce water consumption of concrete mixtures is usage in the concrete composition of low water consumption binders (LWCB) obtained through mechanical-and-chemical activation of components. This allows assumption of that mechanical-and-chemical activation of water resistant plaster binders (WRPB) would allow obtain water resistant low water consumption plaster binder (WRLWCPB) and allow resolve the problem of the binder's usage in winter concreting.

Keywords: research techniques, conclusions, references.

1 Research techniques

In designing of the Water Resistant Low Water Consumption Plaster Binder (WRLWCPB) have been used the following ingredients: the Portland 400 of the

Frost Resistance of Concrete, edited by M.J. Setzer and R. Auberg. Published in 1997 by E & FN Spon, 2–6 Boundary Row, London SE1 8HN, UK. ISBN: 0 419 22900 0.

Shimkent and Semipalatinsk factories, the plaster binder marked G-4A of the «Gypsum» joint-stock company (Zarechny town of the Almaty area), the ashes of the Almaty Heating Electro Station-2 and the Ekibastuz TETS, the superplastificator C-3, the water soluble mount VRP-1. The chemical composition of the ash consisted of the following oxides,%: $SiO_2 = 62.7$; $Al_2O_3 = 27.7$; $Fe_2O_3 = 4.6$; $CaO = 1.7$; $MgO = 0.2$; $K_2O = 0.6$; $Na_2O = 0.1$; $SO_3 = 0.3$; $TiO = 0.7$. The specific surface of the ash was $S = 2100$ m^2/kg; the specific density $\rho = 2180$ kg/m^3; the water-consumption 15%; loses during calcination 1.4%. The ash was low-calcium one and was of acid character according to its basicity modulus.

While developing the technology of WRLWCPB production that meets the requirements of winter concreting the best way, we have studied effect of additional plaster binder grinding, puzzoalanic additives and Portland onto features of the binder under development.

Filled binder system strength is determined by the composition of chemical, physicochemical and physic-mechanical interactions. When mixed particles are fixed in the structural grid through coagulating contacts. Build-up of enough strong contacts in system is possible provided the complex additives surface energy is much higher than that of the binder. This conclusion is based on the thermo-mechanical concept of adhesion that reads that adhesive strength is formed depending on the ratio of the surface energy of an adhesive and that of a substratum.

Increase of surface energy is caused with break of inter-atomic structural connections, for the first place. The essence of that process is the break of the solid material surface layer and the overcoming mutual gravitation forces, i.e. cohesion, according to P.A.Rebinder [1]. The newly formed surface formed after the break of inter-atomic structural connections, have a higher surface energy.

The mechanical energy that immediately influences crystallization and molecular substructure of the particles, is one of the most efficient methods of catalytic influence. That leads to partial dispergation of plaster, cement grains and ash in weak connections, and to mechanodistruction-amorphization-of composition structure elements. In its turn, amorphization contributes to increase (growth) of active centers.

Mechanical treatment was made in a vibration mill of SVM-2 type. The superplastificator content varied from 1.0% to 3.0%.The complex additives content increased equivalently to decrease of plaster binder content from 20% to 80%.

The specific surface of WRLWCPB was registered with UPV-1 device. The basis of UPV-1 device dispersity determing method is evaluation of external specific surface highly dispersed powders from air filtering parameters through the tasted material provided consistent pressure differential at the grain layer. The advantage of this method is elimination of necessity to determine extra contents. Subject to the common Koseni-Karman constant, unity of determing the external specific surface is provided, and besides there is no need to develop standard working specific surfaces. Moreover certification of the latter when made with other methods, does not give universality to the result.

It was found that additional plaster binder grinding in laboratory ball mill is followed by rapid growth of its specific surface. This is followed by rapid increase of the binder's water consumption and as a result, followed by reduction of plaster stone density and stregth. Addition of optimum amount of S-3 superplastificator during grinding allows reduce normal consistency of the plaster binder by more than 30% and obtain a fineporous stone structure with increased strength. Taking into consideration that increase of specific surface of plaster binder contributes to more complete and quicker hydration of the binder accompanied by intensive heat release, the mechanical-and-chemical activation of plaster component of WRLWCPB compositions that is developed for winter concrering is recognized as advisable and efficient [2].

It is known that durability of the Plaster Cement Puzzoalanic Binder (PCPB) depends upon correlation between a Portland and puzzoalanic additives in their composition. The low level of this correlation is provided by high activity of these additives utilized. As the researches have demonstrated, a thorough milling of the low active puzzoalanic additives (including various industrial wastes) allows to sharply increase their activity and thus to decrease a high level of these additives originally demanded in the WRLWCPB composition.

Based on the findings, we suggest a technology to obtain WRLWCPB that includes the following consequent main stages:
1. Preliminary grinding of low-active puzzoalanic additive until the required specific surface rate is obtained
2. Short additional grinding of all components of the binder provided optimum amount of powdered S-3 superplastificator.

The equitation analysis of the multifactor mathematics model regression has determined the following optimum parameters of the WRLWCPB that best of all correspond to the demands of a winter time concreting:
- the plaster content - 55...65%
- the complex additives containing the Portland and ash - 35...45%
- correlation of the Portland to the ash - 1:1...1:1,5
- the quantity of the superplastificator - 2...2,5%
- the unit surface of the WRLWCPB - 900...1000 m^2/kg

The developed composition are characterized by the following qualities:
- the normal density - 25...28%
- the endurance limit under compression after two hours of solidifying in normal conditions - 7,5...9 MPa
- the endurance limit under compression after seven days of solidifying in normal condition - 35...40 MPa
- the endurance limit under offsetting in seven days of solidifying in normal conditions - 9...10,5 MPa
- the softening rate - 0,7...0,8

So, the designed WRLWPCB in comparison to the PCPB allows to decrease more than twice the consumption of the Portland in the cement composition, simultaneously decreasing the forming dampness of the matter in 2...3,5 times, to increase the early endurance in 2...3,5 times, to increase the offset endurance in

1,5...2 times, to increase the mark endurance of the cement in 2... 2,5 times, as well as to increase it's waterproofs.

The further researches studied an impact of the negative temperatures on aquation process, structure and qualities of the WRLWCPB. The PCPB, the plaster binder (PB), have been used as a standard. As the WRLWCPB is characterized by an accelerated period of grasping, together with solidifying water the inhibitor VRP-1 has been used in amount of 0.01% to slow down that period.

The input of the finely milled plaster binder, characterized by the high reactive capability, into the WRLWCPB, has allowed to avoid the negative influence of low temperatures on initial aquation velocity of the WRLWCPB . It is a fact that as the temperature goes down to $0^{\circ}C$, the water demand of the WRLWCPB and PB stays unchanged, while the water demand of PCPB lowers by 1,5...2,0%. The grasping period of the WRLWCPB does not change, the period of the PB and of the PCPB slows down in 1,5 and 2 times accordingly. The temperature falling by $0^{\circ}C$ does not noticeably affect the early endurance of WRLWCPB, PCPB and PB.

The low water demand of the WRLWCPB , the more full and accelerated aquation of the finely milled plaster ingredient in its consistency, accompanied by chemical fastening of higher quantity of water gate with more intensive heat emission, allows to form a fine vesicular structure of the plaster stone and to avoid considerable destruction at early freezing. In this connection in two hours of solidifying at temperature of $-5^{0}...-20^{0}C$ the endurance of WRLWCPB comes out to be in 1,5...2 times higher than of the PB and PCPB. Herewith the higher the content of the plaster ingredient in the binder the higher the early endurance of WRLWCPB and the lesser the influence of negative temperature on it.

Besides that, a sharp decline of the summary vesicularness of the solidified WRLWCPB, with simultaneous redistribution of the pores towards their size decreasing, provides a liquid phase preservation in the matter structure in the negative temperatures, supporting the further aguation of the cement at a high level. It is determined that if WRLWCPB has been solidifying for 28 days at temperatures of $-5...-15^{\circ}C$ its additional endurance increment comes to be accordingly 10...13%, 8...11% of the mark endurance of those cements that solidify normally, and gets higher as level of the complex additives in the cement composition goes up. The endurance increment of PCPB that has large vesicular structure was 3...7% at the $-5^{0}C$, and at the temperatures lower than that was not detected at all.

As the researches have shown the WRLWCPB and the PCPB after they have thawed out are characterized by accelerated aquation. The most intensive endurance growth belongs to those consistencies that has the higher level of complex additives. It has been determined by the method of mercury pore metering that the structure of the stone made of WRLWCPB after 28 days of the further normal solidifying does not considerable differ from the structure of the stone that has been solidifying without freezing and practically does not suffer endurance loss. The endurance loss for PB reaches 15%, for PCPB - 25%, and correlate to the degree of stone structure destroying at early freezing.

Basing on the results of x-ray phase, thermographic and electromicroscopic analyzing tastes it is determined that the phase consistence of the WRLWCPB growth does not qualitatively differ from the growth consistency of the traditional PCPB, the special features of WRLWCPB aquation have also been proved.

2 Conclusions

The optimum compositions for WRLWCPB have been developed to allow to use them in winter time conditions of concreting.

It has been determined that the negative influence of the low temperatures on the initial aquation kinetics of WRLWCPB does down as the unit surface of WRLWCPB increases.

If the level of the complex additive in WRLWCPB increases one can observe the highest endurance growth at a long-run solidifying at negative temperatures.

The complex researches have led to the conclusion that the stone structure has small crystals and has the increased level of pores measured by less that 0.1-0.01 mkm.

3. References

1. Rebinder, P.A. (1966) *Physicochemical mechanics of dispersed structure*, Academy of Sciences of the USSR publishing house, Moscow
2. Veronskaya A.W., Melnichenko S.W. (1992) *The Waterproof Gypsum Binder of Low Water Demand for Winter Time Concreting* , Construction Materials,#5,pp.24-26

Influence of sand on the freeze-thaw resistance of the mortar phase of concrete

C. GIRODET, J.L. BOSC, M. CHABANNET and J. PERA
Unité de Recherche Génie Civil – Matériaux,
Institut National des Sciences Appliquées de Lyon, France

Abstract
From the literature, the influence of aggregates on the freeze-thaw resistance of concrete is not well established. In the present study, the influence of sand on the durability of mortars subjected to a thermo-mechanical loading is investigated.
The following parameters are studied : the nature of sand (siliceous, calcareous, silico-calcareous), its origin (from river or quarry), and its particle size distribution. The mortars are prepared with a cement : sand ratio of 3.375, at the same workability.
The prismatic samples (40 x 40 x 160 mm^3) are subjected both to three-point flexure and freeze-thaw cycles. The rate of loading is 20 % of the 28-day tensile strength. A coefficient of damage is defined and allows the rating of sand characteristics.
The decreasing rating is as follows : the particle size distribution, the origin (sand river behaves better), and finally the nature of the sand (siliceous sand is less efficient). The microstructural investigation of the mortars allows the interpretation of the observed phenomenons.

Keywords :Aggregate, damage, freeze-thaw resistance, microstructure, mortar, sand, tensile strength.

1 Introduction

Publications by many authors show that the freeze-thaw resistance of concrete is influenced by the type of aggregate used [1] to [4]. The coefficient of thermal

Frost Resistance of Concrete, edited by M.J. Setzer and R. Auberg. Published in 1997
by E & FN Spon, 2–6 Boundary Row, London SE1 8HN, UK. ISBN: 0 419 22900 0.

expansion of aggregates is generally lower than that of the cement paste leading to stresses in the material subjected to temperature changes [5], [6]. The size of aggregates influences the properties of the interfacial transition zone [7]. The origin and the porosity of the aggregates may affect the freeze-thaw resistance of concrete [8].

However, the relative influence of the characteristics of the sand present in the mortar phase of concrete remains poorly documented [8]. Therefore, the aim of this study was to elucidate the role played on the freeze-thaw resistance of mortars by the three following parameters : the origin of the sand (from river or quarry), its nature (siliceous, calcareous, silico-calcarerous) and its particle size distribution. In order to accelerate the degradation of mortars, an external flexural load was applied to the mortars during the freeze-thaw cycles [9], [10].

The paper first presents the main characteristics of the sands used in the study, the mixture proportions of mortars and the definition of a damage coefficient. Then, the influence of the parameters considered in the research on the frost durability of mortars is analyzed and a physico-chemical approach is provided to explain some phenomenons.

2 Properties of sands

Usually, porosity and frost resistance of aggregates are the main parameters investigated in the frost durability of concrete. However, other parameters affect this durability :

- the chemical nature of the aggregate (siliceous, calcareous, silico-calcareous), which may induce some interactions between the cement paste and the aggregate,
- the origin of the aggregate (river or quarry), which affects the placing of concrete and consequently its initial microstructure,
- the particle size distribution which, due to the presence of fine particles, contributes to densify the matrix.

Five sands were chosen in the present study [9]. Their main physical and chemical properties are shown in Table 1., while their particle size distribution is presented in Table 2.

Table 1. Physical and chemical properties of sands

Property	Standard sand	Calcareous sands (C)		Silico-calcareous (SC) sands	
Mineralogy	Quartz (100 % SiO₂)	Calcite (98 % CaCO₃)		Mixture of calcite and quartz	
Origin	River	River	Quarry	River	Quarry
Size range	0/2 mm	0/4 mm	0/4 mm	0/4 mm	0/2 mm
Los Angeles		33	33	25	25
Micro-Deval		20	20	8	8
Water absorption (%)	0.09	4.10	4.20	4.50	5.20
Designation	SN	0/4 C$_r$	0/4 C$_c$	0/4 SC$_r$	0/2 SC$_c$

Table 2. Particle size distribution of sands

Sieve opening	Total percentage passing (w_t %)				
(mm)	0/4 C_c	0/4 C_r	0/2 SC_c	0/4 SC_r	SN
0.080	0.7	1.1	0.9	0.3	1.5
0.160	8.3	6.1	5.1	7.1	10.9
0.315	18.8	16.9	21.1	29.5	17.9
0.630	38.1	27.9	41.6	44.3	42.7
1.250	65.6	43.8	66.6	68.3	77.0
2.500	93.6	73.7	98.7	92.4	99.7
5.000	99.9	99.7	99.9	99.7	100.0
Fineness modulus (F.M.)	2.76	3.32	2.67	2.59	2.52

According to the French standard NFP 18-541, all these sands, except sand 0/4 C_r, present a fineness modulus meeting the requirements for use in concrete : $1.80 \leq F.M. \leq 3.20$. The rating of these sands, from the finer to the coarser, is as follows :

$$SN < 0/4\ SC_r < 0/2\ SC_c < 0/4\ C_c < 0/4\ C_r$$

3 Specific mortars and durability factor

The specific mortar used in this study had the following composition : 400 g of ordinary portland cement CPA CEM I 52.5 and 1350 g of sand. The water to cement ratio (W/C) was adjusted to get about the same workability. The properties of fresh mortars and their strengths are given in Table 3. The mixtures designated by a (*) are made of recombined sands. These sands had the same particle size distribution as sand 0/4 C_c.

Table 3. Properties of fresh and hardened mortars

Sand	W/C	Initial spread (mm)	Air content (%)	28-day flexural strength (MPa)	28-day compressive strength (MPa)
SN	0.50	115	4.2	8.8	69.2
0/4 C_c	0.58	110	2.9	10.0	72.7
0/4 C_r	0.58	120	4.5	8.9	64.6
0/4 C^*_r	0.58	115	5.1	8.6	64.6
0/2 SC_c	0.77	135	3.7	7.7	49.9
0/4 SC_r	0.66	115	4.8	7.7	56.2
0/4 SC_r	0.77	160	4.3	7.1	47.1
0/4 SC^*_r	0.77	180	7.1	6.2	42.8

For a given water to cement (W/C = 0.77), the initial workability of the mortar prepared with sand 0/4 SC$_r$ is higher than that of the mortar utilizing sand 0/2 SC$_c$ certainly due to a lower water absorption of the sand (4.5 % instead of 5.2 %).

Prismatic samples of mortars (40 x 40 x 160 mm) were cast and subjected to the same tests as shown in reference [10]. A load equal to 20 % of the average flexural strength of mortar was applied.

A durability factor was defined by the equation :

$$DF = \frac{PN}{M}, \text{where :}$$

P = relative dynamic modulus at N cycles, expressed as P = 1 - D ;

N = number of cycles at which P reaches the specific minimum value P = 0.20 for discontinuing the test, or the specified number of cycles at which the exposure is to be determined whichever is less,

M = specified number of cycles at which the exposure is to be determined.

In this investigation, M is equivalent to approximately 40 cycles.

4 Results and discussion.

4.1 Influence of the mineralogy of the sand

4.1.1 Damage and durability factor

The evolution of the loss in frost resistance defined by (1 - D), with respect to the number of freeze-thaw cycles is given in Fig. 1. for four mortars having about the same workability : 0/4 C$_c$, 0/4 SC$_r$ (W/C = 0.66), 0/2 SN, 0/2 SC$_c$.

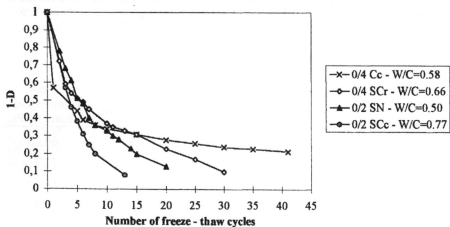

Fig. 1. Influence of the mineralogy of the sand on the freeze-thaw durability of mortars.

The durability factor DF obtained when P = 1 - D = 0.20 is given in Table 4. The DF value of mortar made with the calcareous quarry sand (O/4 C$_c$) is 2 or 2.5 times that of mortars containing either silico-calcareous (0/4 SC$_r$) or siliceous sand (0/2 SN).

Table 4. Influence of the mineralogy of the sand on the durability factor DF

Sand	W/C	Initial spread (mm)	N	DF	DF/DF(SC$_r$)	DF/DF(SN)
0/4C$_c$	0.58	110	40	0.215	2.0	2.5
0/4 SC$_r$	0.66	115	22	0.110	1.0	1.3
0/2 SN	0.50	115	17	0.085	0.8	1.0
0/2 SC$_c$	0.77	135	8	0.040	0.4	0.5

4.1.2 Microstructure

Mortars made with calcareous and silico-calcareous sands are more resistant than those cast with siliceous sand. This result can be explained by the good bond between the particles of calcareous sand and the cementitious matrix, which leads to better strengths as shown in Table 3. As chemical reactions occur between limestone and cement to precipitate carbo-aluminates [11], [12], the strength of the interfacial zone is increased and microcracks appear either in the matrix or throughout the aggregate (Fig. 2). Therefore, the frost resistance of the mortar is linked to the strength of the matrix, as the interface is no longer a zone of weakness.

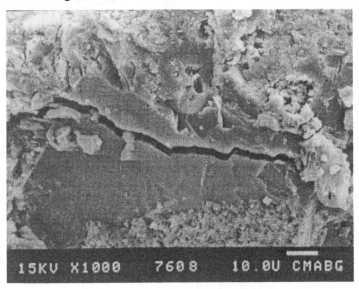

Fig. 2. Propagation of a microcrack throughout a calcareous
aggregate, after freeze-thaw cycles. Sand 0/4 C$_c$.

This result agrees the findings of Bellanger et al [11] : microcracks always appear in the matrix when siliceous or hard calcareous aggregates are used ; when the deformability of the paste is lower than that of the aggregate, these cracks occur in aggregates.

4.2 Influence of the particle size distribution and origin of sand.

Five mortars made with the following sands were compared :
- river sands : 0/4C$_r$, 0/4 SC$_r$, 0/4 C$_r^*$, 0/4 SC$_r$,
- quarry sand : 0/4 C$_c$.

Their properties are given in Table 3. For calcareous sands, the water to cement ratio was 0.58 ; for silico-calcareous sands, it was 0.77.

4.2.1 Damage and durability factor

The residual frost resistance of mortars, P = 1 - D, is shown in Fig. 3.

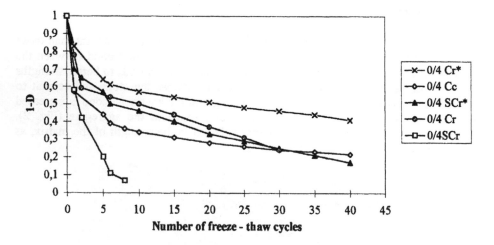

Fig. 3. Freeze-thaw durability of mortars. Influence of the
particle size distribution and origin of sand.

For each mortar, the durability factor obtained for P = 1 - D = 0.20, is given in Table 5. For a same workability, the durability factor of recombined sands is very high : there is an increase of 160 % for sand 0/4 C$_r^*$ and 580 % for sand 0/4 SC$_r^*$, compared to sands 0/4 C$_r$ and 0/4 SC$_r$, respectively. Therefore, the particle size distribution of the sand is a main parameter in the frost resistance of mortar : a fineness modulus of 2.80 is recommanded.

Table 5. Influence of the sand origin and grading on the durability factor DF.

Sand	W/C	Initial spread (mm)	N	DF	DF/DF(C_r)	DF/DF(C_c)	DF/DF(SC_r)
0/4C_r*	0.58	115	40	0.410	2.60	1.90	16.40
0/4 C_c	0.58	110	40	0.215	1.40	1.00	8.60
0/4 SC_r*	0.77	180	34	0.170	1.10	0.80	6.80
0/4 C_r	0.58	120	31	0.155	1.00	0.70	6.20
0/4 SC_r	0.77	160	5	0.025	0.16	0.12	1.00

The influence of the sand origin is less important : DF is enhanced by 90 % for sand 0/4 C_r*, compared to sand 0/4 C_c.

4.2.2. Microstructure

The properties of the porous distribution of mortars before and after freeze-thaw cycles are shown in Table 6. The results obtained point out that the initial porosity of recombined sand (0/4 C_r* and 0/4 SC_r*) are higher than that of natural sand ; this increase is directly linked with a higher air content, as shown in Table 3.

Table 6. Porous distribution of mortars

Sand	State	W/C	Total porosity (%)	Micropores < 0.1 μm (%)	Mesopores (%)	Macropores > 0.6 μm (%)
0/4 C_c	Initial	0.58	12.9	85.7	8.6	5.8
	Final		15.0	55.1	35.1	9.8
0/4C_r*	Initial	0.58	14.5	41.5	52.5	6.0
	Final		14.8	43.6	43.8	12.6
0/4SC_r*	Initial	0.77	19.7	43.8	44.8	11.3
	Final		20.7	42.1	38.1	19.8

The porous distribution is therefore shifted towards mesopores and becomes more homogeneous. Such microstructure ensures a good frost resistance of mortars.

The presence of fine particles in the quarry calcareous sand (0/4 C_c) densifies the matrix and leads to a great initial quantity of micropores (> 80 %) which are transformed into mesopores after the freeze-thaw cycles. The use of this sand gives a lower durability factor than sand 0/4 C_r* and is about the same as that of 0/4 SC_r*.

5 Conclusion

The decreasing influence of some characteristics of sand on the freeze-thaw durability of mortars is as follows :

i) according to the sand mineralogy, the sand grading may increase the durability : DF is enhanced by 160 % for the calcareous sand and 580 % for the silico-calcareous sand ; a fineness modulus of 2.80 is recommended ;
   ii) for a given particle size distribution, river sands perform better than quarry sands,
   iii) calcareous and silico-calcareous sands behave better than siliceous sand.
The microstructural analysis shows that good interfaces between aggregates and paste appear in mortars containing either calcareous or silico-calcareous sands, leading to high 28-day strengths and frost durability.
The modification of the sand grading curve and the choice of river sand increase the air content and make the porous distribution more homogeneous, two factors which enhance the frost durability.

6 References

1. Aubertin, M. (1985). Cryoclastie des granulats minéraux . *Bulletin de Liaison des Laboratoires des Ponts et Chaussées*, Vol. 135, pp. 95-99.
2. Bertrandy, R. (1977). Etude de l'influence des fines sur le facteur durabilité des bétons hydrauliques. *Travaux*, Mai, pp. 43-55.
3. Longuet, M. (1978) Mise au point de méthodes d'appréciation de la résistance au gel des matériaux poreux. *Silicates Industriels*, Vol. 12, n° 39, pp. 275-287.
4. Tourenq, C. (1970). *La gélivité des roches - Applications aux granulats.* Paris : Ministère de l'Equipement et du Logement, Laboratoire Central des Ponts et Chaussées, 54 p., rapport de recherche n° 6.
5. Huovinen, S. (1993). Abrasion of concrete structure by ice. *Cement and Concrete Research*, Vol. 23, pp. 69-82.
6. Sadouki, H., Regnault, P. et Rastagi, P.K. (1988). Analyse numérique et holographique de l'endommagement d'un matériau composite induit par l'incompatibilité de ses phases. *Matériaux et Techniques*, Septembre-Octobre, pp. 27-32.
7. Ping, X., Beaudoin, J.J. et Brousseau, R. (1981). Effect of aggregate size on transition zone properties at the Portland cement paste interface. *Cement and Concrete Research*, Vol. 2, pp. 990-1005.
8. Pigeon, M. (1996). *Microstructure et résistance au gel des ciments et bétons.* Thesis, Paris, 1984, 340 p. (in French).
9. Girodet, C. (1996). *Endommagement des mortiers sous sollicitations thermo-mécaniques-Influence des caractéristiques des matériaux sur la résistance au gel interne des bétons.* Thesis, Lyon, 299 p. (in French).
10. Bosc, J.L., Chabannet, M. et Péra, J. (1996). Comportement au gel interne de bétons sous contraintes mécaniques. *Materials and structures*, Vol. 29, Août-Sept., pp. 395-400.
11. Bellanger, M. Chaouch, M. et Homand, F. (1996). Comportement mécanique des mortiers et bétons de calcaires. *Mines et Carrières, Industrie Minérale*, Mars 1996, pp. 57-61.
12. Bacchiorini, A. (1985). *Interactions physico-chimiques entre l'aluminate monocalcique et différents carbonates au cours de la réaction d'hydratation.* Thesis, Lyon, 215 p.

Investigations on freeze-thaw resistance of recycling concrete

H.K. HILSDORF, R. KOTTAS and H.S. MÜLLER
Institut für Massivbau und Baustofftechnologie,
University of Karlsruhe, Germany

Abstract
The objective of this investigation was to develop acceptance criteria for concrete aggregates which have been obtained by crushing old concrete pavements so that these recycling aggregates are suitable for the production of new concrete pavements. Concretes with water-cement-ratios of 0.38, 0.45 and 0.52 were made from recycling aggregates as well as from natural aggregates. The recycling aggregates were prepared either from the debris of demolished concrete pavements or from laboratory concretes with compressive strengths ranging from 47 MPa to 68 MPa. Air entrainment as well as superplasticizers were employed for the recycling concretes. The recycling concretes were either moist cured or treated with a curing compound. For freeze-thaw testing the CDF-method has been employed. All concretes tested satisfied the criteria for concretes with a high resistance to freezing, thawing and deicing salts. The experiments showed that air entrainment of the old concretes is not a necessary requirement for their use as recycling aggregates. Nevertheless, the air void characteristics of the recycling concrete are dominant parameters for the scaling resistance of such concretes. There is a trend that the extent of scaling decreases with increasing compressive strength of the old concretes from which the recycling aggregates have been obtained.
Keywords: Aggregate freeze-thaw resistance, air entrainment, concrete aggregates, concrete compressive strength, concrete deicing salt resistance, concrete freeze-thaw resistance, recycling concrete.

1 Introduction

The recycling of the debris of demolished concrete structures, particularly as aggregates for recycling concrete is today common practice in many countries, and numerous investigations on the properties of such recycling concretes have been carried out

Frost Resistance of Concrete, edited by M.J. Setzer and R. Auberg. Published in 1997 by E & FN Spon, 2–6 Boundary Row, London SE1 8HN, UK. ISBN: 0 419 22900 0.

and published, e.g. in [1,2,3]. In concrete pavement construction recycling material often is used for the subbase of a pavement, however, there is some hesitation to use recycling aggregates obtained from the old pavement also for the top layer of a new concrete pavement which is directly exposed to traffic and to deicing salts. The primary reason for this is that so far no reliable criteria to be satisfied by the recycling material to ensure durability of the recycling concrete have been established. Therefore, the objective of the investigation described in the following was to establish such criteria.

For conventional aggregates a variety of experimental methods have been developed to check the suitability of a particular aggregate for the production of a concrete with a high resistance to freezing, thawing and deicing salts. Such tests are refered to e.g. in the German specification for concrete aggregates DIN 4226 [4] and described in DIN 52104, Part 1 [5]. In these experiments water saturated coarse aggregates with a diameter > 8mm are exposed to 10 cycles of freezing and thawing in water. The aggregates are considered sufficiently resistant against freezing and thawing if the loss of material after such treatment does not exceed certain limiting values. Since recycling aggregates are in most instances coated with a layer of hydrated cement paste from the old concrete it is at least questionable that the same test methods and the same criteria applicable to conventional aggregates are also suitable for recycling aggregates.

2   Test program

From 4 concrete highways in Germany constructed several decades ago debris of the unreinforced concrete pavements, which had been exposed to traffic as well as to freezing, thawing and deicing salts, were transported to the laboratory and used for the production of recycling aggregates. Concrete cores had been taken from these pavements prior to their demolition in order to determine the physical properties of the old concrete. No records regarding the composition and the initial properties of these pavement concretes were available. Therefore, two additional concretes with water-cement-ratios of 0.55 and 0.65, respectively, and a maximum aggregate size of 22 mm were manufactured in the laboratory. After the laboratory concretes had reached an age of 2 years recycling aggregates were produced from these laboratory concretes.

Mechanical and chemical properties of the recycling aggregates as well as their frost resistance according to the German specification DIN 4226 were determined.

From the recycling aggregates 15 series of recycling concretes and 3 series of concretes with conventional Rhine river gravel as an aggregate were produced with water-cement-ratios of 0.38, 0.45 and 0.52, respectively. Specimens for determining the mechanical properties as well as the freeze-thaw and deicing salt resistance of these concretes were cast.

The resistance of the concretes against freezing, thawing and deicing salts was determined employing the CDF-method [6].

3   Materials

3.1   Concrete aggregates

In the following the recycling aggregates are designated according to their origin:

- PCA: pavement concrete from Alsfeld
- PCAP: pavement concrete from Appenweier
- PCL: pavement concrete from Langwedel
- PCG: pavement concrete from Garlsdorf
- LC65:   laboratory concrete, w/c = 0.65
- LC55:   laboratory concrete, w/c = 0.55
- RRG: Rhine river gravel

The recycling aggregates were crushed and subsequently separated by sieving into aggregate fractions of 4/8 mm, 8/16 mm and 16/22 mm. The fine particles with diameters < 4 mm were discarded. From a sieve analysis of the aggregates after crushing it followed, that about 2/3 of the aggregates had a size in the range of 4 to 32 mm.

Compressive strength f_{cu}, modulus of elasticity E_c, total porosity V_{pc}, chloride content Cl_c and air void characteristics i.e. air content A_c, spacing factor SF and content of micro-air voids with diameters < 0.3 mm L300 of the concretes from which the recycling aggregates have been obtained, are listed in Table 1.

Table 1. Properties of the recycling aggregates

Aggregate	f_{cu}	E_c	V_{pc}	Cl_c	Air void characteristics		
					A_c	SF	L300
	[MPa]	[MPa]	[%]	[%]	[%]	[mm]	[%]
PCA	58	55000	13.7	0.190	-	-	-
PCAP	57	56000	13.3	0.076	2.0	0.40	0.9
PCL	68	45000	11.2	0.107	2.2	0.67	0.6
PCG	47	45000	11.0	0.162	2.3	0.58	0.6
LC65	40	30000	18.8	0.005	1.1	0.50	0.5
LC55	55	34000	16.0	0.005	1.1	0.47	0.5

From Table 1 it follows that with all likelihood none of the concretes from which recycling aggregates were obtained have been manufactured with air entraining agents.

Table 2 lists the results of freezing and thawing tests of the aggregates according to method N as described in DIN 52104, Part 1. In these experiments the aggregates are initially stored for 24 hours in deaired water. Then the aggregates are filled into metal containments and cooled from 20 °C to 0 °C within a period of approx. 150 min. After about 210 min at this temperature the temperature is decreased to -17.5 °C. Following at least 240 min at this temperature the containments are exposed to a temperature of 20 °C. After 10 cycles of freezing and thawing the aggregates are removed from the containments, and the amount of particles with diameters less than half of the lower original particle size is determined. According to the applicable German specifications an aggregate is considered suitable for the top layer of a concrete pavement, if the loss of material after 10 cycles of freezing and thawing as defined above does not exceed 1 percent by mass. In order to check to which extent the rate of deterioration of the recycling aggregates decreases with the number of freeze-thaw-cycles some of the experiments were extended to 21 cycles. As an example the loss of mass of the recycling aggregate PCG is shown in Fig. 1. In Table 2 also the difference of loss of mass after 20 and after 10 cycles of freezing and thawing is listed for all recycling aggregates.

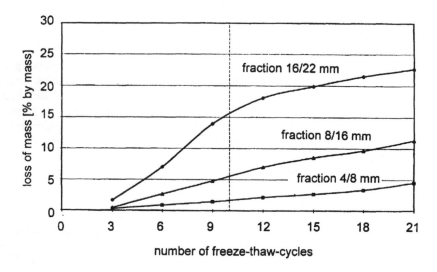

Fig.1. Loss of mass of recycling aggregates PCG during freeze-thaw-testing

Table 2. Results of freeze-thaw-tests of recycling aggregates

Aggregate size [mm]	Type of aggregate						
	PCA	PCAP	PCL	PCG	LC65	LC55	RRG
	Loss after 10 cycles [% by mass]						
4/8	1.5	0.7	2.1	5.3	12.2	2.6	0.6
8/16	2.5	0.8	0.8	1.6	10.2	2.7	0.3
16/22	4.6	0.7	2.4	15.3	18.9	6.7	0.1
	Difference of loss after 10 cycles and after 20 cycles [% by mass]						
4/8	-	1.2	3.5	4.3	8.2	6.0	-
8/16	-	0.4	2.0	2.4	9.2	7.5	-
16/22	-	0.4	3.3	4.6	5.0	9.5	-

From Table 2 it follows that, with exception of aggregate PCAP, the recycling aggregates used in this investigation would not meet the requirement for aggregates with a high resistance to freezing and thawing as listed above. According to [7] also a less stringent requirement may be applied: the loss of deteriorated material after 10 cycles of freezing and thawing may be determined by sieving the material with a 0.71 mm sieve. If this criterion is applied all fractions of the recycling aggregates PCAP, as well as some fractions of the aggregates PCG and PCL may be considered sufficiently frost resistant.

The data in Table 2 as well as in Fig. 1 give no clear indication, that the rate of deterioration of the recycling aggregates decreases with the number of freeze-thaw-cycles.

For all concretes natural sand from the river Rhine has been used for the aggregate fraction 0/4 mm.

3.2 Cement and admixtures

For the recycling concrete a Portland cement CEM 32.5 R has been used. In addition an air entraining agent resulting in air contents of the fresh concrete of 4 to 5 percent has been added. In order to ensure a sufficient workability of fresh concrete a super-plasticizer has been used resulting in a plastic consistency corresponding to a degree of compaction after mixing in the range of 1.19 to 1.08.

4   Concrete Composition

Extensive preliminary tests had been performed in order to determine a suitable concrete composition which also satisfies the requirements generally put forward for pavement concrete. For preliminary experiments most of the recycling aggregates PCA had to be used.

For the actual tests an aggregate gradation in the range A32/B32 according to DIN 1045 adjusted to a maximum aggregate size of 22 mm has been chosen [8]. As an example the composition of the concretes made with the recycling aggregate PCAP is given in Table 3.

Table 3.   Composition of recycling concretes with aggregates PCAP

Compound	Mass [kg/m^3] for w/c of		
	0.52	0.45	0.38
Cement	340	350	360
Water	177	158	137
Sand 0/4 mm	604	618	634
Aggregate 4/8	173	177	181
Aggregate 8/16	345	353	362
Aggregate 16/22	604	618	634

The water added to the mix has been adjusted taking into account the amount of water absorbed by the recycling aggregates.

The amount of air entraining agent was about 0.03 % by mass of the cement for w/c = 0.52 and about 0.08 % for w/c = 0.38. Up to 3.1 % by mass of cement of superplasticizer had to be added to the concrete with w/c = 0.38 in order to obtain the required plastic consistency of the fresh concrete.

5   Production and curing of the concrete specimens

For the production of one test series consisting of specimens for determining the mechanical properties on cubes 200/200/200 mm^3 as well as on cylinders 150/300 mm and for freeze-thaw-testing on prisms 150/150/69 mm^3 two mixes with a volume of 0.375 m^3 were required. After filling the forms the specimens were compacted with internal vibrators.

Table 4.   Mechanical properties of the concretes, curing method A

Aggregate	w/c	f_{cu}	$f_{ct,fl}$	$f_{ct,sp}$	E_c
	-	[MPa]	[MPa]	[MPa]	[MPa]
PCAP	0.52	43	5.8	2.1	29400
	0.45	56	7.6	3.8	32400
	0.38	68	9.2	4.4	35100
PCL	0.52	51	6.5	3.5	30100
	0.45	54	8.2	3.9	31200
	0.38	66	8.6	4.4	33700
PCG	0.52	44	6.7	3.3	29400
	0.45	55	7.8	3.7	32700
	0.38	69	8.3	3.9	33800
LC65	0.52	50	5.4	3.3	27000
	0.45	60	6.6	3.7	28300
	0.38	72	7.6	4.2	33000
LC55	0.52	53	4.5	3.0	28100
	0.45	55	7.1	3.1	27200
	0.38	69	7.6	4.4	30200
RRG	0.52	37	6.2	3.1	28000
	0.45	45	4.6	3.4	31500
	0.38	58	7.8	3.8	34700

Curing method A: After casting the specimens were covered with plastic sheets and demolded after 1 day. Specimens for the determination of the compressive strength as well as of the modulus of elasticity were covered with wet burlap up to an age of 7 days and then stored at 20 °C/65 % rel. humidity until testing. Specimens for the determination of tensile splitting and flexural strength were cured in water up to the time of testing. The specimens for CDF-testing were cured in water at 20 °C and then stored up to an age of 28 days at 20 °C/65 % rel. humidity. Further preconditioning followed the rules of the CDF-method.

Curing method B: Soon after casting and after demolding, respectively, the free surfaces and all other surfaces of the specimens were treated with a curing compound and stored at 20 °C/65 % rel. humidity up to the time of testing. Prior to CDF-testing at an age of 26 days the surfaces subsequently exposed to the deicing salt solution were ground wet in order to remove the curing compound.

6   Results of tests on recycling concretes

6.1   Fresh concrete properties
The air content of the fresh concrete ranged from a minimum value of 3.8 percent to a maximum of about 6 percent. The degree of compaction, V, increased significantly with time indicating a loss of workability irrespective of the type of aggregate used. Typical values of the degree of compaction of concretes with a w/c of 0.45 and 10 min, 30 min and 60 min after casting are $V_{10} = 1.15$; $V_{30} = 1.25$ and $V_{60} = 1.30$.

Table 5.   Total porosity and air void characteristics of the hardened concrete, curing method A

Aggregate	w/c	V_{cp}	A_c	SF	L300
	-	[%]	[%]	[mm]	[%]
PCAP	0.52	22.3	4.8	0.19	2.9
	0.45	19.1	3.3	0.23	1.9
	0.38	22.4	3.5	0.24	1.9
PCL	0.52	20.1	4.2	0.20	2.6
	0.45	20.0	4.0	0.19	2.9
	0.38	20.4	3.0	0.28	1.8
PCG	0.52	22.2	4.2	0.19	2.7
	0.45	20.2	3.9	0.20	2.6
	0.38	16.0	2.5	0.20	1.8
LC65	0.52	22.3	3.5	0.16	2.7
	0.45	19.9	3.2	0.19	2.1
	0.38	22.0	3.6	0.19	1.8
LC55	0.52	22.0	3.7	0.20	2.3
	0.45	21.6	3.9	0.20	2.5
	0.38	21.4	3.5	0.20	2.3
RRG	0.52	17.7	4.7	0.15	3.6
	0.45	18.2	4.7	0.21	2.7
	0.38	15.4	2.9	0.24	1.9

6.2   Hardened concrete

6.2.1   Mechanical properties

In Table 4 the compressive strength, f_{cu}, flexural strength, $f_{ct,fl}$, tensile splitting strength, $f_{ct,sp}$, and modulus of elasticity, E_c, of the concretes at an age of 28 days, cured according to method A, are summarized. From these data it follows that the effect of type of aggregates on the mechanical properties of the concretes is comparatively small, though the compressive strengths of all of the recycling concretes were higher than those of the concretes made with a conventional river gravel.

6.2.2   Total porosity and air void characteristics

Table 5 gives the total porosity V_{cp} as well as the air void characteristics of the hardened concrete, air content A_c, spacing factor SF and content of micro-air voids < 0.3 mm L300.

Despite the fact, that the air content of the fresh concretes was – with one exception – between 4 and 5 percent, low values of A_c have been observed particularly for the concretes with low water-cement-ratios. Also the spacing factor SF of the concretes with w/c = 0.38 exceeded the desired value of 0.20 mm in 3 cases. Though the content of micro-air voids L300 did not fall below the desired value of 1.5 percent it decreased for most of the concretes with decreasing water-cement-ratios. This is due to a frequently observed unfavorable interaction between the air entraining agent and some superplasticizers, refer to e.g. [9].

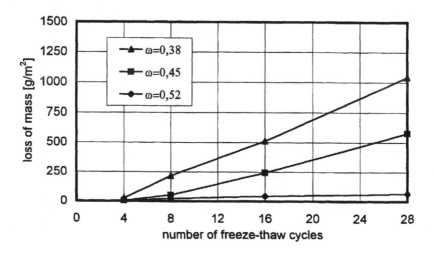

Fig. 2. Loss of mass of recycling concrete, aggregate PCG, during freeze-thaw testing

6.2.3   Resistance against freezing, thawing and deicing salt

The experiments have been carried out in accordance with the CDF-method as described e.g. in [6] except for the specimens cured with a curing compound as delt with in section 5. However, after removal of the curing compounds the CDF-proce-dure has been followed also in these cases. Fig. 2 shows that the loss of material dur-ing freezing and thawing increases linearly with the number of cycles. The unexpected increase of deterioration with decreasing water-cement-ratio is due to the correspond-ing decrease of the content of micro-air voids in Table 5.

The average values of the loss of mass after 28 cycles of freezing and thawing are summarized in Table 6. The coefficients of variation of these values ranged from 5 to 20 % and decreased with increasing loss of mass. From Table 6 it follows that for all types of aggregates the extent of deterioration increased with decreasing water-cement-ratio. This is due to differences in the air void characteristics as pointed out above. There is no clear trend with regard to the effects of the method of curing: for 8 of the 18 types of concrete investigated curing with a curing compound resulted in a lower deterioration than 7 days of water curing. The resistance against freezing, thawing and deicing salts of the recycling concretes was similar to that of the concretes with natural river gravel. According to [6] a concrete may be rated as having a high resistance against freezing, thawing and deicing salts if the loss of mass after 28 cycles of CDF-testing does not exceed a value of 1500 g/m^2. In no case has this value been reached.

Table 6.   Results of CDF-tests: Loss of mass after 28 freeze-thaw-cycles

Aggregate	w/c	Loss of mass	
		Curing A	Curing B
	-	[g/m²]	[g/m²]
PCAP	0.52	33	49
	0.45	198	121
	0.38	389	301
PCL	0.52	47	88
	0.45	111	96
	0.38	551	202
PCG	0.52	61	73
	0.45	579	192
	0.38	1041	759
LC65	0.52	323	631
	0.45	403	550
	0.38	322	808
LC55	0.52	99	282
	0.45	259	359
	0.38	344	524
RRG	0.52	81	154
	0.45	226	107
	0.38	927	297

7 Discussion

In Figs. 3 and 4 relations between the extent of deterioration and the air void characteristics A_c and L300 are presented. An increase of these values results in a decrease of deterioration.

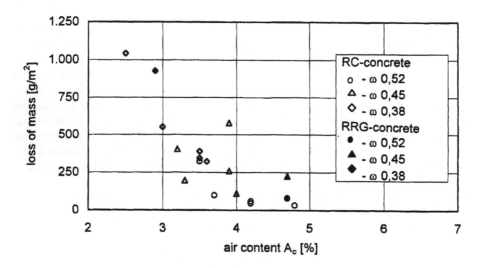

Fig. 3.   Effect of air content of the hardened concrete, A_c, on the scaling resistance of the recycling concretes, curing method A

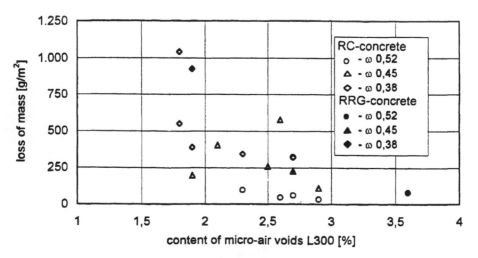

Fig. 4.   Effect of content of micro-air voids, L300, on the scaling resistance of the recycling concretes, curing method A

From Fig. 5 it follows that the deterioration of recycling concretes tends to decrease with increasing compressive strength of the concretes from which the recycling aggregates had been obtained.

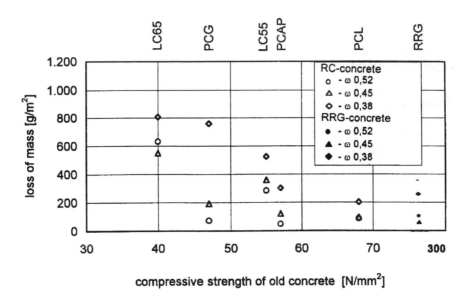

Fig. 5. Scaling resistance of the recycling concretes and compressive strength of the aggregate concrete, curing method B

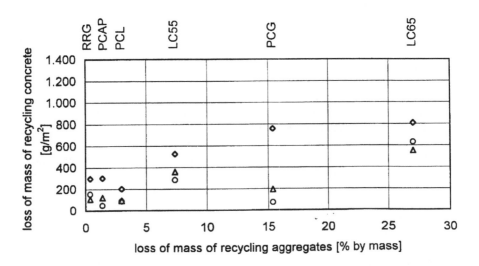

Fig. 6. Scaling resistance of the recycling concretes and frost resistance of the recycling aggregates, curing method B

Fig. 6 shows a correlation between the loss of mass of the recycling concretes and the loss of mass of the recycling aggregates during 10 cycles of freezing and thawing as given in Table 2.

From these data preliminary and rather stringent criteria for the acceptance of an old concrete as recycling material for concretes with a high resistance to freezing, thawing and deicing salts may be a compressive strength of the old concrete of at least 55 MPa and a loss of mass of the recycling aggregate after 10 cycles of freezing and thawing of less than 10 percent by mass.

8   Conclusions

The following conclusions can be drawn from this investigation:
- Recycling concretes may have a frost and deicing salt resistance similar to that of concrete made of conventional aggregates
- Similar to conventional concretes the dominant parameter for sufficient scaling resistance of a recycling concrete is its air void system
- The criteria to evaluate the frost resistance of conventional aggregates are too stringent to be applied to recycling aggregates
- Old concretes are suitable to produce recycling aggregates and concretes with a high scaling resistance even if they do not contain entrained air. However, such concretes should have a compressive strength of at least 55 MPa. This requirement is met by most older pavement concretes.

9   References

1.   Hansen, T.C. (1986) Recycled Aggregates and Recycled Aggregate Concrete, Second State-of-the-Art Report, Developments 1945-1985. Materials and Structures, Vol. 19, No. 111, pp. 201-246
2.   Rahlwes, K. (1991) Recycling von Stahlbeton- und Stahlbetonverbundkonstruktionen – Ansätze zu einer umweltökologischen Bewertung, Deutscher Betontag 1990
3.   Springenschmid, R. and Fleischer, W. (1993) Zur Technologie der Wiederverwendung von altem Straßenbeton, Straße und Autobahn, Vol. 12
4.   DIN 4226 (1983) Zuschlag für Beton, Beuth Verlag GmbH, Berlin
5.   DIN 52104, Part 1 (1982) Prüfung von Naturstein; Frost-Tau-Wechsel-Versuch; Verfahren A bis Q, Beuth Verlag GmbH, Berlin
6.   Setzer, M.J. and Auberg, R. (1995) Freeze-thaw and deicing salt resistance of concrete, testing by the CDF method; CDF resistance limit and evaluation of precision, Materials and Structures, Vol. 28, pp. 16-31
7.   TL Min-StB 94, Technische Lieferbedingung für Mineralstoffe im Straßenbau, Ausgabe 1994, Forschungsgesellschaft für Straßen- und Verkehrswesen, Köln
8.   DIN 1045 (1988) Beton und Stahlbeton; Bemessung und Ausführung, Beuth Verlag GmbH, Berlin
9.   Siebel, E. (1986) Beeinflussung der Luftporenkennwerte bei leicht verarbeitbarem Beton mit Fließmittel für Verkehrsflächen. Ergänzende Beiträge zum 5. Internationalen Betonstraßen-Symposium. Aachen 1986, pp. 231-236

Freeze-thaw resistance of concrete with recycled aggregates

R. DILLMANN
IBPM – Institute of Building Physics and Materials Science,
University of Essen, Essen, Germany

Abstract:
Research was made to find limitations and criteria for the use of recycled coarse and fine aggregates made of hardened concrete with different compressive strength, especially for the freeze-thaw resistance of the new concrete.
Keywords: concrete, recycled aggregates, frost resistance, CF-test, durability

1 Introduction

In Germany there are about 70 million tons[1] of construction material per year, that could be recycled. Due to legal regulations this construction material has to be recycled in a regular way.

Today the usual way of recycling is the „Down-Cycling", that means the re-use of construction material for minor purposes.

It must be the goal to recycle construction material in a way that it could be re-used for the same construction that it was used before.

Especially the problem of the re-use of recycled concrete as aggregates for new concrete must be solved.

There are some investigations 2,3,4,5,6,7,8,9,10 dealing with parts of this problem.

This article is based on the research project „Influence of Recycled Aggregates extracted from Concrete with different Strengths on the Properties of Concrete

The goal of the above mentioned research project was to find limitations and criteria for the use of crushed aggregates and especially crushed sand extracted from concrete.

2 Research Program

For the research recycled aggregates (sand 0/2 mm, coarse aggregates 2/8 mm, 8/16 mm and 16/32 mm) extracted from two different classes of concrete

Frost Resistance of Concrete, edited by M.J. Setzer and R. Auberg. Published in 1997 by E & FN Spon, 2–6 Boundary Row, London SE1 8HN, UK. ISBN: 0 419 22900 0.

strength (β_D < 20 N/mm² and β_D > 50 N/mm²) and a so called „Praxisgemisch" were delivered to the concrete laboratory of the University of Essen.

The „Praxisgemisch" was taken from a recycling plant as it was found there; it consists - besides crushed concrete - of asphalt particles and crushed bricks. i

The crushed concrete 8/16 mm from concrete with a strength > 50 N/mm² had admixtures of about 22 % by weight of asphalt particles.

Concrete of strength classes B 25 and B 45 was made according table 1.

Table 1: Variations of Concrete

Aggregates	Concrete to compare with	β_D < 20 N/mm²	β_D > 50 N/mm²	Praxis-gemisch
100 % natural sand	B 25/B 45	B 25/B 45	B 25/B45	B 25/B45
50 % natural sand 50 % recycled sand	–	B 25/B45	B 25/B45	–
100 % recycled sand	-	B 25	B 25	-

The combined grading curve of the aggregates was up to 2 mm below and between 4 to 32 mm above the grading curve B according DIN 1045.

The concrete composition was as follows:

	Concrete B 25	Concrete B 45
CEM I 32,5 R	280 kg/m³	-
CEM I 42,5 R	-	320 kg/m³
Fly ash	70 kg/m³	50 kg/m³
water/cement-ratio	0,60	0,50
aggregates and plastiziser or super-plastiziser		

For the characterisation of the aggregates the following tests were made:
- composition of delivered aggregates
- bulk density
- density
- water absorption
- percentage of favourable formed aggregates
- organic matter
- resistance against stroke
- resistance against freeze-thaw
- aggregate-crushing-value
- ten-percent-fines-value
- chloride- and sulphate content

The properties of fresh and hardened concrete were tested, the freeze thaw resistance was tested according the CF-Test[11]

---

[i] This project was sponsored by „Forschungsgemeinschaft Transportbeton e.V.; Moers

3 Results

3.1 Aggregates
The results of the aggregates test are shown in the annex.

3.2 Fresh Concrete
The consistency of the fresh concrete with the recycled coarse and fine aggregates was poor, but with plastiziser and super-plastiziser the workability was good.

3.3 Hardened Concrete
The test results of the hardened concrete are shown in table 2 and 3.

Table 2: Concrete class B 25: Variations of test results compared with concrete with

		Compressive strength	Tensile-splitting strength	E-Modulus	Shrinkage	Carbonation
natural aggregates		100	100	100	100	100
	100% natural sand	79	131	77	100	239
β_D<20N/mm²	50% natural sand	74	122	68	100	239
	0% natural sand	62	109	65	140	409
	100% natural sand	74	96	81	94	144
β_D>50 N/mm²	50 % natural sand	65	96	75	97	185
	0% natural sand	56	69	48	135	295
Praxis-gemisch	100% natural sand	82	138	84	125	350

natural aggregates in %.

Table 3: Concrete class B 45: Variations of test results compared with concrete with natural aggregates in %.

		Compressive strength	Tensile-splitting strength	E-Modulus	Shrinkage	Carbonation
natural aggregates		100	100	100	100	100
β_D<20N/mm²	100% natural sand	80	118	71	93	202
	50% natural sand	72	118	65	100	270
β_D>50 N/mm²	100% natural sand	72	128	66	120	122
	50 % natural sand	64	118	64	120	178
Praxis-gemisch	100% natural sand	80	136	76	124	205

The change of the dynamic modulus of elasticity is shown in the annex.

The decrease of the dynamic modulus of elasticity is below 20% except for the two types of concrete where 100 % recycling sand was used. These two types of concrete have failed the freeze-thaw resistance.

The surface scaling after 56 freeze thaw cycles is shown in table 4:

Table 4: Surface scaling after 56 freeze-thaw cycles

		natural sand	surface scaling
		% by weight	g/m²
B 25	natural coarse aggregates	100	82
	β_D<20 N/mm²	100	392
		50	545
		0	1265
	β_D>50 N/mm²	100	255
		50	1645
		0	6200
	Praxisgemisch	100	956
B 45	natural coarse aggregates	100	32
	β_D<20 N/mm²	100	160
		50	214
	β_D>50 N/mm²	100	318
		50	452
	Praxisgemisch	100	1165

Due to the scaling the two concrete types B 25 with crushed aggregates made of concrete with a strength > 50 N/mm² and an amount of crushed sand of 50 and 100 % by weight have failed, if the limit of 1500 g/m² is accepted.

Two aspects are to be considered:

Natural aggregates have a compressive strength of more than 100 N/mm², the compressive strength of the hardened cement paste determines the compressive strength of concrete.

During crushing primarily crushed hardened cement paste with low strength could be found in crushed sand and crushed hardened cement paste with high strength could be found in a higher percentage in coarse aggregates.

The amount of crushed old hardened cement paste in new concrete could be rough estimated:

Table 5: Amount of crushed sand made of concrete in the different size of recycled aggregates and in the new concrete.

	Aggregate	hardened cement paste in crushed aggregates	
		equal distribution	unequal distribution
	% by volume	% by volume	% by volume
0/2 mm	30	9,0	12,0
2/8 mm	15	4,5	4,0
8/16 mm	20	6,0	3,0
16/32 mm	35	10,5	2,0
Total	100	30	21

That means - beside the amount of new hardened cement paste of about 30 % by volume - about 21 % by volume of old hardened cement paste could be found in the new concrete, if crushed sand is used.

This seems to be an amount of porous hardened cement paste that is high enough for not resisting the freeze-thaw cycles.

The results show, that using an amount of 50 % by weight of crushed sand - additionally about 9 % by volume of old hardened cement paste is in the new concrete - the amount of porous hardened old and new cement paste is low enough that the resistance against freeze-thaw cycles is high.

4 Literature

1. Kohler: Recyclingpraxis Baustoffe, Verlag TÜV Rheinland, Köln 1994
2. Forschungsbericht T 86-140 „Aufbereitung von Ziegel- und Betonabbruch durch Zerkleinerung und Klassifizierung in einer Bauschuttaufbereitungsanlage - Nasse Trennung des Holzes aus dem Bauschutt - von Kuhne, Nendza, Ivanyi, Dohmann, Offermann, Heckötter, Lardi, Tränkler, Fachbereich Bauwesen der Universität-Gesamthochschule Essen; gefördert vom Bundesministerium für Forschung und Technologie
3. Müller/Schießl/Kwiasowski: Recyclingzuschlag für Beton - Anwendung von Recyclingmaterialien nach DIN 1045; Beton Heft 8/1996, Seite 473 bis 478
4. Dahms/Brune: Wasseraufnahme und Rohdichte von Betonbruch; Beton Heft 8/1996, Seite 480 bis 486
5. Wesche/Schulz: Beton aus aufbereitetem Altbeton - Technologie und Eigenschaften; Beton Heft 2/1992, Seite64 bis 68 und Heft 3/1992, Seite 108 bis112
6. Wöhnl: Recyclingbeton für Bauteile im Hochbau; Beton Heft 9/1994, Seite 499 bis 503
7. Franke: Recycling von Betondecken im Autobahnbau; Beton Heft 9/1994, Seite 504 bis 509
8. Breitenbücher: Recycling von Frisch- und Festbeton; Beton Heft 9/1994, Seite 510 bis 514
9. RILEM TC 121: Specification for Concrete with Recycled Aggregates; September 1993
10. Springenschmid/Fleischer: Zur Technologie der Wiederverwendung von altem Straßenbeton; Straße und Autobahn Heft 12/1993, Seite 715 bis 718
11. RILEM TC 117 FDC (1995) Test Method for the Freeze-Thaw Resistance of Concrete - Tests with Water (CF) or Sodium Chloride Solution (CDF), Materials & Structures April 1995

Property	Unit	Natural Aggregates				$\beta_D < 20$ N/mm²				$\beta_D > 50$ N/mm²				Praxisgemisch		
		2/8	8/16	16/32	0/2	2/8	8/16	16/32	0/2	2/8	8/16	16/32	0/2	2/8	8/16	16/32
Bulk density	kg/dm³	2,64	2,64	2,64	2,239	2,264	2,234	2,196	2,312	2,382	2,332	2,204	-	2,301	2,297	2,204
Density	kg/dm³	-	-	-	2,507	2,,560	2,554	2,518	2,519	2,522	2,603	2,539	2,531	2,528	2,,623	2,625
Water Absorption	% by weight	-	-	-	6,9	4,7	4,4	4,4	12,7	5,1	3,6	2,5	10,8	4,5	3,7	3,9
Percentage of Favourable Formed Aggregates	% by weight	-	-	-	-	91,6	86,8	92,1	-	90,2	89,9	90,6	-	88,0	94,0	94,7
Resistance against Stroke[1] S_z	-	-	26,5	-	-	-	25,4	-	-	-	23,4	-	-	-	24,3	-
Resistance against Freez-Thaw	% by weight	-	-	-	-	20,1	8,7	6,8	-	14,3	17,1	0,4	-	12,7	10,1	16,8
Aggregate-Crushing-Value[2]	% by weight	-	19,7	-	-	-	27,6	-	-	-	24,7	-	-	-	26,4	-
Ten-Percent-Fines-Value[2]	kN	-	249	-	-	-	108	-	-	-	136	-	-	-	107	-

[1] Tested on aggregates with an size of 8/12,5 mm
[2] Tested on aggregates with a size of 10/14 mm

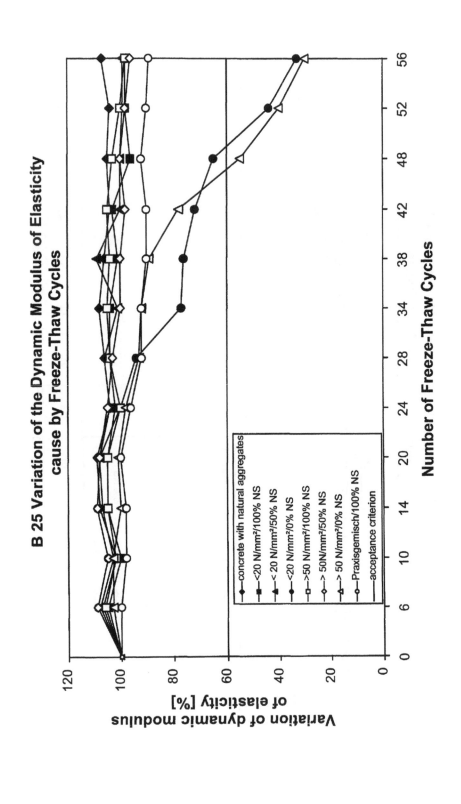

B 25 Variation of the Dynamic Modulus of Elasticity cause by Freeze-Thaw Cycles

Number of Freeze-Thaw Cycles

Variation of dynamic modulus of elasticity [%]

Legend:
- concrete with natural aggregates
- <20 N/mm²/100% NS
- < 20 N/mm²/50% NS
- <20 N/mm²/0% NS
- >50 N/mm²/100% NS
- > 50N/mm²/50% NS
- > 50 N/mm²/0% NS
- Praxisgemisch/100% NS
- acceptance criterion

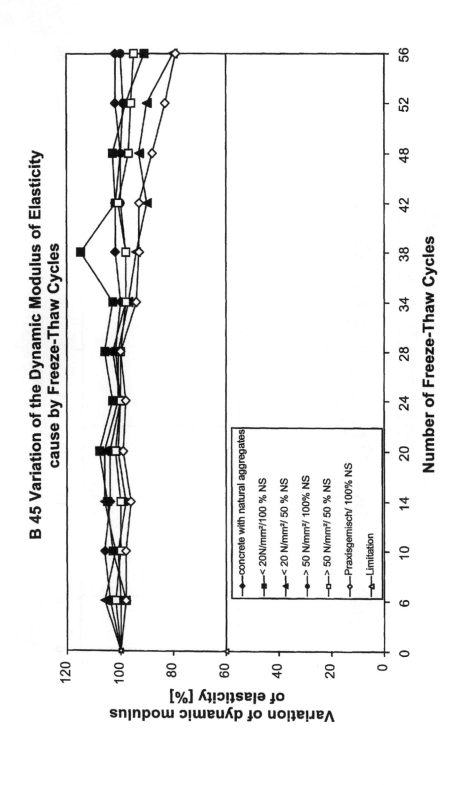

B 45 Variation of the Dynamic Modulus of Elasticity cause by Freeze-Thaw Cycles

Variation of dynamic modulus of elasticity [%]

Number of Freeze-Thaw Cycles

concrete with natural aggregates
< 20N/mm²/100 % NS
< 20 N/mm²/ 50 % NS
> 50 N/mm²/ 100% NS
> 50 N/mm²/ 50 % NS
Praxisgemisch/ 100% NS
Limitation

PART II

Chemical Parameters

Frost resistance with and without deicing salt — a purely physical problem?

J. STARK
Bauhaus-Universität, Weimar, Germany

Abstract
Depending on the type of cement, phase transformations in hardened cement paste during frost or frost and deicing salt attack influence the resistance to freezing and thawing or to frost and deicing salt attack to a different degree and by different mechanisms.

Due to the formation of metastable carbonation products concretes made with blast furnace cement rich in granulated slag (slag content > 50%), and even more so those made with super-sulphated cement (SSC) - with or without air-entraining agents -, exhibit a low resistance to frost and deicing salt attack. A densification of the surface layer may significantly improve the resistance of these concretes to frost and deicing salts.

The resistance of portland cement concretes to frost and deicing salts does not depend on the C_3A content of the cements, but on the amount of new ettringite (AFt) which is formed during frost or frost and deicing salt attack. For portland cement concretes made with air-entraining agents this influence of the type of cement is practically of minor importance. But in terms of high-strength concretes without air-entrainment this chemical aspect of the frost-deicing salt resistance should be taken into consideration.

Keywords: aragonite, blast furnace cement, C_3A content, carbonation, CDF procedure, ettringite, frost-deicing salt resistance, frost resistance, high-strength concrete, initial scaling, monosulphate, phase transformation, portland cement, supersulphated cement, vaterite

Frost Resistance of Concrete, edited by M.J. Setzer and R. Auberg. Published in 1997 by E & FN Spon, 2–6 Boundary Row, London SE1 8HN, UK. ISBN: 0 419 22900 0.

1   Introduction

Generally, frost resistance and resistance to frost and deicing salt attack is understood as a purely physical problem of freezing water in the capillary pore system of the hardened cement paste. With the CDF procedure according to Setzer (Universität Essen) [1] [2] [3] a new procedure for testing the frost-deicing salt resistance of concrete has been available for several years. Due to its high precision this method allows to determine even relatively small differences in the behaviour of different concretes that prove to be significant. Frost resistance and frost-deicing salt resistance of concrete made with cement rich in granulated blast furnace slag as well as with various portland cements was examined.

2   Frost resistance of concrete made with blast furnace cement

After a 28-day standard storage of the cement paste specimens (7 days in water and 21 days in the air at 20°C and 65% relative humidity) the degree of hydration of the portland cements (OPC) studied was between 85 and 90%. Depending on the quality of the granulated blast furnace slag and the fineness of cement, degrees of hydration from 30 to 65% were established in the blast furnace cements (BFC) after the same storage time. The resistance of the concretes to freezing and thawing (+20/-20°C) was determined by means of measuring the dynamic modulus of elasticity the decrease of which implies an internal damage of the concrete.

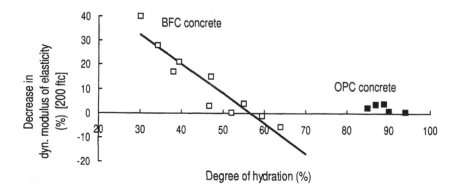

Fig. 1. Dependence of frost resistance on degree of hydration of cements

As may be seen in Fig. 1, there exists a linear relationship between the degree of hydration and the frost resistance of blast furnace cement concrete. The reason for that is the amount of capillary pores which decrease with an increasing degree of hydration. As a result the amount of freezable water in the concrete is reduced. Taking this into consideration, blast furnace cement concrete with a degree of hydration of 50 to 60% will lead to the same good results in frost resistance as a portland cement concrete with a degree of hydration of about 90%. Consequently, blast furnace cement concretes should not be subjected to frost attack too early.

3 Frost-deicing salt resistance of concrete made with blast furnace cement

Compared to the pure frost test (in water) the test for frost-deicing salt resistance according to the CDF procedure in 3% NaCl solution leads to completely different results. As may be seen in Fig. 2, the amount of scaled material after 28 cycles of freezing and thawing is not at all dependent on the degree of hydration of the blast furnace cements.

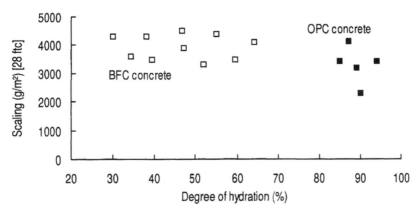

Fig. 2. Dependence of frost-deicing salt resistance on degree of hydration of cements

In this case it seems that the degree of hydration and, consequently, the pore size distribution of the concrete are of minor importance. As further investigations have shown a scaling of a thin carbonated surface layer occurs at the beginning of the frost-deicing salt test. The depth of scaling and the depth of the carbonated layer correlate very well with each other (Fig. 3).

After a 7-day storage of three series of concrete samples under water, the further storage of one series was performed under the conditions required for the CDF procedure (3 weeks in air at 20°C and 65% relative humidity), another series was stored without CO_2 under nitrogen (N_2), and the third one in a 3% CO_2 atmosphere. Those three series of concrete samples were subjected to frost and deicing salt (CDF test). It was proved that the frost-deicing salt resistance of concretes made with cement rich in granulated blast furnace slag (> 50%) is indeed dependent on the state of the carbonated surface layer. The results show (Fig. 4) that storage in CO_2 leads to an extremely high amount of scaling whereas scaling is significantly smaller after storage in nitrogen as compared to air storage.

In all cases the concretes showed a sufficiently high and almost identical air void content (L300 > 1.5%). The spacing factor lies clearly below the limit of 0.20 mm.

The reason for the heavy scaling of concretes made with blast furnace cement is the carbonation of a thin surface layer as well as the formation of the metastable $CaCO_3$ modifications vaterite and aragonite. An explanation for the formation of these $CaCO_3$ modifications, which are extremely unstable in NaCl solution, may be the impediment of calcite crystallisation by Mg^{2+} ions in the pore solution. In this context, the high MgO content of granulated blast furnace slags seems to be decisive. [3-10]

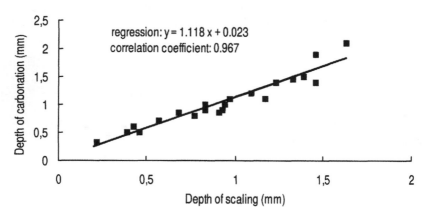

Fig. 3. Relation between depth of scaling and depth of carbonation of various blast furnace cement concretes with and without air-entraining agents

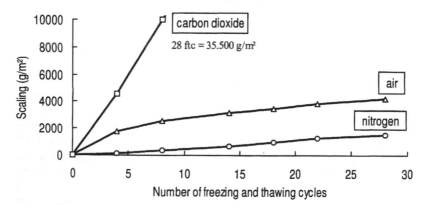

Fig. 4. Scaling curves of a blast furnace cement concrete (65% slag content) without air-entrainment stored under different conditions

Other deicing chemicals, e.g. urea solution or ethylene glycol solution, act in a similar way as NaCl solution.

4 Frost and frost-deicing salt resistance of supersulphated cement concrete

Furthermore, another material rich in granulated slag, namely supersulphated cement (SSC) was included in the investigations. The use of supersulphated cement is no longer permitted in Germany, but as it was required for remedial work it was produced in the laboratory. The supersulphated cement made at our laboratory consists of 85% granulated blast furnace slag with a relatively high alumina content ($\geq$ 15% Al_2O_3), 14% anhydrite, and 1% portland cement. The scaling under pure frost attack as well as under frost-deicing salt attack is extremely high (Fig. 5). [11] [12]

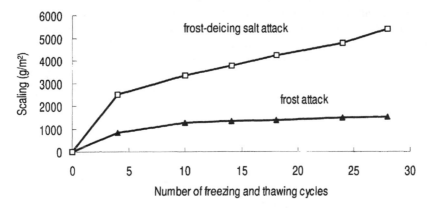

Fig. 5. Scaling curve of supersulphated cement concrete under freeze-thaw attack (CF test) and under frost-deicing salt attack (CDF test)

Examination of the porosity showed that the microstructure of supersulphated cement concrete was relatively coarse at the time of testing. This may be due to the low degree of hydration at the time of testing. Besides, carbonation certainly plays an important role. In supersulphated cement concrete ettringite carbonates earlier and to a larger degree than the C-S-H phases. This reaction results in a coarser microstructure. Products of carbonation are the metastable carbonate modifications aragonite and vaterite. As with blast furnace cement concrete, the initial scaling of supersulphated cement is also extremely heavy, which is due to the instability of the carbonated surface layer of the concrete.

5 Ways of reducing the initial scaling of concretes containing cement rich in granulated slag (blast furnace cement and supersulphated cement)

As is already known from numerous investigations, the frost-deicing salt resistance of blast furnace cement concrete rich in granulated slag cannot satisfactorily be improved by adding air-entraining agents. This holds true even if excellent air void parameters are achieved in the hardened concrete. The addition of air-entraining agents to supersulphated cement concrete will even lead to a further increase in the amount of scaling. Therefore an addition of air-entraining agents to supersulphated cement concrete should be avoided in any case.

There are two ways of improving the frost and frost-deicing salt resistance. On the one hand, it would be desirable to reduce carbonation by technical means. On the other hand, it should be investigated whether the formation of the metastable calcium carbonate modifications may be minimised or even prevented by alteration of the materials used.

As a technical means the use of draining formwork material proved successful. Both in blast furnace cement concrete and in supersulphated cement concrete the microstructure in the surface layer became definitely denser. The depth of carbonation of both these concretes could be reduced extremely. The effects on frost and frost-deicing salt resistance were studied in CF/CDF tests. The frost resistance of supersulphated cement concrete was considerably increased and is now comparable to that of a high-quality concrete. The results for the frost-deicing salt resistance of the two concretes are equally good (Fig. 6). Heavy initial scaling is prevented by using draining formwork material and the amount of scaling is significantly reduced. The total amount of scaling of both concretes is clearly below the acceptance criterion according to CDF.

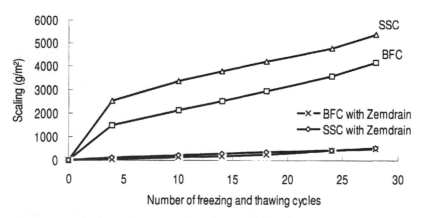

Fig. 6. Influence of draining formwork material on the courses of scaling of blast furnace cement concrete and of supersulphated cement concrete under frost-deicing salt attack (CDF test)

The use of curing agents as another way of improving the frost and frost-deicing salt resistance is being investigated and seems to be promising.

6   Frost-deicing salt resistance of concrete made with portland cement

In various publications the question has repeatedly been discussed whether the C_3A content of cement has an effect on the frost or frost-deicing salt resistance of concrete. It was studied if and in which way the hydration products of C_3A affect the frost-deicing salt resistance of concrete. Fig. 7 shows that for concretes without artificial air voids (i.e. without air-entraining agents) there is no statistically proven correlation between the amount of scaling during the CDF test and the C_3A content. [13-17]

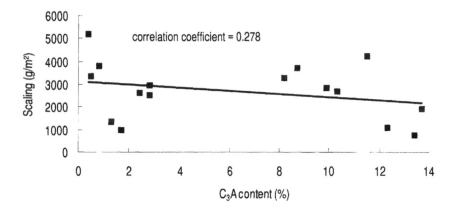

Fig. 7. Correlation between scaling of concretes without air-entrainment during the CDF test and C_3A content of the corresponding cements

Despite that, cements, which show practically no difference in the properties of the hardened cement paste (Table 1), sometimes show great differences in frost-deicing salt resistance. The following influencing factors were tested for correlation to frost or frost-deicing salt resistance of the concrete:

- compressive strength
- degree of hydration
- portion of gel porosity (d < 10 nm)
- capillary porosity (d = 10 nm - 100 µm)
- air void content (d > 100 µm)
- total porosity

Table 1. Correlation coefficients of the relation between frost resistance / frost-deicing salt resistance (FR/FDSR) and various influencing factors

| | Influencing factor | | | | | |
	Compressive strength	Degree of hydration	Portion of gel porosity	Capillary porosity	Air void content	Total porosity
FR	0.053	0.280	0.206	0.210	0.204	0.187
FDSR	0.002	0.156	0.348	0.198	0.008	0.109

The correlation coefficients show that for none of the potential influencing factors a statistically proven correlation exists. Furthermore, the pore size distribution of hardened cement pastes, which show completely different amounts of scaling in the concrete test, is almost identical.

Therefore it is to be questioned whether phase transformations occur in hardened cement paste during frost and deicing salt attack, and if they affect the durability of concrete. X-ray diffraction analyses of monosulphate / ettringite and AFm / AFt phase made from C_3A and C_4AF, respectively, indicate that both during pure frost attack and, particularly, during frost and deicing salt attack monosulphate ($C_3A \cdot Cs \cdot 12H$) and AFm phase (iron-containing monosulphate) are completely or partly transformed into ettringite and AFt phase, respectively (Fig. 8 and 9).

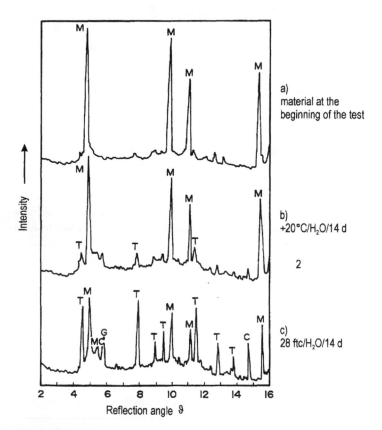

Fig. 8. X-ray diffraction analyses. Monosulphate made from C_3A (M - monosulphate, T - ettringite, G - gypsum, C - calcite, MC - monocarbonate)

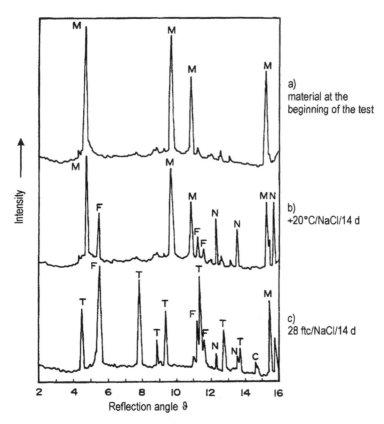

Fig. 9. X-ray diffraction analyses. Monosulphate made from C₃A
(M - monosulphate, F - monochloride, T - ettringite, N - NaCl)

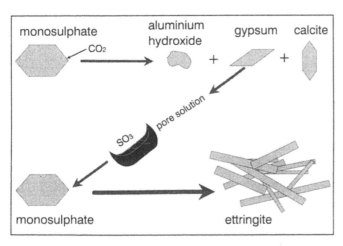

Fig. 10. Mechanism of ettringite formation under frost attack

In contrast to monosulphate and/or AFm phase ettringite and/or AFt phase remain stable under frost attack in water and in NaCl solution. There are various reasons for this depending on the type of attack (pure frost attack or frost and deicing salt attack).

In the case of pure frost attack, monosulphate and/or AFm phase partly decompose as a result of carbonation, so that mobile SO_3 is set free which may react with not yet carbonated monosulphate to form ettringite and/or AFt phase.

I. partial reaction

$2C_3A \cdot Cs \cdot 12H + 6CO_2 + 2H_2O \rightarrow 6CaCO_3 + 2CsH_2 + 4Al(OH)_3 + 16H_2O$ $s = SO_3$

II. partial reaction

$C_3A \cdot Cs \cdot 12H + 2CsH_2 + 16H_2O \rightarrow C_3A \cdot 3Cs \cdot 32H$

Resulting reaction

$2C_3A \cdot Cs \cdot 12H + 6CO_2 + 2H_2O \rightarrow C_3A \cdot 3Cs \cdot 32H + 6CaCO_3 + 4Al(OH)_3$

The bipartite course of reaction is shown in Fig. 10.

Carbonation is of minor importance in the case of frost and deicing salt attack, in which the decomposition of monosulphate and/or AFm phase as a result of the formation of monochloride (Friedel's salt) dominates.

I. partial reaction

$2C_3A \cdot Cs \cdot 12H + 4NaCl + 2Ca(OH)_2 \rightarrow 2C_3A \cdot CaCl_2 \cdot 10H + 2CsH_2 + 4NaOH$

II. partial reaction

$C_3A \cdot Cs \cdot 12H + 2CsH_2 + 16H_2O \rightarrow C_3A \cdot 3Cs \cdot 32H$

Resulting reaction

$2C_3A \cdot Cs \cdot 12H + 4NaCl + 2Ca(OH)_2 + 16H_2O \rightarrow C_3A \cdot 3Cs \cdot 32H + 4NaOH + 2C_3A \cdot CaCl_2 \cdot 10H$

This phase transformation occurring under frost attack in NaCl solution is illustrated in Fig. 11.

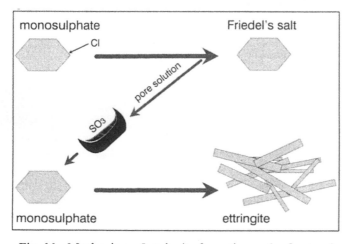

Fig. 11. Mechanism of ettringite formation under frost and deicing salt attack

Obviously, this phase transformation has an influence on the frost and frost-deicing salt resistance of concrete. If the amount of new ettringite formed under frost attack is correlated to the amount of scaling (Fig. 12), a statistically proven correlation becomes apparent in the case of frost and deicing salt attack. The amount of newly formed ettringite was determined by means of a specially developed procedure in which the calcium silicates of the cement are dissolved by salicylic acid extraction while the residue - consisting of C_3A, C_4AF, sulphate, and alkalis - is hydrated and finally subjected to frost and deicing salt attack.

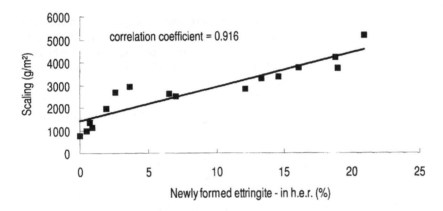

Fig. 12. Correlation between scaling of concrete without artificial air-entrainment and newly formed ettringite in the hydrated extraction residue (h.e.r.) during frost and deicing salt attack (CDF test procedure)

A correlation clearly exists between the amount of monosulphate and/or AFm phase before the frost-deicing salt attack and the amount of newly formed ettringite. As the differences between portland cements with a low (HS) and a high C_3A content (R) show, the amount of monosulphate and/or AFm phase that will be transformed also depends on the iron content of the AFm phase.

Finally, it can be stated that both frost resistance and frost-deicing salt resistance of portland cement concretes are influenced by phase transformations during frost attack. In contrast to the processes observed for blast furnace cement concrete, in the case of portland cement concretes with artificial air voids (air-entrained concretes) these phase transformations are of minor importance. But for high-strength concretes that are produced without adding air-entraining agents, these correlations are of major importance.

7   Frost-deicing salt resistance of high-strength concrete

The term "high-strength concrete" implies that the characteristic feature of these concretes is their particularly high strength. Therefore, requirements on the strength of the concretes B65 and B115 are specified by regulations of DAfStB. However, another favourable property of high-strength concretes, i.e. their particularly high performance, seems to gain increasing importance.

A very important aspect of the performance of concrete is its resistance to frost and deicing salt attack. Whereas normal concrete will achieve the necessary frost-deicing salt resistance by the adding of air-entraining agents and the formation of micro air voids in the hardened concrete, the use of air-entraining agents is impossible in high-strength concrete because the micro air voids would lead to a loss in strength. As was shown by our studies of a white cement, a high-strength concrete with a low water-cement ratio may have a very high frost-deicing salt resistance even without artificial air voids.

The question was whether the frost-deicing salt resistance of high-strength concrete without air-entrainment may be influenced by the composition of the portland cements used. Therefore high-strength concretes were made of two portland cements which in preceding tests had shown a different behaviour of their AFm and/or monosulphate phases under frost and deicing salt attack.

For this analysis concretes were produced which contained the two cements mentioned in Table 2. The 28-day strengths $\beta_D \geq 55$ N/mm^2 , the water-cement ratio w/c varied from 0.50 to 0.30, and superplasticizers and silica fume were used. [18] [19]

Concretes of the same mix proportions but with various types of cement were produced. The cements used showed very different amounts of AFm phase after hydration (see Table 2). Clearly different amounts of new ettringite formed during frost and deicing salt attack were observed as well. In a concrete without air-entrainment this should affect the frost-deicing salt resistance significantly.

It was the aim of these investigations [18] [19] to determine a limiting value of the water-cement ratio which ensures that a high-strength concrete without air-entrainment will clearly fall below the CDF acceptance criterion of $\Delta m = 1,500$ g/m^2 and to find out whether the cement used has a significant influence on these concretes.

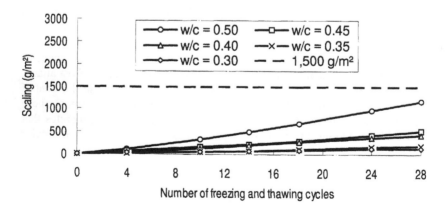

Fig. 13. Frost-deicing salt resistance (CDF test) of concretes made with portland cement OPC No. 1

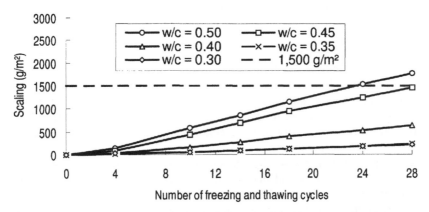

Fig. 14. Frost-deicing salt resistance (CDF test) of concretes made with portland cement OPC No. 2

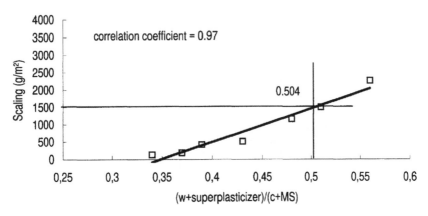

Fig. 15. Amount of scaling in CDF test depending on w/c ratio (OPC No. 1)

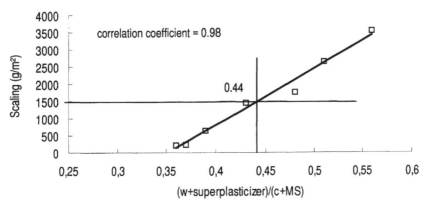

Fig. 16. Amount of scaling in CDF test depending on w/c ratio (OPC No. 2)

As could be expected, the compressive strengths show that with water-cement ratios of 0.50 and 0.45 no high-strength concrete can be produced which corresponds to the regulations of DAfStB.

The pore size distribution of the mortar matrix of these concretes are almost identical. Consequently, differences in the CDF test cannot be explained by varying pore size distributions. The CDF test of the concretes (Fig. 13, 14, 15, and 16) showed, firstly, that high-strength concrete with a water-cement ratio ≤ 0.40 has a high or very high frost-deicing salt resistance. This result was to be expected. The reason is the extremely high density of these concretes which impedes the capillary absorption of the 3% NaCl solution.

With a water-cement ratio ≤ 0.42 the average amount of scaling Δm = 600-800 g/m² after 28 cycles of freezing and thawing remains clearly under the acceptance criterion of 1,500 g/m². Secondly, the test results show a significant influence of cement quality on high-strength concrete without air-entrainment, as already stated in [6]. For instance, as can clearly be seen from a significance test, the amounts of scaling after 28 cycles of freezing and thawing are as follows (Table 2).

Table 2. Amount of AFm after hydration and of AFt after frost-deicing salt attack of the two types of cements used in this study, and scaling of the corresponding concretes (depending on water-cement ratio)

	Amount of monosulphate/ AFm before the CDF test (%)	Amount of new ettringite/AFt formed during CDF test (%)	Scaling (g/m²)	
			w/c = 0.50	w/c = 0.35
OPC No. 1	4.3	2.8	1,162 ± 33	190 ± 50
OPC No. 2	61.0	18.3	1,675 ± 266	233 ± 26

It needs to be emphasised that the frost-deicing salt resistance is always a function of the content of air-entraining agent and - from a chemical viewpoint - of the amount of new AFt formed during the frost and deicing salt attack.

Considering the relatively small amount of high-strength concrete compared to the total amount of concrete produced and taking into account that only a small proportion of it has to have a high frost-deicing salt resistance , it would be advisable to chose the suitable cement for the purpose by means of the suitability tests which - as a rule - are carried through in any case.

The effect of phase transformations in hydrated cement under frost and deicing salt attack in the case of concrete without air-entraining agents should be taken into consideration when producing high-strength concrete.

8 Unsolved questions

After determining the fundamental mechanisms of the frost-deicing salt resistance of portland cement and blast furnace cement concretes the following questions require further clarification:

a) Blast furnace cement concrete

- What role does the MgO content of the granulated blast furnace slag play with regard to the formation of the metastable $CaCO_3$ modifications aragonite and vaterite?
- With regard to the densification of the surface zone, do new curing agents take effect to the same extent as a draining formwork material?
- How does scaling really work: by dissolution and recrystallisation of aragonite and vaterite or due to change in porosity?

b) Portland cement concrete

- What does the amount of monosulphate formed during hydration of different cements depend on? Which factors have an influence on the amount of new AFt formed during frost and deicing salt attack?
- How do the very high amounts of superplasticizers, often employed when high-strength concretes are to be produced, act in this regard?
- What are the practical consequences when air voids become overgrown by ettringite crystals, as often observed in older concretes? (Fig. 17)

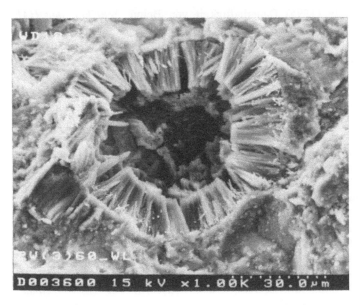

Fig. 17. Ettringite in an air void - orientated growth of needle-shaped crystals from the wall into the pore [20]

- Are these air voids already filled up with water before the growing of ettringite crystals starts and therefore already ineffective?

- The influence of admixtures added to the concrete mixture or to the cement during grinding - as e.g. fly ash, stone dust, and silica fume - has not yet been sufficiently clarified.

9   References

1. Setzer, M., Hartmann, V. (1991) Verbesserung der Frost-Tausalz-Widerstandsprüfung, *Betonwerk + Fertigteil-Technik*, No. 9, pp. 73-82.
2. Setzer, M., Stark, J., Auberg, R., Ludwig, H.-M. (1994) Erfahrungen mit dem CDF-Test zur Prüfung des Frost-Tausalz-Widerstandes an der Uni-GS-Essen und der HAB Weimar, 12. Ibausil, Weimar 1994, Vol. 1, *Wiss. Zeitschrift HAB*, No. 5/6/, pp. 127-131.
3. RILEM TC 117 - FCD (1995) 'Draft Recommendation for Test Method for the Freeze-Thaw Resistance of Concrete. Test with Water (CF) or with Sodium Chloride Solution (CDF).' *Materials and Structures*, Vol. 28, pp. 175-182.
4. Hilsdorf, H. K., Kropp, J., Günther, M. (1984) Karbonatisierung und Dauerhaftigkeit von Beton, *Fortschritte im Konstruktiven Ingenieurbau*, Verlag Ernst & Sohn, Berlin, pp. 89-96.
5. Manns, W., Zeuss, K. (1979) Zur Bedeutung von Zement für den Frost-Taumittel-Widerstand von Beton, *Straße und Autobahn*, Vol. 30, No. 4, pp. 167-173.
6. Stark, J., Wicht. B. (1995) Dauerhaftigkeit von Beton, *Schriften der HAB Weimar-Universität*, No. 100, pp. 1-282.
7. Stark, J., Ludwig, H.-M. (1995) The Influence of the Type of Cement on the Freeze-Thaw and Freeze-Deicing-Salt-Resistance of Concrete, *Intern. Conf. on Concrete Under Severe Conditions 'CONSEC'* Sapporo/Japan, Vol. 1, pp. 245-254.
8. Stark, J., Ludwig, H.-M. (1994) Frost- und Frost-Tausalz-Widerstand von hüttensandreichen HOZ-Betonen; 12. ibausil, Weimar 1994, Vol. 1, *Wiss. Zeitschrift HAB*, No. 5/6/7, pp. 111-117.
9. Stark, J., Ludwig, H.-M. (1996) Freeze-Thaw and Freeze-Deicing-Salt Resistance of Concrete Containing Cement Rich in Granulated Blast Furnace Slag, *ACI Fall Convention*, New Orleans.
10. Stark, J., Ludwig, H.-M. (1997) Freeze-Thaw and Freeze-Deicing Salt Resistance of Concrete Containing Cement Rich in Granulated Blast Furnace Slag, *ACI Journal*.
11. Hilsdorf, H. K., Günter, M. (1986) Einfluß der Nachbehandlung und Zementart auf den Frost-Tausalz-Widerstand von Beton, *Beton und Stahlbetonbau*, Vol. 81, No. 3, pp. 57-62.
12. Stark, J. (1995) Sulfathüttenzement, *Wiss. Zeitschrift der HAB Weimar-Universität*, pp. 1-15.

13. Nobst, P., Ludwig, H.-M., Stark, J. (1994) Reaktionen von Monosulfat und Ettringit bei Frost- bzw. Frost-Tausalz-Beanspruchung; 12. ibausil, Weimar 1994, Vol. 1, *Wiss. Zeitschrift HAB*, No. 5/6/7, pp. 119-125.

14. Stark, J., Ludwig, H.-M. (1995) Zum Frost- und Frost-Tausalz-Widerstand von PZ-Betonen, *Wiss. Zeitschrift HAB*, No. 6/7, pp. 17-35.

15. Stark, J., Ludwig, H.-M. (1994) Frost- und Frost-Tausalz-Widerstand von Beton - ein rein physikalisches Problem?, 12. ibausil, Weimar 1994, Vol. 1, *Wiss. Zeitschrift HAB*, No. 5/6/7, pp. 95-104.

16. Eckart, A., Ludwig, H.-M., Stark, J. (1995) Zur Hydratation der vier Hauptklinkermineralien des Portlandzements, 12. ibausil, Weimar 1994, Vol. 3, pp. 365-379, *Zement-Kalk-Gips* 48, No. 8, pp. 443-452.

17. Stark, J., Ludwig, H.-M. (1996) Die Rolle von Phasenumwandlungen im Zementstein beim Frost- und Frost-Tausalz-Angriff auf Beton, *Zement-Kalk-Gips* 49, No. 11, pp. 648-663.

18. Stark, J., Chelouah, N. (1997) Freeze-deicing salt resistance of high-strength concrete, *8th Intern. Conf. on Mechanics and Technology of Composite Materials*, 29.09.-02.10. 1997, Sofia, Bulgarien.

19. Chelouah, N., Stark, J. (1997) Einfluß von Polymeren und einer saugenden Schalungsbahn auf den Frost-Tausalz-Widerstand von Beton, *13. ibausil*, Weimar.

20. Bollmann, K., Stark, J. (1996) Ettringitbildung im erhärteten Beton und Frost-Tausalz-Widerstand, *Wiss. Zeitschrift Bauhaus-Universität Weimar*, No. 4/5, pp. 9-16.

Influence of C_3A content on frost and scaling resistance

J. STARK and A. ECKART
Bauhaus-Universität, Weimar, Germany
H.-M. LUDWIG
Bauhaus-Universität, Weimar
(currently at Schwenk Zementwerke KG, Karlstadt), Germany

Abstract
The effects of chemical transformation processes on the frost and frost-deicing salt resistance of concrete are by far not of the same order as the physical effects, but they are nevertheless significant. Our investigations showed that monosulphate (AFm phase) is particularly instable and will transform to ettringite (AFt phase) under frost and also under frost-deicing salt attack. This formation of secondary ettringite, which is supported by thermodynamic conditions at low temperatures, may reduce considerably the frost and frost-deicing salt resistance of concretes without air-entrainment.
Keywords: C_3A content, ettringite, frost attack, frost-deicing salt attack, monosulphate, portland cement, phase transformation

1 Introduction

Practical experience of many years has shown that it is possible to produce Portland cement with a high frost and scaling resistance - independent of the types of phases of the cement -, if favourable air void parameters are created by air-entraining agents, if the water-cement ratio is kept low, and if frost and deicing-salt resistant aggregates are employed. On the other hand, however, it is well-known that Portland cement concretes and mortars without air-entrainment which differ only in the type of Portland cement used may vary considerably in their frost and scaling resistance. This is the reason why already for decades attempts have been made to discover a relationship between the types of phases in Portland cements and the frost and scaling resistance of the respective concretes. In the discussions on the influence of cement

Frost Resistance of Concrete, edited by M.J. Setzer and R. Auberg. Published in 1997 by E & FN Spon, 2–6 Boundary Row, London SE1 8HN, UK. ISBN: 0 419 22900 0.

on frost and scaling resistance the evaluation of the clinker phases of Portland cements is of particular interest.

Most publications deal with the influence of the C_3A content on the frost and scaling resistance of concrete. Still there are completely different opinions held on this question.

Some authors hold the view that the C_3A content has a favourable influence on frost and scaling resistance [1] [2] [3]. Other authors conclude from their test results that frost and scaling resistance is adversely affected by the C_3A content [4] [5] [6].

Schrämli [7] compared the varying statements on the C_3A content and supposed that the differences had been caused by different test conditions. In his opinion the influence of the C_3A content on frost and scaling resistance can only be understood if the basic mechanisms are revealed. The C_3A content in itself cannot be a factor directly influencing frost and scaling resistance because there is hardly any non-hydrated C_3A left when the frost and/or frost-deicing salt attack begins. It can only be of importance if there is a direct relationship between the C_3A content of the cement and the physical and/or chemical factors of influence (pore size distribution, degree of hydration etc.). Any correlation of C_3A content and frost and scaling resistance is only an attempt to predict the influence of the cement by means of the clinker phase composition without knowing the cause of a possible influence of the C_3A.

The aim of the investigations described in this paper was to find out whether hydrate phases are formed from C_3A that show a different stability under frost and/or frost-deicing salt attack than hydrate phases from C_4AF. Basic studies were made to find out which phase transformations in calcium sulpho-aluminate and/or calcium sulpho-aluminate ferrite hydrates may occur under frost attack and what the differences between these hydrates are. In a second step the question was considered whether these phase transformations have an effect on the frost and scaling resistance of concrete and which role the C_3A content of the cements may play in this connection.

2 Basic investigations - frost attack

In the basic investigations into frost attack synthesised C_3A in a stoichiometric mixture with gypsum was hydrated to monosulphate and ettringite respectively, stored for 28 days and then subjected to 28 freeze-thaw cycles (+ 20°C / - 20°C). The comparison of these specimens with other non - frozen specimens showed the different stability of monosulphate and trisulphate under the specific humidity and temperature conditions of a frost attack. Whereas ettringite proves very stable, part of the monosulphate is transformed to ettringite during freezing (Fig. 1).

The main cause of these transformation processes is the change in the thermodynamic stability conditions at low temperatures. Compared to monosulphate, there is an increasing probability that trisulphate - expressed as a negative free formation enthalpy - will form with a drop in temperature, so that at low temperatures the formation of trisulphate is supported by thermodynamic conditions.

The XRD-diagram of the non-frozen monosulphate shows that after pre-storage the added gypsum has almost completely been transformed. That means that, prior to freezing, there was no sulphate available for the transformation of monosulphate to ettringite, which is supported by thermodynamic conditions at low temperatures.

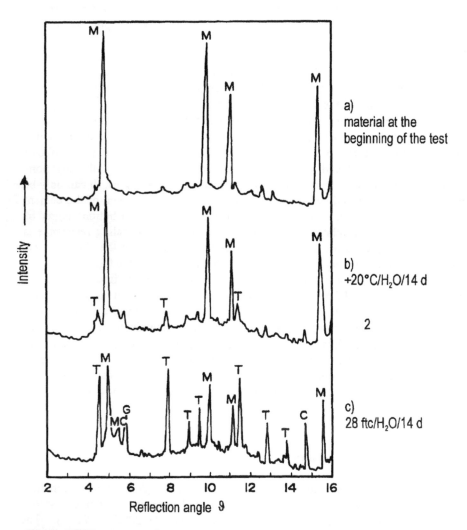

Fig. 1. XRD patterns of monosulphate (frost) - M-monosulphate, E-ettringite, G-gypsum, X-monocarbonate.

The results of the investigations lead to the conclusion that the sulphate required becomes available through the partial decomposition of the monosulphate due to carbonation. The gypsum formed in this process may then form ettringite with the non-carbonated monosulphate. An indication of this reaction mechanism is given by the phase analysis of the reaction products of the carbonation of monosulphate. Both gypsum and calcite (XRD, DTA) as well as aluminium hydroxide (DTA, SEM) could be detected as new formations in addition to ettringite. The reaction proceeds as follows:

$$3\ C_3A \cdot C\overline{S} \cdot 12H + 6CO_2 + 2H_2O \rightarrow C_3A \cdot 3C\overline{S} \cdot 32H + 6CaCO_3 + 4Al(OH)_3$$

The results achieved in an examination of AFm phase made from C_4AF correspond to those of the monosulphate tests. The AFm phase also proved instable under frost attack and was partly transformed to AFt phase, gypsum, calcite and aluminium hydroxide.

The morphology of the secondarily formed trisulphate is similar for C_3A and C_4AF (REM). In both cases needle-shaped trisulphate of comparable size is formed. Taking into account the x-ray results and the thermoanalytical tests, the TG curves of the thermoanalysis allow the quantities of both phases to be estimated by means of the dehydration temperatures which are different for ettringite/AFt phase and monosulphate/AFm phase. The quantities determined by this method (Table 1) confirm the results of the x-ray tests. We see that the rate at which AFm phase is decomposed and AFt phase is secondarily formed is nearly identical to the decomposion of monosulphate and the secondary formation of ettringite in C_3A.

Table 1. Monosulphate (AFm) and ettringite (AFt) contents approximately determined by means of TG curves

	Monosulphate from C₃A in % by weight		Afm phase from C₄AF in % by weight	
	Monosulphate	Ettringite	Afm phase	Aft phase
Initial sample	95.8	0.0	91.6	0.0
28 ftc/H₂O/14 d	42.9	37.4	35.2	37.6

The formation of the possibly expansive secondary ettringite depends on how much SO_3 is produced by the decomposition of the monophases. If the rate of decomposition of the monophases is the same it should be possible by means of the amount of monosulphate before freezing to predict the new formation of ettringite (AFt phase) and also the potential of the cement for a later damage.

3 Basic investigations - frost-deicing salt attack

During the frost-deicing salt test which consisted of 28 freeze-thaw cycles in a 3 % NaCl solution, the monosulphate was totally transformed into Friedel's salt and ettringite (Fig. 2). After 28 freeze-thaw cycles (ftc) no phase transformations were found in the ettringite examined.

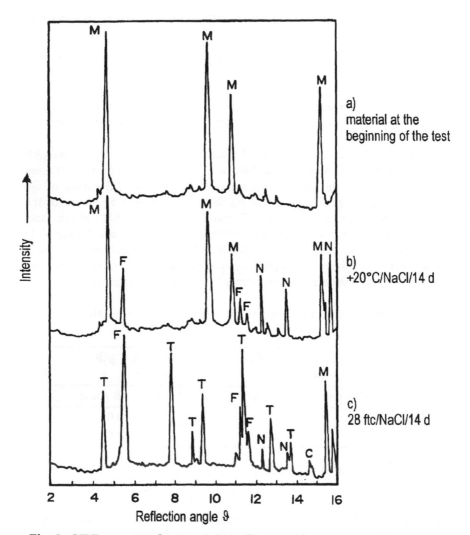

Fig. 2. XRD patterns of monosulphate (frost + salt) - M-monosulphate, E-ettringite, F-Friedel's salt, N-sodium chloride.

Contrary to freezing in pure water the combined frost-deicing salt attack gave no indication of a carbonation of the monosulphate. Obviously the lower solubility of CO_2 in the presence of chlorides leads to a slow-down of the carbonation reaction. It may be concluded that the SO_3 which is required for the formation of ettringite is set free by the partial transformation of monosulphate to monochloride. This hypothesis is supported by investigations by Dorner [8] who found out that in the presence of NaCl solution a transformation of monosulphate to ettringite will take place even if the work is done in an atmosphere of nitrogen and with nitrogen-saturated water.

The transformation processes which occurred in the AFm phase under frost-deicing salt attack are comparable in quality to those in the monosulphate from C_3A. Also here AFm phase was decomposed and AFt phase and Friedel's salt were formed. As under frost attack the secondarily formed trisulphate shows again a needle-shaped morphology. The size of the needles is again similar in both C_3A and C_4AF mixes. The decomposition of AFm phase proceeds, however, much more slowly than that of monosulphate. After 28 ftc only about half of the AFm phase has transformed into other phases (Table 2).

Table 2. Monosulphate (AFm), ettringite (AFt) and monochloride contents approximately determined by means of TG curves

	Monosulphate from C_3A in % by weight			AFm-phase from C_4AF in % by weight		
	Mono-sulphate	Ettringite	Mono-chloride	AFm-phase	AFt-phase	Mono-chloride
Initial sample	95.8	0.0	0.0	91.6	0.0	0.0
28 ftc/ NaCl/14 d	0.0	47.4	42.3	50.7	20.2	18.0

Another source of sulphate than that occurring under frost attack may be assumed to be the reason for the slower decomposition of AFm phase compared to monosulphate. The notoriously bad bonding of chlorides in ferritic phases [9] [10] [11] does not only lead to a reduced formation of monochloride but also to the fact that in comparable periods of time less SO_3 may be provided for the secondary formation of trisulphate than in pure monosulphate from C_3A. The different rate of transformation of monophase under frost-deicing salt attack indicates that by means of the monosulphate amount (AFm amount) before freezing statements on the potential of the cements for a later damage will not be possible without taking the Fe-content of these phases into consideration.

Mixtures of synthetical clinker phases were examined an showed similar results during frost attack in NaCl solution. After 26-day hydration of a mixture of C_3A, C_3S and SO_3 not only monosulphate was formed, but a mixture of both ettringite an monosulphate. After the frost-deicing salt test had been carried out, all monosulphate was transformed into Friedel's salt and ettringite.

28-day hydration of mixture of C_4AF, C_3S and SO_3 led mainly to the formation of AFt. Part of the gypsum and C_4AF was found to be unhydrated. During the frost attack in NaCl solution no changes in phase content were observed.

After a 28-day hydration of a mixture of all four main clinker minerals of portland cement (C_3S, β-C_2S, C_3A and C_4AF) together with gypsum only small amounts of monosulphate compared to ettringite were measured. After frost-deicing salt attack, first of all, the complete decomposition of gypsum and monosulphate is observed, whereas the ettringit content seems to stay the same (Fig. 3).

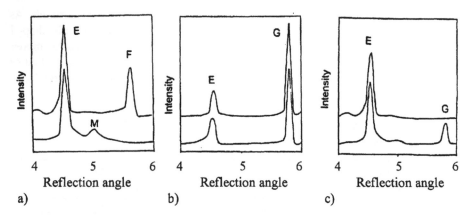

Fig. 3. XRD patterns of mixtures (frost + salt) - M-monosulphate, E-ettringite, F-Friedel's salt, G- gypsum.
a) C_3A, C_3S and SO_3
b) C_4AF, C_3S and SO_3
c) C_3S, β-C_2S, C_3A, C_4AF and SO_3

4 Effects on frost and frost-deicing salt resistance of concrete

The phase transformations of synthetically produced aluminate-containing hydrate phases which had been observed in our basic investigations were found by some other authors also in mortars and concretes [12] [13] [14]. However, the significance of these phenomena for the real damage by frost and/or frost-deicing salts has hardly been evaluated as yet.

Within the context of technological investigations into cement and concrete at the Bauhaus-Universität Weimar it was studied to what extent the new formation of trisulphate may influence the frost and frost-deicing salt resistance of concrete. For this a total of 16 cements with C_3A contents ranging from 0.4 % to 13.7 % were used. With those cements concretes were produced (z = 350 kg/m^3 , w/c ratio = 0.50) with and without artificial air voids, which were then tested by CF test in pure water as well as by CDF test in 3 % NaCl solution [15] for their frost and frost-deicing salt resistance.

It was found that concretes without air voids showed big differences in the degree of damage in both tests which has to be attributed to the different cements because mixtures were kept identical and curing was the same for all concretes tested (Fig.4). Differences in scaling were extreme; under frost-deicing salt attack the amount of scaled material ranged from 770 g/m^2 up to 5212 g/m^2. Even under frost attack alone where normally only minor scaling can be observed different amounts of scaled material occurred (from 97 g/m^2 to 1079 g/m^2) depending on the cement used.

Damages of concretes with artificial air voids proved not to be dependent on the cements used. All of those concretes showed a high frost-deicing salt resistance with very small amounts of scaled material.

Evaluating the results from tests with frost and frost-deicing salt attack no clear relationship exists between the composition of clinker phases in portland cements - in this case C_3A content - and their frost and frost-deicing salt resistance. Physical as well as physical-chemical parameters like the degree of hydration of cement or pore size distribution in the concrete also fail to explain the influence of the different cements in concretes without artificial air voids on their frost and frost-deicing salt resistance. Following hypothesis was formulated: the formation of secondary ettringite, which is supported by thermodynamic conditions at low temperatures, as found in basic investigations, influences damage of concrete.

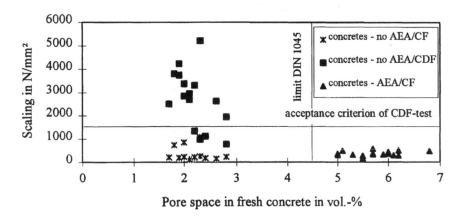

Fig. 4. Amount of scaling of concretes with and without air-entrainment (AEA) in dependence on the pore space in fresh concrete in the frost test (CF) and the frost-deicing salt test (CDF)

After the respective attacks a new formation of ettringite was observed in all concretes, mainly in cracks or at the interface of hydrated cement and aggregate. Obviously some aggregate was forced out of the texture due to the formation of secondary ettringite. A quantification of the newly formed ettringite was not possible because of the diluting effect of the aggregate.

Now it was investigated which aluminate and/or aluminate-ferritic hydrate phases are formed by the 16 cements after 28d hydration and to what extent these hydrate phases are transformed during the CF and/or CDF tests. For this purpose the aluminate-containing clinker phases, the sulphates and the alkalis of the cements were enriched by means of salicylic acid extraction and then hydrated in saturated calcium hydroxide solution. The procedure applied is described in detail in the literature [16] [17]. After 28 d hydration the extractions were subjected to frost and/or frost-deicing salt attack.

The formation of secondary ettringite (AFt formation), which occurred during freezing the hydrated extraction residues in CF and CDF test, was quantified by thermoanalysis and then correlated to the frost and frost-deicing salt resistance of the concretes tested. The results hint towards a close statistical relationship between the phase transformation as observed and the frost and frost-deicing salt resistance of concretes without artificial air voids (Fig. 5).

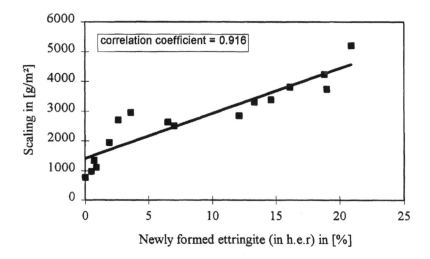

Fig. 5. Amount of scaling of concretes in depence on the formation of secondary ettringite in the frost-deicing salt test (CDF)

Between the formation of new ettringite (new AFt formation) and the amount of scaled concrete material a correlative factor was established of 0.886 under frost attack and of 0.916 under frost-deicing salt attack.

The potential of the cement for a later damage, i.e. the formation of new ettringite (new AFt formation), obviously depends on the amount of monosulphate (AFm phase) existing prior to freezing.   In addition to that the Fe-content of the monosulphate and AFm phase is important under combined frost-deicing salt attack. As it is not yet possible to say how the original cement composition influences the content of monosulphate (AFm-content) prior to freezing nothing can be predicted from the clinker phases concerning the potential of the cement for a later damage. A direct relation between the amount of monosulphate (AFm-amount) prior to freezing and the C_3A-content does not exist.

The aim of future investigations should be to determine the factors influencing the increase in the formation of monosulphate (AFm). Then specific measures may be taken to improve the frost and frost-deicing salt resistance by optimising the cement composition.

The importance of the influence of chemical transformation processes is shown in Fig. 6, where the concretes made of two different cements without air-entrainment are compared to air-entrained concrete. It may be seen that these phenomena play a minor role compared to the physical influencing factors such as air-entrainment.

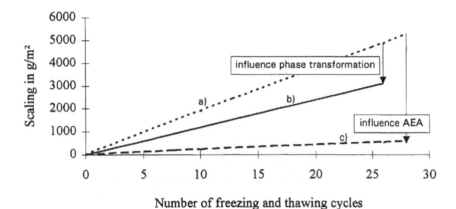

Number of freezing and thawing cycles

Fig. 6. Effect of physical and chemical factors of influence on the frost-deicing salt resistance (CDF test)
a) concrete with cement A (high amount of newly-formed ettringite)
b) concrete with cement B (small amount of newly-formed ettringite)
c) concrete with cement A and with air-entraining agent (AEA)

The results show that it is possible to produce concretes with a very high frost and frost-deicing salt resistance, independent of the clinker composition. The new findings may, however, be of importance for the future optimisation of concrete with regard to high requirements on its frost or frost-deicing salt resistance. Furthermore phase transformations, as observed in latest investigations, play an important part when the damage of high performance concrete without artificial air voids under frost and frost-deicing salt attack is concerned.

5 References

1. Joosting, R. (1968) Über den Einfluss der Portlandzementmarke auf die Frostbeständigkeit des Betons und des Mörtels. *Schweizerische Bauzeitung*, Vol. 86, No. 49, pp. 881-883.
2. Stark, J. and Stürmer, S. (1991) The Influence of the C_3A-Content on the Durability at Freezing and Thawing. *Report TVBM - 3048*, RILEM TC 117, Lund, pp. 155-166.
3. Rasmussen, T.H. (1985) Long-Term Durability of Concrete. *Nordic Concrete Research*, Vol. 4, pp. 159-178.
4. Jackson, F.H. (1955) Long-Time Study of Cement Performance in Concrete. *ACI - Journal*, Vol. 27, pp. 159-193.
5. Rose, K., Hope, B.B. and Ip, A.K.C. (1989) Statistical Analysis of Strength and Durability of Concrete made with Different Cements. *Cement and Concrete Research*, Vol. 19, pp. 476-486.

6.   Peterson, P.E. (1993) Scaling Resistance Tests of Concrete. Experience from Practical Use in Sweden. *Proceedings, International Workshop on Freeze-Thaw and Deicing Salt Scaling Resistance of Concrete*, Quebec, pp. 249-259.

7.   Schrämli, W. (1978) An Attempt to Assess Beneficial and Detrimental Effects of Aluminate in the Cement on Concrete Performance. *World Cement Technology*, Vol. 9, No. 3, pp. 35-42 and No. 4, pp. 75-80.

8.   Dorner, H. and Rippstain, D. (1986) Einwirkung wäßriger Natriumchlorid-lösungen auf Monosulfat. *Tonindustriezeitung*, Vol. 110, No. 6, pp. 383-386.

9.   Neroth, G. (1985) Zur Bindung von Chlorid in Zementstein. *Baustoffe '85*, Bauverlag, Wiesbaden/Berlin, pp. 140-147.

10.   Smolczyk, H.-G. (1984) Flüssigkeit in den Poren des Betons - Zusammensetzung und Transportvorgänge in der flüssigen Phase des Zementsteins.*Beton - Information*, Vol. 24, No. 1, pp. 3-11.

11.   Lukas, W. (1983) Zur Frage der Chloridbindung und -korrosion von Stahl bei nachträglicher Einwirkung von Chlorid auf Beton. *Proc., Internationales Kolloqium Werkstoffwissenschaften und Bausanierung*, Esslingen, pp. 171-173.

12.   Kukko, H. (1993) Frost Effects on the Microstructure of High Strength Concrete, and Methods for Their Analysis. *Diss.*, Helsinki University of Technology.

13.   Illgner, R. and Bubig, M. (1991) Alterung des Betons durch Veränderung physiko - chemischer Gleichgewichte unter Nutzungsbedingungen. *Proc., 11. Internationale Baustoff- und Silikattagung, Weimar*, Vol. I, pp. 220-231.

14.   Ghorab, H.Y. and Ludwig, U. (1981) Modellversuche zur Klärung von Schadensursachen an wärmebehandelten Betonfertigteilen. Teil I: Zur Stabilität von Monophasen und Ettringiten, *Tonindustriezeitung*, Vol. 105, No. 9, pp. 634-640.

15.   RILEM TC 117-FDC, (1995) Draft Recommendation for Test Method for the Freeze-Thaw Resistance of Concrete. Test with Water (CF) or with Sodium Chloride Solution (CDF). *Materials and Structures*, Vol. 28, pp. 175-182.

16.   Stark, J., Müller, A. and Stürmer, S. (1994) Hydration of C_3A and $C_2(A,F)$ - Separated from Sulphate-Resisting and White Portland Cement - under Conditions of Normal Hardening and Heat Treatment. *Cement and Concrete Research*, Vol. 24, pp.503-513.

17.   Stark, J. and Ludwig, H.-M. (1992) Effects of Low Temperature and Freeze-Thaw Cycles on the Stability of Hydration Products. *Proc., 9th International Congress on the Chemistry of Cement*, New Dehli, Vol. IV, pp. 3-9.

Influence of cement type on resistance against freezing and thawing, with or without deicing chemicals, of cement mortar

U. BALTERS and U. LUDWIG
Institut für Gesteinshüttenkunde, RWTH Aachen, Germany

Abstract
The intention of the investigations was to find out the resistance of different types and qualities of cements in standard mortars against freeze thaw cycles (FTC) in water and in NaCl-, $CaCl_2$- and urea- containing solutions.
With respect to the thaw media water the results show expansion together with cracking and decrease of strength nearly without exception. The largest average number of FTC were reached in water. In this case it could be assumed that the mortar specimens freeze from outside to the center. Hydraulic pressure is the cause of the destruction of the entire structure.
The NaCl-solution causes mainly loss of weight with scaling of the surface and with the lowest average number of FTC. Chloride and enrichments of chloride could only be analysed in the surface regions of the specimens. In contrast to those in $CaCl_2$-solution losses of weight as well as expansion could be measured. The damages started after more cycles. Chloride and chloride enrichments could be measured in the surface region and in the center of the specimens which had been treated for a longer period.
The thaw medium urea takes a middle position regarding the type and intensity of the damage.
It can be assumed that the specimens freeze in solutions with deicing additions preferably from the center to the surface causing their scaling and less expansion accompanied by strenght decrease.
The degree of the damages due to the deicing media increases in the following row:

$$Water < CaCl_2 < Urea < NaCl$$

Frost Resistance of Concrete, edited by M.J. Setzer and R. Auberg. Published in 1997 by E & FN Spon, 2–6 Boundary Row, London SE1 8HN, UK. ISBN: 0 419 22900 0.

This is valid for the 1 % solution as well as for the 3 % solution. If looking at the investigations regarding the different cement types, it can be shown that the portland composite cements have a good resistance against freezing and thawing in comparison with the portland cements. At this it should be considered that the composite cements need a longer subsequent wet treatment before exposed to frost action. Composite cements behave similar with respect to the cause of the damage in the thaw media. But the pure portland cements show greater differences with regard to the cause of damages. Correlations can be shown between a high specific heat of hydration and a high K_2O-content within the same kind and quality of cement and a lower resistance against freezing and thawing. An important influence on the resistance against freezing and thawing has the structure of the hydrated cement. Specimens with a dense and more complete hydrated structure achieve a good resistance against freezing and thawing.

Keywords: Frost resistivity, deicing salts, types and qualities of cements, expansion, weight loss, resonance frequency, SEM- and micro-analysis

1. Introduction

To keep concrete roadways free of ice considerable amounts of anorganic deicing salt i.e. NaCl, $CaCl_2$ and $MgCl_2$ are used. Organic deicing mediums like urea, glycole or alcoholderivates are used on airports. Deicing mediums containing chloride cannot be used due to their corrosive effects.

The physical attacks which damage or destroy the microstructure of concrete are well known. In earlier literature the frost and frost deicing damage were explained by physical effects. To chemical and mineralogical processes only little attention were paid. Nevertheless nowadays the idea is, that the damage is an interaction of physical and chemical effects. The damages of the frost respectively frost deicing attack causes the following effects [1]:

- scaling of the mortar layers at the surface
- pop-outs above the aggregate ("pop-outs")
- map-, d-cracking
- complete destruction of the hardened cement paste
- destruction of the aggregates

The scaling of the surface can be a result of the hydraulic pressure in the structure, which depends on the degree of water saturation in the concrete layers and on the gradients of the thaw medium inside the structure [2].

Type and quality of the used cements have an important influence on the durability and corrossion resistance of an concrete building while exposed to frost and deicing salt. Increasingly not only portland cements and blast furnace slag cements but also composite cements are used. Nevertheless there are many different opinions about the influence of latent hydraulic or pozzolanic additives with respect to the frost-thaw cycle (FTC) resistance [3, 4, 5, 6].

2.   Frost thaw investigation

2.1   Type of cements

Experimental frost-thaw investigations were carried out with ordinary Portland cements as well as with blended cements. Table 1 show the types and the qualities of the cements.

Quantity	cements	admixtures [wt.-%]	
2	PZ 55	-	
1	PZ 55 - NA (HS)	-	
2	PZ 45 F	-	
1	PZ 45 L - NA (HS)	-	
3	PZ 35 F	-	
3	PZ 45 F - HS	-	
3	EPZ 35 F	27 / 25 / 29	blast furnace slag
2	HOZ 35 L	50 / 46	"
2	FPZ 45 L	20 / 26	fly ash
2	PKZ 35 F	17 / 18	lime stone flour
1	TrZ 35 L	25	trass
2	PÖZ 35 F/E	n.b.	oil shale

Table 1   Types and qualities of used cements

2.2   Experimental work

Mortar prisms were produced with the cements described above according to German standard DIN 1164 or to EN 196-1. The prisms were sized 4x4x16 cm^3 and stored for 24 hours at 20°C and 95 % relative humidity (r.h.). Thereafter the prisms were stored in water of 20°C for 27 days and afterwards at 80 % r.h. and 20°C for 7 days, respectively at 10°C for 24 hours. During the storage the samples were physically and technically examined. Afterwards the prisms were frozen in water, NaCl-, CaCl$_2$ and Urea-solution. The concentration of the solutions were prepared with 1 and 3 %. One FTC lasted 10 hours and was in the bounds of +10 to -20°C. The criterions of destruction were 30 percent loss of weight or 3 mm/m expansion. Only half of the cements were examined with solutions of 1% deicing additions. It could be shown that the degree of damage in 1%-solutions lies between that in water- and that in 3%-solutions. During the examinations the mass, length and resonance frequency were measured after defined cycles. When the FTC had been finished physical-technical as well as chemical-mineralogical investigations on flakes, surfaces and centers of the specimen were carried out.

3.   Results of the physical and technical investigations

3.1   Change of mass, length and resonance frequency after freeze thaw cycles in water and in 3 %- solutions

Table 2 show the results of the change of mass and length due to FTC in water and in 3% solutions. The average, the maximum and the minimum numbers of FTC, which lead either to a loss of mass of 30 % or to an increase of length of 3 mm/m are given beside. The values are listed for the total-, the portland- and the composite cements. The thaw mediums are water and 3% NaCl-, $CaCl_2$- and Urea solution. The last column shows which percentage of the tested cements had a loss of mass and/or a change of length in the respective thaw medium. The table is to point out possible coherence between the types of cement and the thaw medium.

	$\emptyset$ FTC	$FTC_{max.}$	$FTC_{min.}$	Δm	criterion Δl [%]	$\Delta m + \Delta l$
Water						
Cements (1-26)	145,9	471	38	-	88,5	-
PC (1-12)	137,4	362	38	-	91,7	-
Composite (13-26)	153,1	471	49	-	85,7	-
NaCl 3 %						
Cements (1-26)	66,9	170	29	80,8	11,5	7,7
PC (1-12)	58,6	170	29	83,3	-	16,7
Composite. (13-26)	74,0	137	18	78,6	21,4	-
CaCl₂ 3 %						
Cements (1-26)	130,1	380	24	61,5	34,6	3,8
PC (1-12)	132,7	380	35	50,0	41,7	8,3
Composite. (13-26)	127,9	272	24	71,4	28,6	-
Urea 3 %						
Cements (1-26)	102,7	226	26	69,2	19,2	11,5
PC (1-12)	90,3	181	43	66,7	8,3	25,0
Composite (13-26)	113,2	226	26	71,4	21,4	-

Table 2   Average number of FTC by reaching of Δm = 30 wt.-% and/or Δl = 3 mm/m in the different thaw media

A comparison of the average number of FTC, which could be reached in water and in the respective thaw medium, show that the cements 1-26 reached the largest number with 146 FTC in water. When looking at portland- and composite cements separately it can be noticed, that in this study the composite cements reached with 153 FTC a higher number than the portland cements. The type of damage which occurs after freeze thaw cycles in water is uniform for all investigated cements: these are expansions resulting in a loss in strength. The examinations of three cements were stopped after 190 FTC, because no appreciable increase of the length or decrease of the resonance frequency could be measured. - The cements frozen in 3% NaCl solution reached only a smaler number of FTC. The loss of mass was the criterion for most of the cements to break off the FTC. The behaviour of portland and composite cements in this thaw media was similar. - In the 3 % $CaCl_2$ solution the cement motars behaved different. With an average of 130 FTC more cycles could be reached in this solution than in the 3 % NaCl solution. Also the behaviour of the damage of the portland and composite cements was different. Nearly the same number of mortar specimens made with portland cements were damaged by loss of weight as well as by expansion. The relation of specimens with composite cements which were damaged by loss of weight to those which were damaged by expansion was 71 to 29.

The average number of FTC in the urea solution takes a medium position between the NaCl and $CaCl_2$ solution. Approximately 67 % of the portland cements showed loss of weight and 25 % of the cements showed loss of weight as well as expansions.

Although most of the investigated cements reached first the criterion loss of weight, some cements and cement groups showed expansions in the range between 1,4 and 3 mm/m in all of the used thaw media. Particularly the portland cements of the quality 35 F and 45 F reached distinct expansions. The blast furnace slag cements behave very similar with regard to the type of damage. One fly ash cement showed expansions and only little loss of weight, the other reacted with high mass losses after a few number of FTC.

3.2 Influence of physical-technical parameters on the damage by freeze thaw cycles

The majority of the cements which have loss of mass as well as distinct expansions in all thaw media showed within their quality classes a correlation between the heat of hydration and the length change. The total heat, which was reached after the third respectively the fifth day, and the second maximum of the emitted specific heat were considered as a reference parameter. The cements with the highest heat of hydration in their classes showed especially high expansions in the deicing media.

The reason could be an extremly high and an early rate of hydration together with an unfavourable formation of the cement stone structure. The two portland cements 35 F and 45 F (No. 5 and 8) are good examples of the above mentioned correlation. Moreover they have a high specific surface, an increased water demand and a high early strength. These parameters influences a fast and high heat of hydration. The cements marked in table 3 shows a correlation between the expansion and the listed physical and technical parameters.

Cements	No.	dl/FTC *100	surface area [cm²/g]	heat of hydration [J/g] 5d	spec. heat [mW/g] 2.max.	water demand. [wt.-%]	Setting initial [h,min]	final [h,min]	compr. strength [N/mm²] 2d
PZ 55	1	1,49	6100	406	5,34	31,0	1,30	2,40	46,0
PZ 55 NA	2	0,63	5210	300	3,22	29,0	2,40	4,15	31,1
PZ 55	3	0,89	5170	438	4,33	32,0	1,30	2,20	51,2
PZ 45 F	4	0,46	3730	255	3,16	27,0	2,30	3,30	26,8
"	5	6,24	4910	473	4,33	31,0	2,35	3,15	40,8
PZ 45 L-NA	6	5,55	3350	350	2,35	25,0	3,55	5,15	21,2
PZ 35 F	7	0,26	3000	241	2,40	25,0	2,30	3,08	23,1
"	8	5,07	3520	260	2,74	27,0	1,45	2,55	24,7
"	9	1,75	2860	196	1,39	24,0	1,50	2,50	18,7
PZ 45 F-HS	10	6,61	3660	434	3,61	27,0	4,02	4,38	28,4
"	11	5,74	3400	268	2,26	27,0	3,25	4,40	25,9
"	12	0,50	4130	268	2,84	25,0	2,55	4,20	23,1
EPZ 35 F	13	1,83	3330	342	2,21	27,0	2,15	3,25	18,1
"	14	0,55	3230	333	1,70	27,0	4,00	5,17	10,7
"	15	0,54	3050	222	1,73	24,0	2,50	3,55	12,1
HOZ 35 L	16	1,93	4550	376	1,87	25,0	2,45	3,45	12,5
"	17	4,21	4120	351	2,26	29,0	2,20	3,25	14,8
HOZ 45 L	18	2,62	4470	187	2,00	28,0	3,20	4,35	18,1
FAZ 35 F	19	3,13	3280	294	1,48	28,0	3,50	5,25	20,0
"	20	1,11	4090	262	2,02	28,0	2,10	3,35	21,5
"	21	3,09	3700	435	1,82	28,0	2,55	3,50	26,2
PKZ 35 F	22	1,27	4810	337	2,50	25,0	2,50	4,20	19,1
"	23	-0,05	4930	169	2,30	26,0	2,40	4,10	22,8
TrZ 35 L	24	0,42	6080	285	2,07	34,0	3,35	5,25	20,0
PÖZ 35 F	25	1,65	4850	300	2,61	28,0	3,35	4,20	17,1
PÖZ 35 E	26	0,18	4520	310	3,14	30,0	3,30	4,44	25,7

Table 3   Correlation between the expansion and the physical-technical data

4.   Results of the chemical-mineralogical investigations

4.1   SEM-examinations

The microstructure of the hydrated cement samples and of the samples which were treated with FTC were examined with the scanning electron microscope (SEM). For that polished sections of sectional areas of the specimens were used. These sectional areas cover the surface of the specimens as well as the center of the specimens. Besides point and area analyses were carried out with an energy-dispersive system to obtain information about the chemistry of the cement stone before and after the treatment with thaw media and to look on possible local enrichments of chloride.

After FTC in water the cements with a stable matrix, a uniform distribution of micropores and nearly completed hydration showed no crack formation, i.e. the portland cement No 4 (figure 1) and the portland limestone cement No 23 (figure 4). In comparison the microstructure of the portland cement No 5 (figure 2) and the fly ash cement No 20 (figure 3), which hydrated to a smaller extend, showed a lot of partially some μm width cracks. The cracks were unfilled. No crack formation could be observed at the stable cements after the FTC in solution with deicing agents. On the other hand the structure of the portland cement No 5 showed even after the FTC in the thaw media solutions a distinct formation of cracks. This cement had a strong expansion already after a few FTC. The polished sections of the cement stone from the mortar prisms with the fly ash cement, which had a high loss of weight already after a few FTC, showed a spongy like structure.

At specimens treated in NaCl-solution chloride could only be detected in the surface area. On the other hand chloride could be measured also in the center of the test specimens after a longer treatment with FTC in $CaCl_2$ solution. The chloride content at the surface was higher than in the center of the test specimens. The diffusion coefficient of chloride ions from a solution of $CaCl_2$ was found to be two times higher than that of a NaCl solution. The sodium of the NaCl-solution could not be measured. In the surface region enrichments of chlorides could be found in calciumaluminat-ferrithydrates and on clinker particles, preferably on the interface of C_2S particles.

4.2 Influence of the chemical composition on the loss of weight and the expansion

As expected a correlation between the results of the chemical investigations and the loss of weight was not found. Again only those cements of the same quality showed correspondence in their chemistry, which had in addition to their loss of weight expansions with nearly all thaw media. Most of these cements possess a relativly high K_2O-content. Figure 5 show the tendency that higher expansions are reached with an increasing K_2O content. The portland cements with a high sulfate resistance respectively with a low alkali content were not taken into account. The K_2O content of the fly ash- and trass- cements was assumed without the portion of the fly ash respectively the trass, because these admixtures have a high alkali content themselves. K_2O, partially incorporated into C_3A, raises the early hydration and by this the early strength. Those cements with a high K_2O content, i.e. the portland cements No. 3, 5, 8 and the portland blast furnace slag cement 13 as well as the fly ash cement No. 21, have a high 2-days-compressive strength.

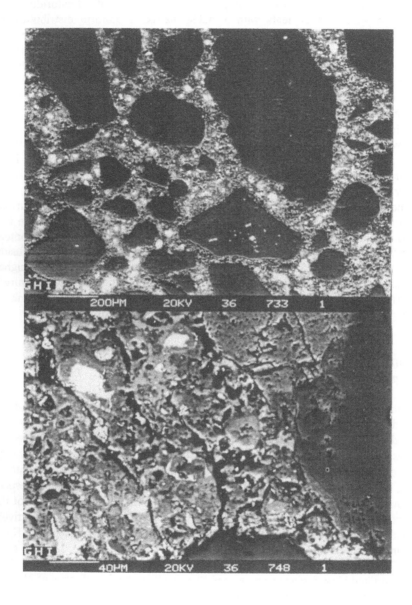

Figure 1   SEM-photos of mortar with PZ 35 F (4),
            upper photo: 27 d water storage, surface of the specimen,
            bottom photo: 362 FTC in water, center of the specimen.

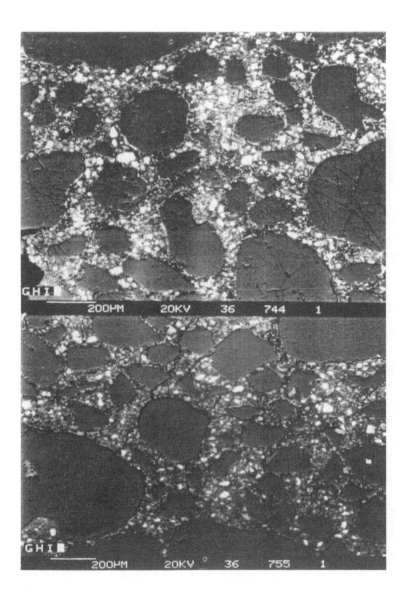

Figure 2   SEM-photos of mortar with **PZ 35 F (5)**,
upper photo: 27 d water storage, center of the specimen,
bottom photo: 48 FTC in water, center of the specimen.

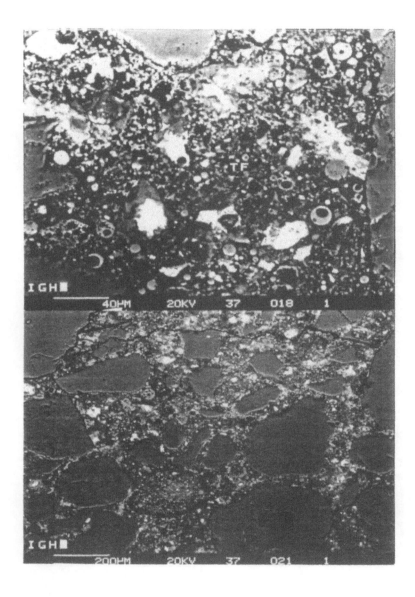

Figure 3   SEM-photos of mortar with FAZ 35 F (20),
             upper photo: 27 d water storage, center of the specimen,
             bottom photo: 67 FTC in water, surface of the specimen.

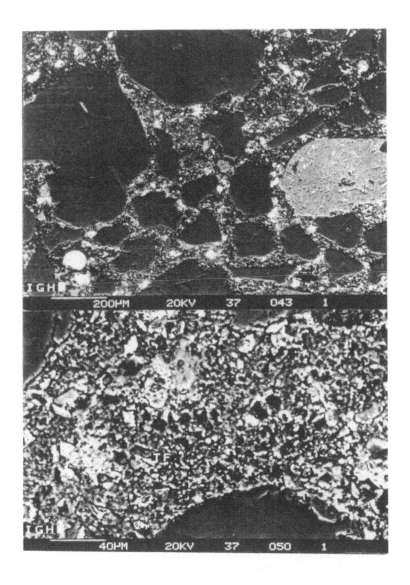

Figure 4   SEM-photos of mortar with PKZ 35 F (23),
            upper photo: 27 d water storage, surface of the specimen,
            bottom photo: 471 FTC in water, surface of the specimen.

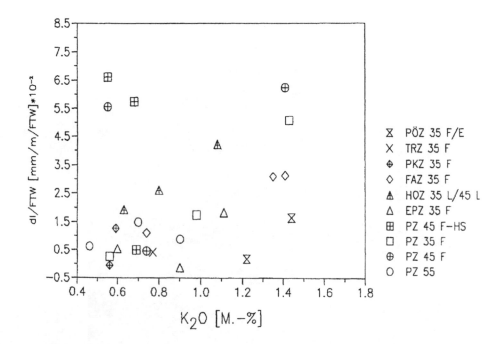

Figure 5   Influence of the K_2O content of the cements on the change of length
during FTC

5. References

1   Glock, G.: Frost-Taumittel-Widerstand von Betonzuschlagstoffen und Beton,
Diss. RWTH Aachen (1991), S. 8

2   Browne, F.P., Cady, P.D.: Deicer scaling mechanism in Concrete, ACI, Spec.
Pupl. 47-6, 1975, S. 103 u. 110-112

3   Bonzel, J., Siebel, E.: Neuere Untersuchungen über den Frost-Tausalz-
Widerstand von Beton, beton 6 (1977), Heft 6, S. 237-244

4   Schorr, K.: Frost-Tausalz-Widerstand von Zementstein aus verschiedenen
Zementarten und mit unterschiedlichen Beimengungen von Flugaschen,
Betonwerk + Fertigteil-Technik, Heft 2 (1983), S. 96-100

5   Fagerlund, G.: Influence of slag cements on the frost-resistance of the green
concrete, Swedish Cement and Concrete, Research Institute, Stockholm, CBI
3.83 (1983), S. 53-54

6   Schießl, P., Härtl, R.: Steinkohlenflugaschen im Beton, beton 12 (1993), S. 644-
648

Freeze-deicing salt resistance of concretes containing cement rich in slag

J. STARK
Bauhaus-Universität, Weimar, Germany
H.-M. LUDWIG
Schwenk Zementwerke KG, Karlstadt, Germany

Abstract
Whereas blast-furnace cement concretes have proved successful in structures subjected to freeze-thaw attack, their use in structures subjected to freeze-deicing salt attack is still a problem. The present paper discusses this problem. It was found that the freeze-deicing salt resistance of blast-furnace cement concretes is closely related to the carbonation of the surface area. The carbonation of blast-furnace cement concretes does not only lead to an increase in capillary porosity but also to metastable calcium carbonates soluble in NaCl. Based on the results of investigations, ways of improving the freeze-deicing salt resistance of blast-furnace cement concretes are proposed.
Keywords: Blast furnace slag, calcit, carbonation, freeze-deicing salt resistance, porosity, slag cement.

1 Introduction

In the assessment of the freeze-deicing salt resistance of concretes made of cements containing granulated blast-furnace slag a number of questions is still unanswered.

The uncertainties in the use of these cements have been caused by test results indicating that with blast-furnace cement concretes both the air-entrainment into fresh concrete and the effectiveness of air voids in the hardened concrete are problematic. To obtain the required fresh air content the amount of air-entraining agents in blast-furnace cement concretes rich in granulated slag has to be much greater than in portland cement concretes [1],[2]. But even a sufficient content of air voids in the hardened concrete does often not lead to an improvement of the freeze-deicing salt resistance of blast-furnace cement concretes rich in granulated slag

Frost Resistance of Concrete, edited by M.J. Setzer and R. Auberg. Published in 1997 by E & FN Spon, 2–6 Boundary Row, London SE1 8HN, UK. ISBN: 0 419 22900 0.

[3],[4],[5],[6]. Only few authors achieved results that indicate that the freeze-deicing salt resistance of blast-furnace cement concrete may be improved by air-entrainment in the same way as that of portland cement concrete [7],[8].

In general, the lower effectiveness of air voids in blast-furnace cement concretes is explained by the fact that because of the very dense microstructure and its very low permeability the water can hardly reach the expansion spaces, even if the spacing factors are small [4],[9]. In contrast to this hypothesis experience has shown that the effect of air-entraining agents in blast-furnace cement concrete may be improved by longer curing and the resulting denser microstructure [4]. In an attempt to explain the different effect of air-voids on portland and blast-furnace cement concretes Brodersen [10] refers to the fact that the gradients of moisture and chlorides in blast-furnace cement concretes are steeper because of the very dense microstructure in the core. Therefore greater stresses may occur near the surface of these concretes. But Brodersen himself points out that this cannot be a satisfactory explanation of the low effectiveness of air-entraining agents in blast-furnace cement concretes, since according to this statement also blast-furnace cement concretes without air-entrainment should behave in a more unfavourable way than the respective portland cement concretes. According to his own investigations, however, this is not the case. This means that at the moment a satisfying explanation of the different behaviour of air-entrained portland and blast-furnace cement concretes under freeze-deicing salt attack cannot be given.

In the opinion of some authors the low effectiveness of air-entraining agents in blast-furnace cement concretes rich in granulated slag is of secondary importance, as blast-furnace cements are able to develop a very dense microstructure and so a high freeze-deicing salt resistance may be developed even withour air-entraining agents [11],[12]. On the other hand, it was found in several studies that the use of systems rich in granulated slag without air-entraining agents reduced the freeze-deicing salt resistance in comparison to portland cement [13],[14].

The controversial opinions on the freeze-deicing salt resistance of blast-furnace cement concretes with and without air-entrainment initiated the systematic investigation into this problem. In recent years specific research into the causes of these phenomena has been carried out at the F.A. Finger Institute of Building Materials at the Bauhaus University Weimar. This paper presents the most important results of the research work.

2 Investigations into the freeze-deicing salt resistance of blast-furnace cement concretes

2.1 Basic materials and production of concrete
In addition to blast-furnace cements with varying contents of granulated slag a portland cement and a portland composite cement were included in the test programme for comparison. The cements used in the tests for freeze-deicing salt resistance come from the current production of various German cement factories and are listed in Table 1.

Table1. Cements used

Trade name of cements	Slag content	Short name of resp. Concretes
CEM I 32,5 R	0 %	P 0
CEM II/B-S 32,5	25 %	E 25
CEM III/A 32,5	45 %	H 45
CEM III/A 32,5	55 %	H 55
CEM III/A 32,5	65 %	H 65
CEM III/B 32,5 NW/HS	75 %	H 75

The mixture proportions are given in Table 2. To achieve the required fresh air content of 5.5 % ± 0.5 % in the air-entrained concretes the proportion of air-entraining agent had to be increased with rising granulated slag content. While for concrete P 0 the addition of 2.0 ml air-entraining agent per kg cement was sufficient to achieve the required pore space, 3.5 ml / kg cement were necessary for concrete H 75.

Table 2. Composition of the concretes

Concrete	Concrete without aea	air-entrained concrete
Cement content (kg/m^3)	350	350
Water content (kg/m^3)	175	175
w/z ratio	0,50	0,50
Air-entraining agent	without	with (target: 5.5 % air)
Aggregate type	sandy gravel	sandy gravel
Grading/ maximum particle size	A/B 16	A/B 16
Aggregate content (kg/m^3)	1830	1725

2.2. Freeze-deicing salt resistance of the concretes
The freeze-deicing salt resistance of the concretes was determined by the CDF test acc. to [15] in 3 percent NaCl solution. The total amounts of scaling after 28 freeze-thaw cycles are represented in Table 3. It may be seen that with identical curing and mixture proportions of all concretes the amount of scaling differs considerably depending on the cements used.

The amounts of scaling of the concretes without air-entrainment ranged from 2185 to 5337 g/m^2. Concretes made of cements with a medium granulated slag content showed significantly lower amounts of scaling than those made of portland cements or cements rich in granulated slag (granulated slag content $X_{GS} \geq 65$ %). Obviously, the optimum freeze-deicing salt resistance of concretes without air-entrainment can be achieved with a medium granulated slag content ($X_{GS} = 25 - 55$ %).

The amounts of scaling of air-entrained concretes ranged from 375 g/m^2 to 4875 g/m^2. The freeze-deicing salt resistance decreases with rising granulated slag content in the concretes. Due to the addition of air-entraining agents the concretes made of cements containing ≤ 45% granulated slag are below the CDF limiting value (scaling after 28 freeze-thaw cycles ≤ 1500 g/m^2) and may be classified as concretes having a high freeze-deicing salt resistance. On the other hand, concretes made of cements with a higher granulated slag content (55 %, 65 %, 75 %) show amounts of scaling higher than 1500g/m^2 despite the entrainment of artificial air voids.

Table 3. Amounts of scaling of the concretes after 28 freeze-thaw cycles

		Amount of scaling after 28 cycles (g/m^2)	
Concrete	Slag content of cement (%)	Concretes without air-entrainment	Concretes with air-entrainment
P 0	0	3750	375
E 25	25	2185	609
H 45	45	2648	1140
H 55	55	3096	2140
H 65	65	4321	3630
H 75	75	5337	4875

A comparison of the concretes with and without air-entrainment proves that the effect of air-entraining agents on the improvement of the freeze-deicing salt resistance decreases considerably when the granulated slag content is increased. Whereas the amount of scaling of the portland cement concrete could be reduced by 90 %, the air-entrained concrete made of blast-furnace cement 35 L-NW/HS (X_{GS} = 75 %) showed hardly any improvement of the freeze-deicing salt resistance related to the 0-concrete.

2.3 Analysis of courses of scaling

Not only the total amounts of scaling after 28 freeze-thaw cycles (FTC) but also the courses of scaling of the concretes differ greatly. In principle the courses of scaling of the tested concretes may be divided into three different groups (see also Fig. 1):

1. Progressive scaling behaviour with little initial scaling
2. Linear scaling behaviour
3. Degressive scaling behaviour with heavy initial scaling

For all concretes with degressive course of scaling the curve of scaling shows a distinct point of discontinuity between the initial scaling in the first 8 freeze-thaw cycles and the further course of scaling. The classification of the concretes according to the courses of scaling was made after comparison of the rates of scaling (g/m^2/ FTC) during the freeze-thaw cycles 0-8 and 8-28.

Generally, it may be stated that independent of the air void content of the concretes a degressive scaling behaviour with heavy initial scaling can be observed with concretes made of cements with a granulated slag content ≥ 55%. With rising granulated slag content the intensity of initial scaling increases. The heavy initial scaling of blast-furnace cement concretes rich in granulated slag which occurs only in the first 4-8 freeze-thaw cycles is responsible for the relatively high amounts of scaling of these concretes. So the initial scaling of the concretes H 65 and H 75 is so heavy that independent of the air void content the limiting value of the CDF test of 1500 g/m^2 is already exceeded after 4 freeze-thaw cycles, although the rate of scaling in the further course of the freeze-deicing salt attack is relatively small.

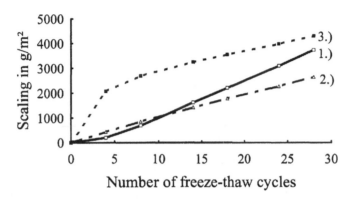

Fig. 1. Progressive (1. - example P 0), linear (2. - example H 45) and degressive (3. - example H 65) scaling behaviour of concretes without air-entrainment

3 Tests for influence of carbonation

It has been stated that the freeze-deicing salt resistance of blast-furnace cement concretes rich in granulated slag is adversely affected by heavy initial scaling. Now the question arises what the causes of this initial effect are. As initial scaling occurs in the surface layer of the concretes, an influence by carbonation may be supposed.

3.1. Correlation between depth of carbonation and depth of scaling at the point of discontinuity of the curve

To find out whether the heavy initial scaling of blast-furnace cement concretes ($X_{GS} \geq$ 55%) is related to carbonation the depth of carbonation prior to freezing was compared to the depth of scaling at the point of discontinuity of the curve of scaling.

The depth of carbonation at the point of discontinuity of the curve of scaling was determined by means of the amount of scaling and the bulk density of the concrete. To simplify the matter it was assumed that under freeze-deicing salt attack the scaling proceeds evenly over the entire area.

A reliable statement, however, could not be made because of the uncertainties in the calculation of the depth of scaling of only 6 concretes. For this reason blast-furnace cement concretes from other test series, whose depth of carbonation and course of scaling were already known, were included in the analysis. The result of this analysis is shown in Fig. 2. The relationship assumed between depth of carbonation and depth of scaling at the point of discontinuity was confirmed. There is a statistically proven relationship between the two phenomena (correlation coefficient = 0.938).

From these results it may be concluded that heavy initial scaling occurs only in the carbonated surface layer, whereas the noncarbonated core of the blast-furnace cement concretes rich in granulated slag offers sufficient resistance to freeze-deicing salt attack. The decrease in the rate of scaling (point of discontinuity of the curve) occurs at the moment when the damage proceeds from the carbonated to the noncarbonated zone of the concrete.

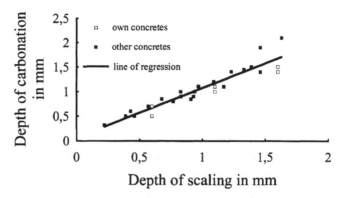

Fig. 2. Relationship between depth of initial scaling and depth of carbonation of blast-furnace cement concretes with and without air-entraining agents

3.2. Study of different kinds of prestorage

Obviously, carbonation has a strong influence on the initial scaling of blast-furnace cement concretes. This would mean that with concretes stored in low CO_2 concentration the scaling under freezing should be lower and with concretes stored in a higher CO_2 concentration the damage should be greater. To verify this assumption concrete H 65 (X_{GS} = 65 %) was once more produced without air-entraining agents. After one weeks' storage in water the specimens were separated and stored for three weeks under different ambient conditions at 20° C and 65 % rel. humidity. According to the CDF test specifications some of the specimens were stored either in nitrogen without CO_2 or in a carbonation chamber in increased CO_2 concentration (3 % CO_2). At the beginning of the CDF test the following depths of carbonation were measured:

- nitrogen storage   - 0.3 mm
- air storage   - 1.5 mm
- CO_2 storage   - 12.5 mm

As was to be expected the results of the CDF test differed greatly depending on the kind of prestorage (Fig. 3).

Air storage
After 28 freeze-thaw cycles the average scaling of the air-stored concrete specimens amounted to about 4200 g/m^2. The course of scaling confirmed the statement that the progressive initial scaling during the first 4 to 8 freeze-thaw cycles is followed by a linear course of damage of a lower intensity.

CO_2 storage
With an amount of about 35.000 g/m^2 after 28 freeze-thaw cycles the concretes stored in increased CO_2 concentration showed the worst freeze-deicing salt resistance of all concretes. The typical point of discontinuity in the curve of scaling is shifted to later freeze-thaw cycles. It occurs after the 24th freeze-thaw cycle with an amount of scaling of about 30.000 g $/m^2$ which for a bulk density of 2.35 kg/m^3 would correspond to a depth of scaling of 12.8 mm (depth of carbonation = 12.5 mm).

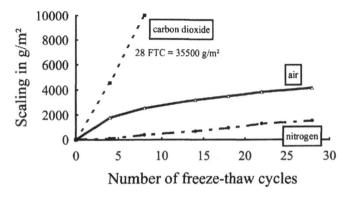

Number of freeze-thaw cycles

Fig. 3. Courses of scaling of concrete H 65 without air-entrainment stored under different conditions

Nitrogen storage
With an amount of scaling of about 1500 g/m^2 the concretes stored in nitrogen exhibited the best freeze-deicing salt resistance. In contrast to the air-stored specimens these concretes showed no progressive initial scaling in the CDF test and the course of scaling was linear from the beginning. The increase in scaling corresponded closely to that of the air-stored specimens in their second phase of damage.

Similar tests were subsequently made on blast-furnace cement concretes with air-entraining agents. Also in this case the specimens stored in nitrogen showed no initial scaling at all.

4 Research into causes of influence of carbonation

The carbonated layer of blast-furnace cement concretes rich in granulated slag had a very low resistance to freeze-deicing salt attack which resulted in heavy initial scaling in the freeze-deicing salt test. Now the question arose what the causes of the adverse effect of carbonation were. First the physical factors of influence in different layers of the concrete were analysed.

4.1 Physical factors of influence
In the investigation of the physical factors of influence in concretes without air-entraining agents the main emphasis was put on pore size distribution. In concretes with air-entraining agents the air void parameters of the hardened concrete were analysed.
    The 28 d compressive strengths of the concretes differed only slightly as all cements used were Grade 32.5 and the mixture proportions were identical.

Concretes without air-entraining agents
For the analysis of pore size distribution the separate examination of carbonated and noncarbonated layers seemed to be advisable because of the findings on the influence

of carbonation on freeze-deicing salt resistance. However, for minimum depths of carbonation of 0.5 mm the exact mechanical separation of these layers was technically not feasible. Starting from the surface of the concrete a surface layer of 0 to 2 mm and a core zone of 3 to 20 mm was analysed in the porosity tests. In the interpretation of the results it has to be noted that in the surface layer the effects from carbonated and noncarbonated layer will overlap.

The pore size distributions were determined and the pore sizes were classified acc. to Romberg [16] into:

- gel pores $r = < 10$ nm
- capillary pores $r = 10$ nm - 100 μm
- air pores $r = > 100$ μm

In the evaluation of the pore sizes with regard to freeze-deicing salt resistance special attention must be given to the capillary pores. Under the climatic conditions of Central Europe it is almost exclusively the capillary water that freezes, so that the proportion of capillary pores greatly influences the amount of ice formation. Moreover, the transport of water or vapour from the surrounding atmosphere into the interior of the concrete takes place mainly through the capillary pore system (influence on water content of the concrete).

In Fig. 4 the capillary porosity of the core zone in dependence of the granulated slag content of the cements is compared to that of the partially carbonated surface layer.

The partial carbonation of this layer causes a slight densification of the microstructure in portland cement and portland blast-furnace cement concrete which can be recognised in the small decrease in capillary pore space. On the other hand, the microstructure of blast-furnace cement concretes made of cements containing ≤ 55% granulated slag becomes definitely coarser by carbonation. The increase in capillary porosity compared to the noncarbonated core is the more significant the higher the granulated slag content of the cement is. With blast-furnace cement Grade 35 L-NW/HS (X_{GS} = 75%), which has the highest content of granulated slag, the capillary porosity due to carbonation increases absolutely by 6 % by volume.

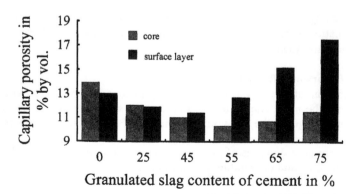

Fig. 4. Capillary porosity of surface layer and core in dependence of the granulated slag content of the cements

The answer to the question for the causes of the heavy initial scaling of blast-furnace cement concretes rich in granulated slag is, however, somewhat contradictory. For instance, in spite of its coarser microstructure blast-furnace cement concrete H 65 (X_{GS} = 55%) shows a lower capillary porosity in the carbonated layer than portland cement concrete. The rate of scaling in the surface layer of this blast-furnace cement concrete is, however, 177.8 g/m^2, that is about twice the rate of portland cement concrete (rate of scaling 86.5 g/m^2/FTC).

Air-entrained concretes
Experience shows that concrete may be protected against damage due to freeze-deicing salt attack by an appropriate air void system. If the system of artificial air voids in the concrete is adeaquate the "natural" porosity in the range of gel and capillary pores is of minor importance in the freeze-deicing salt resistance of the concrete. The results of the CDF test which showed that air-entrainment improved the freeze-deicing salt resistance of blast-furnace cement concretes ($X_{GS} \geq 55\%$) only to a small degree led to the conclusion that the air void system achieved in these concretes was less effective than that in portland cement concrete or blast-furnace cement concrete with a medium or small granulated slag content. By means of automatic image analysis the air void parameters of all concretes were determined acc. to [18]. The results are represented together with the total amounts of scaling after 28 freeze-thaw cycles in Table 4.

Table 4. Total amount of scaling and air void parameters of the concretes

Concrete	amount of scaling after 28 FTC (g/m^2)	Total air void content (% by vol.)	micro air void content < 300 μm (% by vol.)	spacing factor (mm)
P 0	375	4,8	3,2	0,12
E 25	609	5,3	3,4	0,10
H 45	1140	4,7	3,0	0,13
H 55	2140	5,0	3,9	0,10
H 65	3630	5,7	4,1	0,08
H 75	4875	6,2	4,3	0,07

The test results show that the pore space achieved in the fresh concrete can also be found in a similar quantity in the hardened concrete (total air void content), independent of the granulated slag content. That means that the stability of artificial air voids is not influenced by the granulated slag content in the cement. Compared to portland cement the air void parameters become even better when the granulated slag content is increased.

For instance, the air void system of concrete H 75 may be considered almost ideal with respect to freeze-deicing salt resistance. The limiting values for a favourable air void system (air void spacing factor of 0.07 mm and a micro air void content of 4.3 % by vol.) have been adhered to so that a high freeze-deicing resistance should be expected. Nevertheless the total amount of scaling of this air-entrained concrete was very high (4875 g/m^2). Compared to the 0-concrete without air-entrainment the amount of scaling was only reduced by 1 %. Similarly, the concretes H 65 and H 75 showed a great discrepancy between favourable air void parameters and high amount of scaling. Even after the addition of air-entraining agents to blast-furnace cement concretes rich in granulated slag a two-phase course of scaling with high initial scaling was observed.

It may be concluded that neither the relatively high total amount of scaling nor the heavy initial scaling of the blast-furnace cement concretes ($X_{GS} \geq 55\%$) may be attributed to an insufficient air void system. The addition of air-entraining agents led to highly favourable air void parameters in all layers of these concretes.

4.2 Products of carbonation

As already mentioned, carbonation effects the microstructure of portland cement and blast-furnace cement concretes in different ways. The coarser microstructure due to carbonation of blast-furnace cement concretes rich in granulated slag, which has already been reported by other authors, is certainly an important factor of influence on the heavy initial scaling of these concretes. However, the analysis of the present test results shows that the coarser microstructure due to carbonation cannot be the only cause of the low freeze-deicing salt resistance of the carbonated layer of these concretes, particularly since almost no reduction in initial scaling could be achieved by the addition of air-entraining agents. So it had to be found out whether under freeze-deicing salt attack differences occur between portland cement and blast-furnace cement concretes as to the types of carbonate phases and their stability.

The investigations into the types of carbonate phases formed depending on the granulated slag content of the cement were made on hydrated cement prisms. The prisms (4 x 4 x 16 mm) were made of the same cements as were used in the production of the concretes. After production (water/cement ratio = 0.35) the prisms were stored in water for 7 days. Unlike the concrete specimens the prisms were then not stored in the air but for three weeks in 1 percent CO_2 atmosphere (20° C/ 65% rel. hum.). Acc. to Knöfel et al. [19] a 1 percent CO_2 concentration ensures in contrast to higher concentrations that the process of carbonation is accelerated but that no other carbonate phases will form than in the normal CO_2 concentration of the air (about 0.03% by vol.). The aim of the modified prestorage was to achieve a carbonate layer sufficiently thick for the analysis (depth of carbonation ≥ 2 mm) and in this way to provide the conditions for a clear separation of carbonated and noncarbonated layer during the preparation of the specimens.

After 28 days' storage the composition of the carbonated layer was examined by radiography. Fig. 5 shows the results in dependence of the granulated slag content of the cements. It can be seen that up to a granulated slag content of about 45 % the carbonated layer consists exclusively of calcite. With higher granulated slag contents an increasing amount of the metastable calcium carbonate modifications aragonite and vaterite may be observed. Subsequent tests on concrete indicated a similar tendency, but the disturbing effect of the aggregate prevented a semi-quantitative x-ray analysis.

Research into the causes of the different types of phases in dependence of the granulated slag content was not carried out in the present study. There is some indication that the stabilisation of the usually metastable forms of calcium carbonate in cements rich in granulated slag may be caused by the magnesium oxide in the slag. Further research into this problem is required in the future.

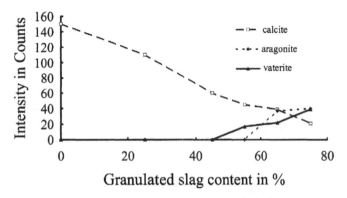

Fig. 5. CaCO₃ modifications in carbonated hydrated cement in dependence of the granulated slag content (detection peaks: calcite $d = 0.303$ nm; aragonite $d = 0.340$ nm; vaterite $d = 0.358$ nm)

After the types of carbonate phases had been determined the hydrated cement prisms were subjected to the CDF test. In principle the resulting course of scaling corresponded to the respective concretes, that means, also here heavy initial scaling was observed for granulated slag contents ≥ 55 % in the first 4 to 8 freeze-thaw cycles.

Some of the specimens made of the cements H 55, H 65 and H 75 were taken out of the CDF test already after the 4th freeze-thaw cycle. At this moment the damage was still confined to the carbonated layer so that the analysis of the surface layer provided information on the stability of the metastable carbonate phases in the carbonated hydrated cement.

The comparison of the x-ray diffraction diagrammes of the carbonated surface layer prior to freezing and after the 4th freeze-thaw cycle shows that already after this short freezing time a large proportion of the metastable calcium carbonates has transformed into stable calcite. (Fig. 6).

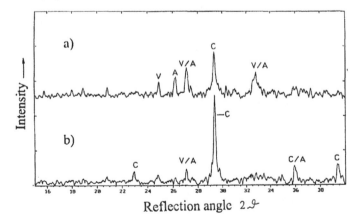

Fig. 6. X-ray diffraction diagramme hydrated cement H 65 a) before freezing and b) after the 4th freeze-thaw cycle (C - calcite, A - aragonite, V - vaterite)

The scanning electron micrographs prove that the microstructure loosens in this process. Figures 7 and 8 show the interface of nonhydrated granulated slag and carbonated hydrated blast-furnace cement before and after freezing. Smaller vaterite spheroids along with well-crystallised aragonite needles may be recognised in the carbonated zone before freezing.

During the freeze-deicing salt attack these metastable calcium carbonate modifications were completely dissolved. The dense microstructure which had formed by the various types of carbonates before the attack was clearly destabilised during the dissolution process.

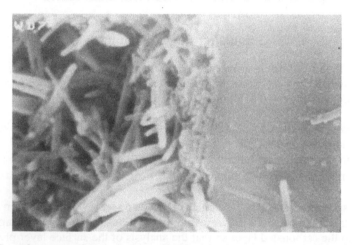

Fig. 7. Interface granulated slag / carbonated hydrated cement before freeze-deicing salt attack (20.000-fold enlargement)

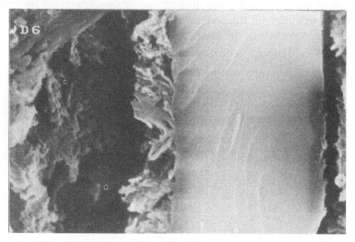

Fig. 8. Interface granulated slag / carbonated hydrated cement after freeze-deicing salt attack (20.000-fold enlargement)

As a result of the dissolution of the metastable calcium carbonates in 3 percent NaCl solution badly crystalline calcite is formed in situ which will quickly scale off under frost attack.

The investigations into the types of carbonates and their stability are an important starting point for the explanation of the heavy initial scaling of blast-furnace cement concretes rich in granulated slag. Particularly the adverse effect of a usually favourable system of artificial air voids in the carbonated layer of these concretes can now be better understood. As the existing expansion space plays no role in the dissolution of the metastable carbonates and as the depth of carbonation can hardly be reduced by air-entrainment, heavy initial scaling will also occur in air-entrained blast-furnace cement concretes rich in granulated slag.

Parallel measurements were made to study the scaling behaviour of blast-furnace cement concretes in other media than the 3 percent NaCl solution used in the CDF test (e.g. distilled water, alcoholic solutions, urea solutions). The results confirm the importance of metastable carbonates for initial scaling. The solubility of vaterite and aragonite in water is lower than in diluted chloride solutions. Therefore, under pure frost attack (CF test) a point of discontinuity in the curve indicating the transition from the carbonated to the noncarbonated zone is hardly to be seen, although in both cases the same increased portion of capillary pores due to carbonation was found in the surface layer.

5 Conclusion and ways of reducing initial scaling

The present study proves that the heavy initial scaling of blast-furnace cement concretes with a granulated slag content $\geq 55\%$ is due to the fact that the freeze-deicing salt resistance of the carbonated surface layer is lower than that of the core concrete. With lower granulated slag contents no increased initial scaling is observed.

Properly produced blast-furnace cement concretes with a lower content of granulated slag ($X_{GS} \leq 45\%$) have a high freeze-deicing salt resistance and without air-entrainment they are better than the respective portland cement concretes.

From the investigations follows that the low freeze-deicing salt resistance of the carbonated layer of blast-furnace cement concretes rich in granulated slag has two essential causes. One cause is that in contrast to portland cement concretes the carbonation of blast-furnace cement concrete rich in granulated slag leads to a coarser microstructure and an increase in capillary pore space. The other cause is that in blast-furnace cement concrete containing $\geq 55\%$ granulated slag along with calcite also considerable amounts of metastable vaterite and aragonite are formed due to carbonation. The instability of these carbonate phases under combined freeze-thaw and freeze-deicing salt attack results in a lower freeze-deicing salt resistance of the carbonate layer.

Based on these findings it was tried to find suitable ways of reducing the initial scaling of blast-furnace cement concretes rich in granulated slag and of improving the freeze-deicing salt resistance of these concretes. Two principal ways seem to be practicable: On the one hand, suitable technical measures may be taken to minimise the carbonation of the concrete in general and thus to impede the development of a coarser microstructure and to reduce the formation of metastable carbonates. For this purpose draining formwork material may be recommended, which densifies the surface by draining off the superflous mixing water so that the progress of carbonation is slowed down [20],[21]. On the other hand, certain substances might be used to influence the carbonation of the blast-furnace cement concrete at least to such an extent that it resembles that of portland cement concrete and in this way to reduce the development of a coarser microstructure and /or the formation of metastable carbonates. For instance, the addition of calcium hydroxide to the concrete might be a possibility. The effects of both measures on initial scaling were tested on concrete H 65 without air-entrainment. The courses of scaling are shown in Fig. 9.

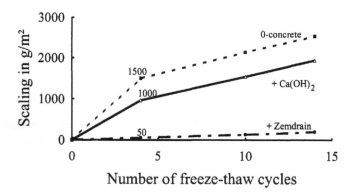

Fig. 9. Course of scaling of concrete H 65 without air-entrainment: without additional measures, after addition of 10 % $Ca(OH)_2$ (related to the cement) to the concrete, and after using the formwork material Zemdrain

The comparison to the 0-concrete shows that initial scaling was reduced by both measures. By the addition of 10 % $Ca(OH)_2$ (related to the cement content) to the concrete initial scaling was reduced from 1500 to 1000 g/m^2. The complete prevention of initial scaling was, however, not possible by this measure. The typical two-phase course of damage remained. The draining formwork material Zemdrain had a much greater effect. Initial scaling disappeared completely (reduction from 1500 to 50 g/m^2) and the course of damage became linear with a low total amount of scaling (500 g/m^2). It must be emphasised that in this case the addition of an air-entraining agent is not required.

These first attempts show that under consideration of the new findings initial scaling may be reduced by appropriate measures and the freeze-deicing salt resistance of blast-furnace cement concretes rich in granulated slag may be increased. Further ways of reducing initial scaling should be studied in the future. For reasons of economy emphasis should be put on material measures.

6 References

1. Haegermann, B. (1987) Zum Einfluß der Nachbehandlung und der Lagerung auf die Betonqualität unter besonderer Berücksichtigung des Frost-Tausalz-Widerstands. *Dissertation*, Bergische Universität-Gesamthochschule Wuppertal.

2. Virtanen, J.(1984) Mineral by-products and freeze-thaw resistance of concrete. *Nordic Concrete Research,*Vol. 3, pp. 191 - 208.

3. Bonzel, J. and Siebel, E. (1977) Neuere Untersuchungen über den Frost-Tausalz-Widerstand von Beton. *Beton*, Vol.27, No. 4, pp. 153 - 157 and No. 5, pp. 205 - 211 and No. 6, pp. 237 - 244.

4. Hilsdorf, H.K. and Günther, M. (1986) Einfluß der Nachbehandlung und Zementart auf den Frost-Tausalz-Widerstand von Beton. *Beton- und Stahlbetonbau*, Vol. 8, No. 3, pp. 57 - 62.

5. Haegermann, B. (1988) Einfluß der Nachbehandlung und der Lagerung auf die Betonqualität unter besonderer Berücksichtigung des Frost-Tausalz-Widerstands. *Beton-Informationen*, Vol. 28, No.1, pp. 3 - 9.

6. Hartmann, V. (1993) Optimierung und Kalibrierung der Frost-Tausalz-Prüfung von Beton - CDF-Test. *Dissertation*, Universität-Gesamthochschule Essen.

7. Blunk, G. and Brodersen, H.A. (1980) Zum Widerstand von Beton gegenüber Harnstoff und Frost. *Straße und Autobahn*, Vol. 31, No. 3, pp. 119 - 131.

8. Luther, M.D. (1994) Scaling resistance of ground granulated blast furnace slag concretes. *Proceedings*, 3rd International CANMET/ACI-Conference "Durability of Concrete", Nice, pp. 47 - 64.

9. Weigler, H. and Karl, S.(1989) *Beton. Arten - Herstellung - Eigenschaften*, Ernst & Sohn Verlag, Berlin.

10. Brodersen, H.A. (1978) Zum Frost-Tausalz-Widerstand von Beton und dessen Prüfung im Labor. *Beton-Informationen*, Vol. 18, No. 3, pp. 26 - 35.

11. Moritz, H.and Vinkeloe, R. (1989) Erhebungen an Kläranlagen. *Beton-Information*, Vol. 29, No. 2, pp. 21 - 23.

12. Geiseler, J. and Lang, E. (1994) Dauerhaftigkeit von Hochofenzementbetonen ohne künstliche Luftporen. *Wissenschaftliche Zeitschrift HAB Weimar*, Vol. 40, No. 5/6/7, pp. 105 - 110.

13. DIN 19569/ Teil 1. (1987) *Kläranlagen - Baugrundsätze für Bauwerke und technische Ausrüstung - Allgemeine Baugrundsätze.*

14. Malhotra, V.M., Carette, G.G. and Bremner, T.W. (1987) Durability of granulated blast-furnace slag concrete in marine environment. *Proceedings*, International Workshop on Granulated Blast-Furnace Slag in Concrete, Toronto, pp. 171 - 201.

15. RILEM TC 117-FDC. (1997) CDF-Test - Prüfverfahren des Frost-Tau-Widerstands von Beton- Prüfung mit Taumittel-Lösung (CDF) - RILEM Recommendation. *Betonwerk + Fertigteil-Technik*, Vol. 63, No.4, pp. 100 - 106.

16. Romberg, H. (1978) Zementsteinproben und Betoneigenschaften. *Beton - Information*, Vol. 18, No. 6, pp. 269 - 277.

17. Bier, T.A. (1988) Karbonatisierung und Realkalisierung von Zementstein und Beton. *Dissertation*, Universität Karlsruhe.

18. Arbeitsausschuß DIN 1048. (1991) Prüfung von Beton - Empfehlungen und Hinweise als Ergänzung zu DIN 1048. *Deutscher Ausschuß für Stahlbeton*, No. 422.

19. Knöfel, D. and Eßer, G.(1983) Einfluß unterschiedlicher Kohlendioxidkonzentrationen auf Zementmörtel. *Proceedings*, 1. Int. Kolloqium Werkstoffwissenschaften und Bausanierung, Esslingen, pp. 1408 - 1418.

20.   Karl, J.-H. and Solacolu, C. (1993) Verbesserung der Betonrandzone. Wirkung und Einsatzgrenzen der saugenden Schalungsbahn. *Beton*, Vol. 43, No. 5, pp. 222 - 225.
21.   Vißmann, H.-W. (1994) Qualität der Betondeckung. Dichte der Betonrandzone. *Beton*, Vol. 44, No. 5, pp. 260 - 264.

Frost and frost-deicing salt resistance of supersulphated cement concrete

U. KNAACK and J. STARK
Bauhaus-Universität, Weimar, Germany

Abstract
The present paper reports on investigations made into a cement very rich in granulated blast-furnace slag (>75% granulated blast-furnace slag) and chemically activated by sulphates: supersulphated cement. The frost and frost-deicing salt resistance of supersulphated cement concrete is very low. Tests have shown that just as with blast-furnace cement concrete the carbonation of supersulphated cement concrete leads to a coarser microstructure and to the formation of metastable $CaCO_3$ modifications. Due to the strong tendency of ettringite towards carbonation the rate of carbonation is, however, even higher than that of blast-furnace cement concrete.

Various ways of improving the frost-deicing salt resistance, such as the improvement of curing, the reduction of the water-to-cement ratio, the use of a curing agent as well as the use of draining formwork material have been explored. The required improvement of the frost-deicing salt resistance could only be achieved by means of the draining formwork material.

Keywords: carbonation, curing, draining formwork material, frost resistance, frost-deicing salt resistance, supersulphated cement concrete

1 Introduction

Numerous studies [1] [2] [3] have shown that blast-furnace cement concrete (blast-furnace slag content > 55%) has a good frost resistance, but that its frost-deicing salt resistance is sometimes low and the initial scaling is heavy. The low frost-deicing salt resistance is caused by carbonation. Carbonation leads to a considerably coarser microstructure and to the formation of metastable $CaCO_3$ modifications which under

Frost Resistance of Concrete, edited by M.J. Setzer and R. Auberg. Published in 1997 by E & FN Spon, 2–6 Boundary Row, London SE1 8HN, UK. ISBN: 0 419 22900 0.

frost-deicing salt attack will behave unstable. [1]

The chief objective of our investigations was to study a system even richer in granulated blast-furnace slag: supersulphated cement. This was due to the fact that hardly any test results on the durability of supersulphated cement concrete were available, particularly on the resistance to frost and frost-deicing salt attack. Supersulphated cement was used in Germany mainly in the fifties. Since 1970 the use of supersulphated cement has no longer been allowed by the building authorities. It is necessary to deal with this binding material/concrete as structures made of supersulphated cement (e.g. dams) require proper remedial work.

2 Characteristics of supersulphated cement

Supersulphated cement consists of 75 to 85 % highly basic granulated blast-furnace slag, with an Al_2O_3 content $\geq$ 15%. In contrast to blast-furnace cement its hydraulic properties are activated by sulphates. Therefore, 12 to 18% anhydrite has to be added. In order to achieve the optimum pH value (pH = 11.5 - 12) required for the hardening of supersulphated cement it contains about 1 - 5% portland cement. Along with the C-S-H phases ettringite is an essential product of hydration. Furthermore, monosulphate, C-A-H phases and $Al(OH)_3$ are formed.[4]

The following composition of supersulphated cement was found to be most suitable for small-scale production:

· 85.0% granulated blast-furnace slag (Al_2O_3 = 16.5%)
· 13.5% anhydrite
· 1.5% portland cement

Carbonation has a great effect on the quality of supersulphated cement concrete. Just as in the case of blast-furnace cement concrete carbonation leads to a coarser microstructure and to the formation of metastable $CaCO_3$ modifications (aragonite, vaterite). The rate of carbonation is, however, higher than that of blast-furnace cement. One reason is that the tendency of ettringite towards carbonation is greater than that of the C-S-H phases. Besides, there is no $Ca(OH)_2$ that could impede carbonation.

Table 1 shows the depths of carbonation measured after 28 days storage (7 days under water / 21 days at 20°C and 65% relative humidity) and after an accelerated carbonation (7 days under water / 21 days in an atmosphere containing 3% CO_2 at 75% relative humidity).

The carbonation of supersulphated cement concrete leads to a reduction in compressive strength and is cause of the dusting of set concrete surfaces.

Table 1. Depths of carbonation of blast-furnace cement concretes rich in granulated slag and of supersulphated cement concretes

	standard storage (7 days water/ 21 days 20°C-65%r.h.)	'CO$_2$-storage' (7 days water/ 21 days 3% CO$_2$-atmosphere-75%r.h.)
blast-furnace cement concrete (75% b.f.slag)	1.5 ... 3.0	8.0
supersulphated cement concrete	4.0 ... 4.5	12.0

3 Frost and frost-deicing salt resistance

The frost and frost-deicing salt resistance of supersulphated cement concrete was determined by CF/CDF tests [5]. The results were compared to those of blast-furnace cement concretes rich in granulated slag.

In contrast to blast-furnace cement concrete supersulphated cement concrete has a very low frost resistance (Fig. 1)

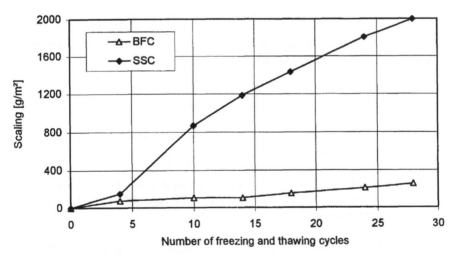

Fig. 1. Comparison of blast-furnace cement concrete and supersulphated cement concrete under pure frost attack (CF test)

One cause of the low frost resistance may be the low degree of hydration at the time of testing (age of concrete - 28 days). The capillary porosity of the microstructure of the supersulphated cement concrete is still rather high. Furthermore, the storage under water for one week is too short to achieve a better frost resistance of supersulphated cement concrete. Certainly, carbonation which causes dusting and scaling of the concrete surface plays also an important role.

Not only the frost resistance but also the frost-deicing salt resistance of supersulphated cement concrete is lower than that of blast-furnace cement concrete (Fig. 2). The courses of scaling of both concretes are similar. The heavy initial scaling is followed by a period of lower intensity of scaling. This course of scaling is caused by carbonation. As described in chapter 2, the carbonation of supersulphated cement will also lead to a coarser microstructure and to the formation of metastable $CaCO_3$ modifications.

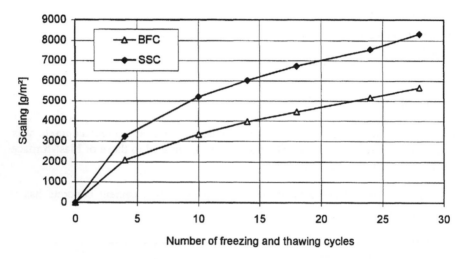

Fig. 2. Comparison of blast-furnace cement concrete and supersulphated cement concrete under frost-deicing salt attack (CDF test)

4 Ways of improving the frost-deicing salt resistance

A way of improving the frost-deicing salt resistance is to impede carbonation as far as possible. But the entrainment of artificial air voids which proved so successful in portland cement concrete will not lead to any significant improvement neither in blast-furnace cement concrete rich in granulated slag nor in supersulphated cement concrete. On the contrary, it had to be stated that the frost-deicing salt resistance of supersulphated cement concrete became even worse.

So, in the first place, the effect of the following <u>changes in curing</u> was studied:

· 23 days under water (20°C) / 5 days in the laboratory (20°C - 65% relative humidity)
· 23 days under water (35°C) / 5 days in the laboratory (20°C - 65% relative humidity)
· 7 days in saturated $Ca(OH)_2$ solution / 21 days in the laboratory (20°C - 65% relative humidity)
· 23 days in saturated $Ca(OH)_2$ solution / 5 days in the laboratory (20°C - 65% relative humidity)

The storage in saturated $Ca(OH)_2$ solution was chosen in accordance with ASTM C 672-92 and ASTM C 511-93.

The courses of scaling are represented in Fig. 3.

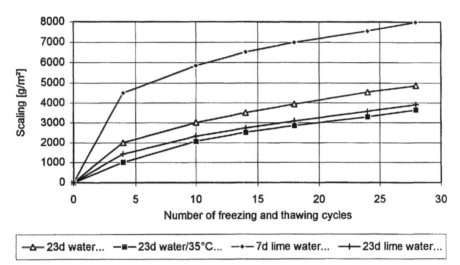

Fig. 3. Scaling of supersulphated cement concrete after different methods of pre-storage

By the improvement of curing the amount of scaling could be reduced, in some cases to 50% of the amount of scaling after standard storage. A significant improvement of the microstructure could not be observed. That means that the improvement was only achieved by the impediment to carbonation due to the moist storage and not by a denser microstructure.

The following depths of carbonation of the concretes (age 28 d) were measured:

· 23 days under water (20°C) / 5 days 20°C - 65% r.h.	0.5 mm
· 23 days under water (35°C) / 5 days 20°C - 65% r.h.	1.0 mm
· 7 days saturated $Ca(OH)_2$ solution / 21 days 20°C - 65% r.h.	2.5 mm
· 23 days saturated $Ca(OH)_2$ solution / 5 days 20°C - 65% r.h.	0.5 mm

It had to be concluded that by changes in curing it was not possible to produce a supersulphated cement concrete with a high frost-deicing salt resistance.

Other ways of reducing the initial scaling of supersulphated cement concrete had to be considered.

A denser microstructure may be achieved by a low water-to-cement ratio. As the complete hydration of supersulphated cement requires a high water content the maximum reduction of the water-to-cement ratio had to be limited to 0.4.

Superplasticisers made from melamine resin or naphthalene sulfonate were employed to achieve the consistency required.

The effect of a <u>curing agent</u> was also studied. Aliphatic paraffin-wax emulsion protects the material from drying out and impedes carbonation.

In the third place the influence of a <u>draining formwork material</u> on the frost-deicing salt resistance was examined. It is a spinning-fibre material made of thermally continuous polypropylene fibre.

Fig. 4 shows the effect of this material on the frost-deicing resistance.

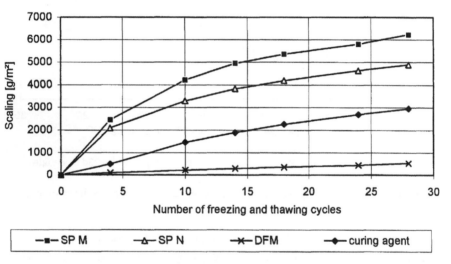

Fig. 4. Scaling of supersulphated cement concrete with superplasticisers made from melamine resin (M) or naphthalene sulfonate (N)/(w/c = 0.4) and with a curing agent and draining formwork material (DFM)

A comparison of the courses of scaling shows that the required improvement of the frost-deicing salt resistance can only be achieved by the use of the draining formwork material.

It is true that the reduction in the water-to-cement ratio and the addition of superplasticisers results in an increase in compressive strength and the formation of a slightly denser microstructure, but a sufficient improvement of the frost-deicing salt resistance could not be achieved. The depth of carbonation at the beginning of the frost-deicing salt attack was 3.0 mm.

The use of a curing compound could also not prevent carbonation (1.5 mm at the beginning of frost attack). Although a significant reduction in scaling has been achieved the supersulphated cement concrete modified by these measures cannot be classified as frost-deicing salt resistant.

In the remedial work of a dam in Thuringia the resistance of concrete to pure frost attack was of particular importance. For this reason the effect of draining formwork material on the frost resistance of supersulphated cement concrete was investigated.

Also in this case a very high frost resistance of the concrete could be achieved by the use of draining formwork material. Whereas the total scaling of 'normal' super-sulphated cement concrete (w/c = 0.5; cement content = 370 kg/m^3) amounted to 1997 g/m^2, the amount of scaling was reduced to 64 g/m^2 when the formwork material was used. Consequently, a section of the dam was concreted using the formwork material. After one year's service life no damage to the concrete elements produced in this way has been observed.

5 References

1.   Ludwig, H.-M. (1996) *Zur Rolle von Phasenumwandlungen bei der Frost- und Frost-Tausalz-Belastung von Beton,* Dissertation an der Hochschule für Architektur und Bauwesen Weimar.
2.   Hilsdorf, H. K. and Günter, M. (1986) Einfluß der Nachbehandlung und Zementart auf den Frost-Tausalz-Widerstand von Beton. *Beton und Stahlbetonbau,* Vol.81, No.3. pp.57-62.
3.   Lang, E. (1996) Untersuchungen zur Erhöhung des Frost-Tausalz-Widerstandes von HOZ-Betonen. *Report des Forschungsinstitutes,* No.1. pp.2-3.
4.   Stark, J. (1995) Sulfathüttenzement. *Wissenschaftliche Zeitschrift HAB Weimar,* Vol. 41, No. 6/7. pp.7-15.
5.   RILEM TC 117-FDC (1995) Draft Recommendation for Test Method for the Freeze-Thaw Resistance of Concrete: Test with Water (CF) or with Sodium Chloride Solution (CDF). *Materials and Structures,* Vol.28, pp.175-182.

Sorption of chlorides on hydrated cements and C₃S pastes

O. WOWRA and M.J. SETZER
IBPM – Institute of Building Physics and Materials Science,
University of Essen, Essen, Germany

Abstract

A concentration dependent adsorption of chloride ions for hydrated cements and C_3S pastes was detected. The capacity of adsorption is mainly determined by the specific surface, the pH value and the associated cation. Chloride adsorption is obviously preceded by a sorption of calcium ions. The adsorption of Ca^{2+} ions increases the positive charge on the surface of the used materials and induces further adsorption of chloride.

Keywords: deicing agents, chloride adsorption, calcium-induced adsorption

1 Introduction

It has been known for a long time that chloride ions lead to corrosion in reinforced concrete under the presence of oxygen and water. On the other hand chlorides have been used as hydration accelerators. Chlorides penetrate hardened concrete since they are used as deicing agents. Therefore, research work has been done very early on binding of chloride ions. According to well-known works of Richartz [1] chloride ions during hydration are principally bound by the aluminates of cement paste. The mechanism of chloride binding on hardened cement paste is not clarified. According to Binder the aluminate phases are the chief part of the chloride sorption [9]. However, results of Arya et al. show that C_3A content has little effect on the binding of chlorides introduced after 2 days of hydration [2]. The experimental results of Tang und Nilsson [3] show that the chloride binding capacity of concrete strongly depends on the content of CSH gel in the concrete. These results are confirmed by Beaudoin et al. [6]. The authors examined the chloride content of CSH-phases after the action of a 3,8% $CaCl_2$ solution.

Frost Resistance of Concrete, edited by M.J. Setzer and R. Auberg. Published in 1997 by E & FN Spon, 2–6 Boundary Row, London SE1 8HN, UK. ISBN: 0 419 22900 0.

Chlorides of deicing salts which penetrate hydrated cement or concrete may also be adsorbed. The sorption characteristics of the cement paste surface probably also influences the freeze thaw and deicing salt resistance in addition to the corrosion. Exact knowledge about the sorption capacity of hydrated cement and of the CSH phase under the influence of different concentrations and temperatures is lacking up to now.

2 Materials and Methods

The chemical compositions and the physical properties of the materials used in this study are listed in Table 1. The chemical analyses were carried out with AA spectrophotometry and ion chromatography. For the hydration of the binders deionized water was used. Dilutions were prepared with ultrapure water with a specific conductivity less than 18 μScm^{-1}.

2.1 Preparation of binders

If chloride solutions attract hydrated cement it always should be taken into account that there will be diffusion effects. To minimise such effects bottle hydrated compounds with a water-solid ratio of 3 were used in the following experiments [4]. The hydration of the cements were carried out in polyethylene bottles under continuous rotation for one year at room temperature. The pure cement minerals were cured for 3 months under the same conditions. To avoid carbonation the samples were produced with degassed water.

Tab.1 Compositions of the studied cements and calcium silicates [w-%]

	C_3S	C_3S B	CEM I 32.5 R	CEM III 32.5 NW/NA
CaO	69.29	77.9	62.4	48.20
SiO_2	26.44	19.88	19.1	27.03
Al_2O_3	0.30	0.80	4.91	7.75
Fe_2O_3	n.b.	0.18	2.20	1.25
MgO	0.53	0.67	1.50	6.04
K_2O	0.16	0.12	0.90	0.37
Cl	<0.01	<0.01	<0.01	<0.01
SO_3	<0.01	0.11	4.30	8.74

Two cements a CEM I 32.5 R (OPC) and a CEM III 32.5 NW/NA (SC) with different contents of aluminate were chosen for the experiments. Further two calcium silicate hydrates (CSH and CSH B) with a different C/S ratio were used. Characteristics of the samples are given in Tables 2 an 3. X-Ray diffraction (Siemens

D5000) and IR spectra (Bruker IFS 55) of the samples were made. After the hydration of the C_3S a large absorption band around $975 cm^{-1}$ was observed in the IR spectra. The surface areas were measured by nitrogen adsorption and calculated with the ESW method [5]. Density values were determined pycnometrically with heptan.

Tab. 2 Properties of cements

cement	CEM III 32.5 NW/NA	CEM I 32.5 R
loss on ignition [%]	1.35	1.78
relative density [kg/dm³]	2.9022	3.0354

Tab. 3 Properties of hydrated binders

binder	C_3S	C_3S B	CEM III 32.5 NW/NA	CEM I 32.5 R
loss on ignition [%]	21.37	22.7	24.56	35.00
specific surface [g/m²]	7.58	15.72	19.84	11.47

2.2 Experimental procedures

Chloride sorption was measured at various concentrations using NaCl and $CaCl_2$ solutions. About 3 g of sample was put into a polyethylene vessel before 30 ml of pre-tempered salt solution were added. The vessels were covered and stored at 273 K or 293 K. In preliminary experiments the sorption equilibrium could be reached in 14 days. Nevertheless, all samples were stored for 28 days. At the end of the storage period, the solutions were separated by filtration (< 5 µm). The concentration of Cl⁻ was determined by ion chromatography.

The amount of the Cl⁻ sorbed (c_{ads}) was calculated from the difference between the added solution (c_o) and the equilibrium solution (c_e). The content of the sorbed Cl⁻ was calculated by the following equation:

$$n = c_{ads} \cdot \frac{V_{Lsg}}{M_{Chlorid} \cdot m_{ads}} \qquad (2)$$

where:

n = sorbed chloride content [mmol/g]
$M_{Chlorid}$ = molar weight of chloride (35.453) [g/mol]
m_{ads} = weight of dry sample [g]
c_{ads} = sorbed amount of chloride [g]
V_{Lsg} = volume of solution [mL]

To observe the influence of competing anion, the sorption of Cl⁻ from $CaCl_2$ solution with C_3S paste was also measured in the presence of SO_4^{2-} ions. For that purpose 30 mg $CaSO_4$ (705 mg SO_4^{2-}/L) were added to each vessel.

3 Results and Discussion

Chloride sorption isotherms for CEM I 32.5 R and CEM III 32.5 NW/NA with NaCl and $CaCl_2$ are shown in Figures 1 and 2. The sorption behaviour of both cements is very similar with one exception. The sorption capacity of CEM I 32.5 R in presence of $CaCl_2$ solution is mostly higher than that of CEM III 32.5 NW/NA. As shown in Table 3, it is important to note that the specific surface area of the two hydrated cements are rather different. When the content of bound chloride would be expressed by specific surface or aluminate content of the two materials there will be the same difference between the two isotherms. Therefore, it can not be distinguished clearly from which effect the sorption capacity primarily depends. For both cements the following could be observed:

- A higher sorption capacity could be observed when Cl⁻ was added as $CaCl_2$ instead of NaCl.

- The chloride sorption increased with temperature in the presence of $CaCl_2$.

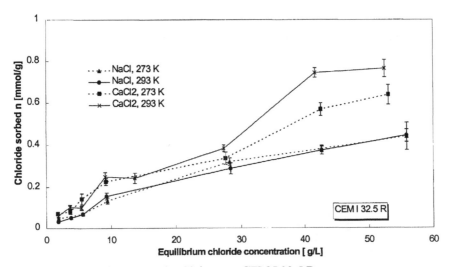

Figure 1 Sorption isotherms for Cl⁻ ions on CEM I 32.5 R

The sorption isotherms for the C_3S paste are shown in Figure 3. The amount of sorbed Cl⁻ depends unequivocally with the concentration of the solution. There is also a significant influence of the associated cation. The sorption capacity of the C_3S paste is, under equal conditions, in all cases lower than that of the hydrated cements. It is important to note that the smallest specific surface area (7.58 m²/g) was measured for this material (Table 3).

Figure 4 presents the results obtained when the CaCl$_2$ concentration in solution was varied at constant SO$_4^{2-}$ concentration. The sorption experiments showed that sorption was lower in the presence of SO$_4^{2-}$. Increasing the temperature leads to a higher sorption of Cl$^-$ ions.

The results indicate that there is a limited number of sites for Cl$^-$ sorption. The quantity of sites probably increases with increasing temperature of the system.

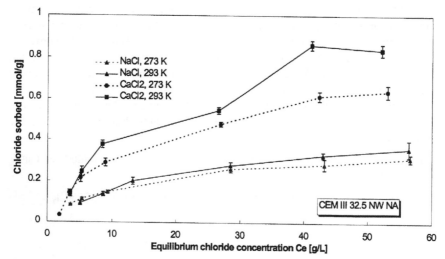

Figure 2 Sorption isotherms for Cl$^-$ ions on CEM III 32.5 NW NA

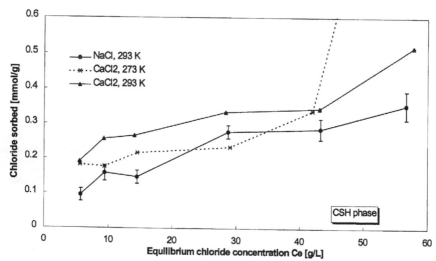

Figure 3 Sorption isotherms for Cl$^-$ ions on C$_3$S paste

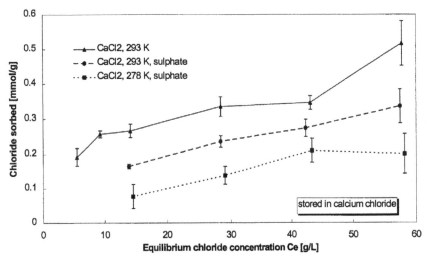

Figure 4 Sorption isotherms for Cl⁻ ions on C₃S past

The sorption isotherms of the two types of C₃S pastes with $CaCl_2$ solution are shown in Figure 5. The materials had varying C/S ratio and specific surface areas. It can be seen that there is a good relationship between the amount of sorbed Cl⁻ and the specific surface areas as shown in Table 3.

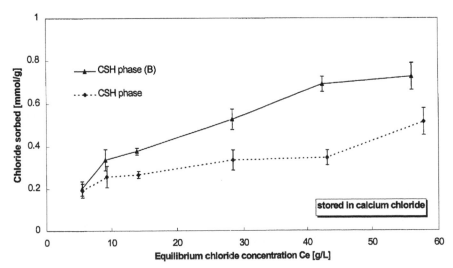

Figure 5 Sorption isotherms for Cl⁻ ions on C₃S and C₃S B paste

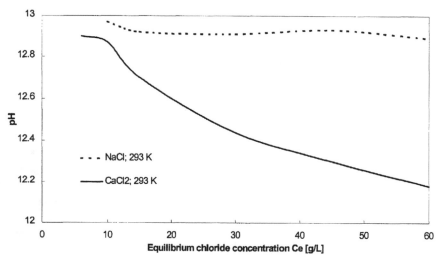

Figure 6 pH values for C₃S paste suspension at various salt concentrations

The solutions of $CaCl_2$ and NaCl are neutral. Figure 6 demonstrates graphically the evaluation of pH values for C_3S paste suspensions with various salt concentrations . Contrary to NaCl which does not influence the pore solution, the influence of the Ca^{2+} ions leads to a decline of the pH-values.

4 Conclusion and Interpretation

Although it was observed by other workers [9] that chlorides in deicing salts react with the alumina compounds in hardened cement to Friedel's salt, the results show that the aluminate content has only little effect on the Cl^- binding.

The main influence was that of the specific surface areas, the chloride concentration and the associated cation [7]. A lower Cl^- sorption was detected in the presence of SO_4^{2-} ions. These results support the interpretation that the sorption can only take place at a limited number of reactive surface sites or in a diffuse ion swarm. The Cl^- sorption with calcium chloride increased with increasing in temperature. The sorption of Ca^{2+} ions lowers the pH values.

This is a strong evidence that Ca^{2+} ions are able to penetrate the compact part of the ionic double-layer or are bound to the solid surface. Different authors obtained calcium-induced anion adsorptions on ionic surfaces in aqueous solutions [10,11]. There are $-OH^{2+}$, $-OH$ and O^- groups on an ionic surface in aqueous solution, the relative amounts depending on pH. At high pH values, where the surface is negatively charged, the Ca^{2+} ions are adsorbed and reverse the surface charge of the solid phase. Different measurements show that Ca^{2+} ions are able to convert the charge polarity of $\gamma\text{-}Al_2O_3$, SiO_2, Fe_2O_3 and other minerals [8,10]. No such effects could be observed with monovalent electrolytes like NaCl.

Reactive surface sites for hydrated cement could be defects in the surface structure or silanol and siloxan groups. The decreasing pH values in the presence of $CaCl_2$ solution, could be a result of a ligand exchange reaction on the surface, which could be described by the formation of an inner sphere complex . An example for such a reaction can be described by the stoichiometric formula:

$$SiOH + Ca^{2+} \rightleftharpoons SiOCa^+ + H^+$$

A reverse of surface charge may lead to a concentration of counter ions (Cl⁻ and OH⁻) in the Stern or the Gouy layer. In accordance with the triple layer model, the adsorbed anions are assigned to outer-sphere complexes and the diffuse ion swarm [13].

5 References

1. Richartz, W. (1969) *Zement-Kalk-Gips*, **22**

2. Arya, C., Buenfeld, N.R. and Newman, J.B.(1990) *Cem. Concr. Re.*, **20**

3. Tang L. and Nilsson, L.O.(1993) *Cem. Concr. Res.*, **23**

4. Brunauer, S., Kantro, D.L. and Copeland, L.E. (1958*) J. Am. Chem. Soc.*, **80**

5. Adolphs, J. and Setzer, M. J. (1996*) J. Colloid Interface Sci.*, **180**

6. Beaudoin, J.J., Ramachandran, V.S. and Feldman, R.F. (1990) *Cem. Concr. Res.*, **20**

7. Wowra, O. and Setzer, M.J. (1997) *13ᵗʰ International Conference on Building Materials*, Weimar (in the press)

8. Breeuwsma, A. and Lyklema, J. (1973) *J. Colloid Interface Sci.*, **44**

9. Binder, G. (1993) *Zement Kalk Gips*, **12**

10. Bolan, N.S., Syers, J.K. and Summer, M.E. (1993) *Soil Sci. Soc. Amer. Proc.*, **57**

11. Kinniburgh, D.G., Syers, J.K. and Jackson, M.L. (1975) *Soil Sci. Soc. Amer. Proc.*, **39**

13. Davies, J.A., James, R. O. and Leckie J.O. (1978) *J. Colloid Interface Sci.*, **63**

PART III

Physical Parameters and Testing

Basis of testing the freeze-thaw resistance: surface and internal deterioration

M.J. SETZER
IBPM – Institute of Building Physics and Materials Science,
University of Essen, Essen, Germany

Abstract

Appropriate testing of freeze thaw resistance requires a sufficient understanding of the macroscopic and microscopic mechanisms leading to frost damage under freeze thaw and deicing salt attack. Macroscopic and microscopic effects such as abnormal freezing of the structured and pre-structured gel water, depression of freezing point by dissolved salts and by surface interaction, supercooling and nucleation, transport mechanisms due to temperature gradients, and different thermal expansion of water and solid and the formation of micro-ice lenses connected with gel shrinkage is described. Macroscopic and microscopic for damage are described. On this basis the essential consequences for freeze thaw testing are outlined. The test procedures treated in RILEM TC 117 FDC „Freeze Thaw and Deicing Salt Resistance“ are shortly discussed. The essential aim of any testing, the precision was meanwhile assessed for CDF test which is therefore recommended by RILEM. Future work especially with respect to internal damage is proposed.
Keywords: Test Procedures, Freeze thaw resistance, internal damage, basic models

1 Introduction

Testing the freeze thaw resistance of concrete is no simple task. Since two contradictions have to be solved. Firstly, the long term behaviour should be measured within a time as short as possible. Secondly, the broad variety in modifications of frost attack shall be covered by a test procedure which must, on the other hand, be restrictive in the relevant test parameters to achieve a sufficient precision. Nevertheless, practical experience not at least of RILEM TC 117 FDC „Freeze Thaw and Deicing Resistance of Concrete“ shows that a fair compromise is possible. To reach this aim some at least

Frost Resistance of Concrete, edited by M.J. Setzer and R. Auberg. Published in 1997
by E & FN Spon, 2–6 Boundary Row, London SE1 8HN, UK. ISBN: 0 419 22900 0.

semi-quantitative knowledge is necessary about the mechanisms generating the frost damage. A short survey of these mechanisms will therefore be given first in this contribution.

On the other hand a procedure to test both freeze thaw and freeze thaw and deicing salt resistance of concrete can be separated into several parts:

1. The methodology how freeze thaw and deicing salt attack is generated in the laboratory. In this part the relevant test parameters and test conditions must be fixed in such a way that the precision, that means both repeatability and reproducibility is sufficiently good as documented in ISO 5725.

2. The type of damage and its measurement have to be defined. At the time being there are two major types of damage:

    2.1. *The scaling of material from the surface.* This is the mostly used kind of measurement since weighing of segregated material can be done relatively easy.

    2.2. *Internal damage of the concrete.* This damage precedes this scaling. Its progress is much deeper. It leads to loss of strength, elastic modulus, i.e. of the mechanical properties for which concrete is designed.

2 Mechanisms leading to frost damage

2.1 Pore water - basic considerations

This part of the article shall give only a comprehensive survey of the different mechanisms discussed in literature. To understand the correlation between these mechanisms and the freeze thaw testing one must keep in mind that during the test procedure the specimens are subjected to freeze thaw cycles and that the specimens are allowed during these cycles to suck up the test solution. Therefore, we must be aware that we have a dynamic system where the thermodynamic equilibrium is never reached in a macroscopic scale and only restricted in a microscopic scale. We have a combined heat and moisture transport to and from the outer surface and within the microstructure of the concrete.

In addition, the pore system is rather heterogeneous where these transport phenomena take place. The reason for these observations is that the pore size distribution ranges from some nanometres to some millimetres, that means, its range covers more than 6 orders of magnitude. The physical and chemical behaviour in this system is not only subjected to macroscopic bulk phenomena but also to submicroscopic surface physics and chemistry. Therefore, we have to distinguish between macroscopic and microscopic effects and to take into account that frost action is a dynamic process with stable and metastable states. To separate the different effects I distinguished between gel pores, capillaries and coarse pores. In gel pores surface physics are at least not negligible or even dominant. In capillaries water behaves mostly macroscopic. However, due to interfaces capillary suction takes place.

Table 1. Classification of pores

Name	R_H	Characteristics	Type of pore water
Coarse		Empty	
Macro Capillaries	<1 mm	Suction: Instantly,	Macroscopic
Meso Capillaries	<30 µm	" minutes to weeks,	bulk
Micro Capillaries	<1 µm	" no macroscopic equilibrium	water
Meso gel pores	<30 nm	Transition from bulk to surface physics	Prestructured condensate
Micro gel pores	<1 nm	Surface Physics	Structured surface water

Besides, the pore system water itself has some specific physical behaviour which should be kept in mind. Due to the high polarity of the water molecule the crystal lattice of ice is governed by hydrogen bonds. Every water molecule in ice is surrounded by four other molecules. There is a tetrahedral arrangement of the water molecules. Above melting point this tetrahedral arrangement is still found in liquid water. Only 15 % of the hydrogen bonds are broken. However, in contrast to ice every water molecule is now surrounded by five or more other molecules with a decreased binding strength. That means, the arrangement is denser and the density of water is increased. Due to this we have the well-known expansion of ice by 9 vol.-%.

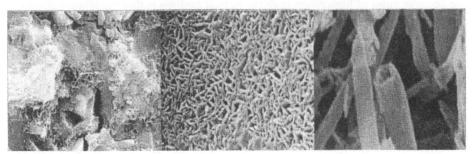

Fig. a: 5000x ⊢————⊣5 µm   Fig. b: 30 000x ⊢——⊣ 0.5 µm Fig. c: 100 000x ⊢⊣ 50 nm

Figure 1. Raster Electron Microscopy of hardened cement paste. Fig. A meso-capillaries:, fig. B: micro-capillaries and fig C: meso-gelpores (by Kreuzfelder IBPM Uni. Essen).

Furthermore, water is an excellent solvent. The solved ions decrease the freezing point. However, in contrast to liquid water the ice crystal is not able to solve substantial amounts of the solved molecules in water. Therefore, by ice formation the concentration in the unfrozen solvent is increased and an osmotic pressure is generated in a pore system.

2.1.1 Water in gel pores

In contrast to the capillary water the water in gel pores is substantially affected by the surface interaction. This phenomenon is apparent already by the fact that this water is kept even at relative humidity below saturation point. We distinguish between two types of water:

2.1.1.1 Adsorbed water

In the immediate vicinity of the surface, i.e. some molecular layers of water, the water is structured by surface forces (structured water). Its properties deviate from bulk water markedly. It is as a rule non-freezable. Its adsorption has been described by BET theory or in a more appropriate form by ESW theory [2]. This adsorption is dominant until 50 % relative humidity.

2.1.1.2 Condensed water

Above 50 % relative humidity condensation sets in which can be described in a first approximation by Kelvin equation. A somewhat improved form can be given by thermodynamic approaches [42,40]. This condensed water is still remarkably prestructured and somewhat different from bulk water.

The total region, where sorption and capillary condensation takes place is called the hygroscopic region since the water uptake takes place by condensation below saturation pressure. The pore size where this hygroscopic phenomena are dominant is restricted to the gel pores that means to a region of hydraulic radius below approximately 30 nm.

2.2 Depression of freezing point by surface interaction

The decisive point of frost action in micro-porous systems is that the phase transitions of water both from fluid to vapour and from fluid to ice are substantially changed. Similar as with dissolved ions the freezing point is depressed by surface interaction. There are different models to describe this depression of freezing point by surface forces. Surface tension is in any case the decisive parameter. In an own thermodynamic model [45,40] it was taken into account that between the internal surface of the solid matrix and the pore ice there is an unfrozen adsorbed water layer of some molecules thickness. Before freezing there is the surface energy (or surface tension) between the solid and the pore-water $\gamma_{s,w}$. After freezing this is changed to the surface energy between solid and ice $\gamma_{s,i}$ with an adsorbed layer in between. The difference between both is the decisive energy depressing the freezing point as can be seen from thermodynamics (s_w and s_i are the molar entropies of pore water and pore-ice, v_w and v_i the molar volumes, V_{pore} is the pore volume, n the number of moles in the pore and $R_{H,C}$ the hydraulic radius - i.e. the ratio of volume to surface - of the ice crystal formed.

$$(s_w - s_i)dT = (v_w - v_i)dp + (\gamma_{s,w} - \gamma_{s,i})\frac{V_{pore}}{n}d\left(\frac{1}{R_{H,c}}\right) \qquad (1)$$

In a first approximation the entropy change can be described by the molar enthalpy of fusion H_0

$$(s_w - s_i) \approx \frac{H_0}{T}$$

and the surface energy change by the surface tension of ice in water $\gamma_{s,w}$

$$\left(\gamma_{s,w} - \gamma_{s,i}\right) \approx \gamma_{w,i}$$

By this approximation with macroscopic parameters a correlation is found nearly identical to a radius freezing point approximation by Brun et.al. However, the use of macroscopic values gives only a fair guess.

$$\frac{H_0}{T} dT = -\gamma_{w,i} v_i d\left(\frac{1}{R_{H,c}}\right)$$

$$\ln\left(\frac{T}{T_0}\right) = \frac{\gamma_{w,i} v_i}{H_0 R_{H,c}};$$

$$\Delta T = \left(T_0 - T\right) = \frac{\gamma_{w,i} v_i}{H_0 R_{H,c}} T_0 \approx \frac{32 nm * K}{R_{H,c}}$$

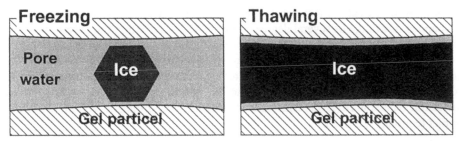

Fig. 2. Different sizes of ice nucleus when freezing sets in and of the melting ice particle with different hydraulic radius.

By this surface interaction a depression of freezing point in gel pores down to -40°C is possible. As shown in figure 2 the ice nucleus when freezing starts has a smaller hydraulic radius than the final ice particle has when melting sets in. This is the reason for a significant freeze thaw hysteresis [44]. Since both capillary condensation and depression of freezing point depend from pore radius there is a correlation between depression of freezing point and vapour pressure where condensation sets in. It can be shown that a slight drying to relative humidity near to 60 % removes all water which is freezable until - 20°C. This effect has been used to find adequate starting conditions for freeze thaw testing.

It should be kept in mind that due to hydration and with decreasing water cement ratio the amount of gel pores increases while capillary pores decrease, i.e. the ratio of freezable to non-freezable water increases.

In addition, it can be shown, that the formed ice in the bigger pores has a lower chemical potential than the unfrozen pore water. Due to this a pressure p is generated which results in a water flux from gel pores to the bigger ice pores. By this effect remarkable pressure differences up to 20 or even 40 MPa can be generated between pores of different size (hydraulic radius).

$$p = \frac{\gamma_{s,i} - \gamma_{s,w}}{R_H}$$

It can be shown that surface effects and other disturbances e.g. by dissolved salt can be dealt independently i.e. the effects superimpose additive.

2.3 Supercooling and nucleation

Supercooling must clearly be separated from the depression of freezing point by dissolved salts and by surface interaction in small pores. Supercooling is a statistical process. It is caused since on freezing a small nucleus has to be formed first. Like in depression of freezing point such a nucleus is not stable to due water ice surface interaction until either the temperature is low enough or the nucleus is big enough. Due to statistical reasons the size of the ice embryos is limited. However, in contrast to depression of freezing point the ice nucleus grows freely after it has reached its critical size. As soon as there exists one centre of nucleation all the water in the vicinity will freeze as well. Therefore, supercooling and nucleation is a statistic process which is can happen in a range of several degrees centigrade. By definition the precise value of depression of freezing point is not predictable precisely. We distinguish between heterogeneous and homogeneous nucleation. While in homogeneous nucleation the ice embryos are formed freely in water, in heterogeneous nucleation ice formation starts from surface defects such as roughness etc. Following Bigg [6] The mean value of nucleation temperature T_u depends in a logarithmic way from the volume V of the droplet (constant A and B).

$$\ln V = A + BT_u$$

In a first approximation in concrete the pore volume can be taken as this droplet volume. In hardened cement paste the supercooling is observed between -10 °C and -15 °C. In concrete the degree of supercooling is between -1 °C and -4 °C. This is reasonable since in the pore size distribution of concrete some larger pores are more probable.

Essential with supercooling is, that when ice formation starts the melting enthalpy is set free. Due to this the specimen is heated up within some nsec [10]. It is not possible to remove this generated heat. Therefore, after supercooling we have quasi adiabatic states. If the amount of freezable water is sufficiently high and if supercooling is not very pronounced the specimen heats up to a temperature where the normal freezing point of pore water would have been without supercooling. The rest of the macroscopic water then freezes with removal of its heat of fusion. Only with very high degrees of supercooling and with very small amounts of freezable water equilibrium temperature is not reached.

2.4 Transport mechanisms

We can distinguish between several transport mechanisms of water.

2.4.1 Water flux due to temperature gradient

Since freezing and thawing is coupled with a heat flux from and to the specimen respectively, we have a dynamic non-equilibrium process. With heat supply and removal water and dissolved salts are transported as well. At a temperature gradient a transport of water is generated from the warm to the cold side due to generated potential difference. Therefore, there is a tendency for a water transport out of the specimen when cooling and into the specimen when heating.

2.4.2 Water flux due to different thermal expansion

A second water transport is generated due to the different coefficients of thermal expansion of water on one side and of solid matrix and ice on the other side. The coefficient of thermal expansion is approximately ten times higher. Since there is a hysteresis between freezing and thawing, i.e. that on cooling freezing commences at lower temperatures than at melting. Since water contracts more than the solid this effect leads to a uptake of water with lowering temperature and to a squeeze out of water with increasing temperature.

2.4.3 Micro-ice-lens formation

Finally, there is a water transport within the microstructure of hardened cement paste and concrete. As shown above due to surface forces and non-freezable water a pressure difference is generated between the unfrozen gel pore water and the ice in the capillary pores. Since this pressure difference is remarkable and restricted to very small distances a dramatic water flux from smaller to bigger pores is generated in the microstructure when freezing starts. The different cases depending from boundary conditions are described in [43].

3 Models of damage

In the following chapter only these damages are taken into account, which are observed at normal freezing temperatures, i.e. the temperatures until -20 °C. It should be noted that there are special freezing phenomena between -20 °C and -60 °C in the condensed water and between -60 and -120°C in the surface layer.

The most detrimental point of frost damage is definitely the expansion of ice by 9 %. However, due to the special pore size distribution of concrete certain effects are superimposed. Basically, we have to distinguish between two groups of damaging. One group deals with macroscopic stresses and phenomena such as concentration gradients, temperature gradients and different thermal behaviour of aggregates and matrix. The other group deals with damages due to microscopic effects and stresses in the matrix of hardened cement paste and water itself.

3.1 Macroscopic models for damage.

Snyder, Blümel and Springenschmid [47,7] explain freeze thaw and deicing salt damage by the gradients of deicing salt concentration and temperature. The gradients can be different in such a way that the outer layer of the concrete freezes due to a sufficient low temperature and the inner part of the concrete due to a reduced concentration of salt and by this a normal freezing point. Between both there remains an unfrozen layer. When the temperature is lowered then this layer is freezing as well and the ice is breaking up the outer shell. There are certain boundary conditions which have to be fulfilled for this mode. In reality they are observed seldom.

Another model is by Rösli and Harnik [39] who point out that if an ice or snow layer on top of the concrete is molten by deicing salts the temperature is suddenly lowered due to the heat of fusion. By the drastic temperature gradient near the external surface stress is generated in this region which can exceed the tensile strength of concrete. However, in practical tests these temperature gradients could not be observed.

Podvalnyi [22] points out that due to the different thermal expansion of aggregates and cement matrix stresses are generated around the aggregate particles which can surpass the tensile strength.

Another macroscopic phenomenon has been observed here in Essen [3,4]. As a rule the outer layer of concrete is carbonated. Due to this and depending from the type of the cement the physical pore structure and the type of hydrates is modified. Therefore, the carbonated layer will scale off differently from the not carbonated core of the concrete. Depending from the type of concrete and depending whether deicing chemicals are used or not an increased primary scaling or a retarded primary scaling is observed.

3.2 Microscopic models

For the microscopic models the following basic phenomena are used:
- The 9 % volume exchange of ice when freezing osmotic and crystallisation phenomena.
- Depression of freezing point and unfreezable water due to surface forces combined with pressure differences.
- Supercooling effects and certain temperature changes.

In the capillary pore region ice formation and the volume expansion is an essential effect. Following Powers a hydraulic pressure is generated leading to a water migration which is possible due to the unfrozen water. If there is no sufficient room for expansion this pressure leads to damage [23,24]. Powers calculates by this the spacing factor. Helmuth [14] and Danielsson [9] improved the theory. Fagerlund describes some model scenario in this context which are published in the conference proceedings [12]. In any of these cases the essential part is whether there is sufficient room left for the transport and expansion of forming ice. There exists a critical degree of saturation above which the internal damage is generated.

An approach similar to Fagerlund and Powers has been formulated, calculated and measured by Erbaydar [10]. He also uses the freezing of water in capillary pores and the hindered water transport for explanation. In addition, he calculates the generated pressure using different models such as the semi-closed container or the inverse problem of a water flux into an empty larger pore in the concrete. He especially deals with the supercooling phenomena which he measured extensively. During freezing of supercooled water a increase of temperature up to 7 K is generated due to the heat of fusion. It is so fast (some 100 nsec) that an adiabatic condition can be assumed. Nevertheless, even under these extreme fast conditions only 15 % of the expansion expected from calculation can be found in measurement until an the observed value reaches 1 ‰. Above this the calculated and measured values are equal.

3.3 Submicroscopic surface models.

Independent from all the above effects surface interaction in the gel pores cannot be neglected. As explained above this surface interaction leads to a depression of freezing point. Water in the gel pores is unfrozen.

Everett and Haynes [11] take as a starting point that a curved surface between ice and unfrozen water is generated at the entrance area of a gel pore to a capillary pore. Due to the surface tension a low pressure is generated which attracts the unfrozen water of the gel pore to the ice.

Litvan [18] uses a similar difference in potential between water and ice. However, in his model water is evaporated and after this condensed on the ice surface.

3.4 Model of micro ice lenses and artificial saturation

As stated above freezing point is depressed by surface interaction. Therefore, the pore ice in the meso gel pores freezes gradually with falling temperature in the smaller pores. At the same time pressure differences are generated. The pressure is the lower the smaller the pores are even if they are ice filled. The pressure is generated independent from time. However the consequences for the stress in the matrix depends from water transport. This transport capacity of course is reduced if substantial amounts of water is frozen even in the bigger gel pores. Whether there is expansion or contraction is related to the boundary conditions especially to the fact whether the water is transported fast enough in the pore system with respect to ice formation. This has been the basis of my considerations long ago [40,45]. In [43] these differences have been substantiated with respect to pore size, degree of saturation, time and temperature.

For a repeated freeze thaw cycle the behaviour of a specimen can be described sufficiently precise by the following model of micro ice lens formation. Starting point is a specimen which has sucked up water to degree of saturation near the final value of capillary suction. Then during freeze thaw cycles the following steps can be separated:

1. Between +20 °C and -3 °C depending from the degree of supercooling: Water is only sucked up due to the different thermal expansion of solid matrix and water. It is pressed out slightly to the outer and cooler surface by the temperature gradient. Both effects should be reversible after cooling and heating.
2. Start of Freezing after the supercooling period: Between -0 °C and ca. -3 °C.
   2.1 If the meso- and macro capillaries are not critically saturated a sudden contraction is observed [48]. This is generated by the low pressure in the smaller pores resulting from surface forces.

PZ   WC = 0.45   RH.= 100 %   PZ   WC = 0.80   RH.= 100 %

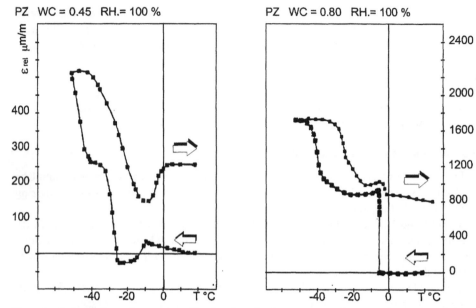

Figure 3.   Relative thermal expansion of hardened portland cement paste (the expansion of dry solid is subtracted). Fig A: w/c = 0,8 and fig B: w/c = 0,45 [43].

2.2 If the pore system is critically saturated an expansion is found due to the expansion of ice by 9 vol.% as described by other models.

3. After supercooling until approximately -12 °C: Water is transported from the unfrozen meso-gelpores to the existing ice. An increase of the volume of the bigger pores is generated due to this process and the hindered transport to free space. Smaller pores are gradually freezing and expanding since they are saturated. This leads to the irreversible expansion after a freeze thaw cycle as e.g. described in [48,43].

4. At 20 °C a remarkable amount of the water in even the meso-gelpores is frozen. Transport is only possible in the very small gelpores and in the adsorbed unfrozen surface layer. It is, therefore extremely restricted and slow. However it can be observed. Ice formation leads to expansion.

5. Between -20 °C and ca. -10 °C ice is practically not melting due to the freeze-thaw hysteresis. Thermal expansion is not severely affected by melting phenomena.

6. Between -10 °C and 0 °C ice in the meso gelpores starts to melt. Not an contraction is found as would be expected by the decreasing of volume when pore water is formed but an expansion [48]. I explain this by the transport of the now mobile water to micro ice lenses.

7. In concrete with increased amount of bigger ice filled capillaries a contraction near bulk melting point is found.

There is a remarkable difference in experiments where no outer water supply is provided and experiments - as normally in frost testing - where this supply is given. With an outer supply an artificial saturation of the pore system is generated which is much more effective than capillary suction. Even larger pores such as macro capillaries which are as a rule empty within concrete and course pores are filled by this phenomenon. It has been even observed that entrained air pores are filled due to this [10,5]. Therefore, the specimens are gradually saturated and the volume expansion becomes increasingly dominant. The degree of saturation reaches values is far above the degree obtainable by isothermal capillary suction.

4 Freeze thaw testing - Prerequisites

1. As mentioned above I distinguish between the attack of frost and deicing salts on one side and the measuring of the damage on the other side. From the above considerations several essential pre-requisites can be defined for testing the freeze thaw and freeze thaw and deicing salt attack. The condition of the specimens when frost testing starts has to be defined sufficiently. RILEM TC 117 FDC decided that this starting point should be a concrete dried out as under natural conditions, i.e. to a relative humidity of 60 % where no freezable water until - 20°C is detected. At least an outer layer of the specimens should fulfil these conditions [34].
2. Since the degree of saturation is essential for the test result the test liquid should have the opportunity to be sucked in sufficiently well before freeze thaw cycle starts. As a compromise we used a suction time of 7 days which in fact leads to a degree of saturation similar to outdoor exposure [29].
3. For the suction period we discussed whether we should use water or test solution. If water is used, the uptake of the deicing salt is retarded and mostly due to the artificial suction by frost action during the freeze thaw cycles. In addition, an unpredictable diffusion of the salt into the already sucked in water has to be taken into account. Since this diffusion process is insufficiently known and completely uncertain in its effect, RILEM TC 117 FDC decided to use as a rule even for the pre-suction period at room temperature a test solution [29].
4. As outlined above, freeze thaw testing is a highly dynamic process where moisture heat and salt transport is generated both macroscopically and microscopically. Therefore, it is essential to have well-defined boundary conditions. RILEM TC 117 FDC therefore decided to stick firstly and strictly to uniaxial water and heat transport. The heat transport should be well-defined over a certain surface of attack. Following the very first decisions of RILEM TC 117 FDC we decided, that water uptake should be by capillary suction even during the freeze thaw cycles. To achieve this, it was necessary to invert the specimens; the surface of attack is now positioned at the bottom and not at the top. This leads to a well-defined degree of saturation both during the suction process in pre-storage and during freeze thaw cycles. All the other surfaces of the specimens besides the surface of attack have to be sufficiently thermally insulated to reach a uniaxial heat flux [29].
5. The temperature regime is of high importance. Therefore, a temperature cycle was defined which should be met at a fixed point outside the test container with high precision [29].
6. For a precise temperature profile during freezing and thawing two pre-requisites have to be fulfilled. Outside the test arrangement a precise heat transport must be guaranteed. Within the test arrangement the different parameters leading to modifications of the temperature gradient must be precisely defined. The most

essential point is the geometric arrangement starting from the surface outside the test liquid over test liquid until surface of attack and specimen. Very important in this context is that water consumes extraordinarily high amounts of heat during fusion which is equivalent to the heat consumed for a temperature change of water by 80 °C. Therefore, a change of the thickness of the liquid layer on top or as in CDF test below the test surface by 1 mm changes the complete temperature profile during the freeze thaw cycle.

If these test parameters are sufficiently well-defined a freeze thaw and deicing salt attack is achieved with high precision as proved in [41,46].

5 Tests discussed in RILEM TC 117 FDC

5.1 CDF and CF test

5.1.1 CDF test [25]

With the CDF test the freeze thaw and deicing salt resistance of concrete is tested. It bases primarily on the discussions of the RILEM TC 117 FDC. It adopts precisely its decisions. The prerequisites as outlined above are consequently taken into account. As deterioration the scaling of material is measured.

Besides these steps in the final CDF test the procedure has been refined. The specimens are now each stored in one test container which is not only an advantage in handling but also from a physical point of view. The containers are immersed in a liquid temperature bath. A liquid has a 4000 times higher heat capacity than the air and is therefore much better suited as a medium for heat transport. The air which is surrounding the specimens on all sides but the test surface acts as thermal insulation. Due to the inverted arrangement the thickness of the test liquid between the test surface and the bottom of the test container is fixed with high precision. This is of high importance since during freezing most heat is consumed in this layer. Under a dynamic freeze-thaw cycle this geometric aspect is essential for the test procedure.

Some details are found in [41]

5.1.2 CF test

Using the same method of freeze thaw attack CF test measures the freeze thaw resistance of concrete with water. Both the draft of RILEM recommendation for CDF and CF test have been published [38]. However, since for CDF test the precision data have been available and proved by RILEM TC 117 FDC, only the CDF test is a final recommendation.

Furthermore, freeze thaw testing with water showed that internal damage should be taken into account in this case since it is more dominant than in testing with deicing salt attack. Meanwhile an improved procedure is available [5]. First precision data are published. [46]

5.2 Cube test and Slab test

There are two further test procedures which should be checked by RILEM TC 117 FDC following a decision in Québéc: Cube test and Slab test. Both are submitted as CEN tests, a draft is published in [37].

5.2.1 Slab test

In slab test the test liquid is on top of the specimen contained in the specially designed sealing at the rim of the specimen. In Slab test all the other surfaces are thermally insulated. On the top a plastic foil is placed to prevent the evaporation of the test liquid. For freeze-thaw cycles the specimens are stored in an air-cooled chamber. However, the temperature control of the chamber is not very sophisticated to reduce costs.

Tests with the slab tests showed the following results:

All measurements proved that it is not easy to keep the freeze-thaw cycle within the given regime although, this regime allows a rather broad bandwidth in temperature. This findings are consistent with data from other laboratories.

A member of RILEM TC, Mr. Studer, measured the temperature profile within the test specimens in different depths [26]. Analysing these data it could be deduced, that although the lateral and bottom of the specimens are thermally insulated the specimens are freezing from behind. As stated above this can be explained by two facts: 1. The layer of test liquid on top has a very high heat consumption during freezing. 2. Between this liquid and the surrounding air there is a layer of non-ventilated air due to the protection against evaporation. This non-ventilated air layer acts as an additional thermal insulation.

5.2.2 Cube test

Cube test is completely different from the other procedures discussed above. In Cube test the complete specimens are immersed in the test liquid and subjected in a special test compartment to freezing and thawing. The details are described in the draft of the procedure. [37]

Due to this complete immersion of the specimens the damage and scaling process depends on geometrical and thermal boundary conditions which are specific for this test.

5.2.3 Other test procedures.

There are quite a lot of different procedures used for freeze-thaw and deicing salt testing. However, the committee decided not to take them into account. It should be mentioned that with respect to ASTM 666 there exists a decision of the committee that this procedure should not be recommended in the future [34].

5.3 Assessment of the precision of the tests

Round robin tests (both national and international) to assess precision following ISO 5725 have been performed for all three procedures - CDF test, cube test and slab test. However, only the CDF test reached a satisfying result [41]. Round robin results available to the RILEM TC 117 FDC for CF test (the freeze-thaw test with water and without deicing agent), for Cube test, and for the Slab test were not convincing. Therefore RILEM TC 117 FDC decided at this time to recommend only the CDF test for future testing. All the other tests have to prove their precision first. The data have to be analyzed by RILEM TC 117 FDC.

6 Applications and further work to be done

Setzer and Auberg [46,3,4] as well as Stark and Ludwig [20] used the precision of test procedures to analyse the influences on damage of concrete with Portland cement and high slag cement.

The precision of CDF test allows to separate several phenomena which could not be separated up to now. At first three typical slopes can be distinguished in accordance to results of other test procedures. A constant increase and progressive slope and a degressive slope. However, in CDF test it proves that both progressive and progressive slope can be separated in two parts. An initial scaling and a later continuous scaling. The later scaling has linear progress. The initial scaling is highly linked to the depth of carbonation. Depending on the carbonated layer the scaling is rather rapid as for instance high slag cement with extreme slag contents. The scaling of the carbonated layer is reduced if this layer is becoming denser by carbonation such as in many Portland cements. The phenomenon that a carbonated layer is weathering differently can be attributed both to the changed pore size distribution and cement unstable carbonation products in high slag. Using CDF test it was possible first to distinguish between chemical and physical influences.

7 Task for future work

There are two aspects for future work: Improvement of the test procedure and study and analysis of the phenomena leading to the damage.

In the test procedures it can be efficient to modify the boundary conditions. As already mentioned in other future testing the test surface can be modified. It may involve the investigation of cut surfaces for mixture evaluations using the CDF and CF procedures and the investigation of cast surfaces in the slab test. Such Modifications cannot be excluded. However, the modifications must be well described. It must be checked and documented how they affect the results.

Any changes of the principals of the method, especially, how both the freeze-thaw and the deicing agent attack are applied have to be analysed with even more care.

Most essential for the test procedure will be the measuring of the internal damage. Some promising work is found in [46].

Since apparently the carbonation plays a rather dominant role for the initial scaling the age, the curing method and time and the further storage conditions of concrete structures have to be checked under this aspect. In addition, the measurement of internal damage has to be improved with emphasis. It is apparent that for a concrete structure where the surface its appearance and its service life plays a dominant role, such as in road structure, the scaling has to be measured. However, for structures where a structural strength is essential internal damage has to be observed.

Testing with water instead of deicing salt solution shows that for pure freeze-thaw testing internal damage is of extremely high interest. However, the composition of „pure" tap water changes the results by a factor of three (Setzer Auberg in[34]) [3].

If the precision of a test procedure is sufficiently high the materials properties which lead or prevent damage can be analysed. Application CDF test shows quite clearly, that it is possible to separate chemical, physical and structural influences. The different types of cement, admixtures and additives should be studied.

8 References

1. Adolphs, J. (1994): Thermodynamische Beschreibung der Sorption. Dis. In Forschungsberichte aus dem Fachbereich Bauwesen der Univ. GH Essen, Heft 61
2. Adolphs, J.; Setzer, M.J. (1996): A model to describe adsorption isotherms. J. Colloid Interface Sci. **180**, p.70
3. Auberg, R. (1997): Zuverlässige Bestimmung des Frost- und Frost-Tausalz-Widerstands. Dis. Univ. Essen
4. Auberg, R.; Setzer, M.J.: Chloride gradient, change of pore structure and freeze-thaw resistance of concrete. Betonwerk und Fertigteiltechnik to be published
5. Auberg, R.; Setzer, M.J.: Influence of water uptake during freezing and thawing. In Proc. International RILEM workshop on Resistance of concrete to freezing and thawing. Essen
6. Bigg, E.K. (1953): The supercooling of water. Proc. Phys. Soc. **B66**, p. 688
7. Blümel, O.W.; Springenschmid, R. (1970): Grundlagen und Praxis der Herstellung und Überwachung von Luftporenbeton. Straße und Tiefbau **24**, p. 92
8. Brun, M.; Lallemand, A.; Quinson, J.-F.; Eyraud, C. (1977): A new method for the simultaneous determination of the size and shape of pores: the thermoporometry. Thermochimca Acta **21**, p. 59
9. Danielsson, U.; Wastesson, A. (1958): The Frost Resistance of Cement Paste as Influenced by Surface-Active Agents. Swedish Cement and Concrete Research Institute, Proceedings Nr. 30, Stockholm
10. Erbaydar, S. (1986): Eisbildung, Volumendilatationen und Wassertransport im Gefüge von Beton bei Temperaturen bis -60 °C. Dis. TU München
11. Everett, D. H.; Haynes, J. M. (1975): The Thermodynamics of Fluid Interfaces in a Porous Medium. Part II. Capillary Condensation, Surface Area and Pore Size Distribution Determination, and Hysteresis. Z. f. Physikalische Chemie Neue Folge, 97, p. 301-312
12. Fagerlund, G.: Internal frost attack - State of the art. In Proc. International RILEM workshop on Resistance of concrete to freezing and thawing.
13. Graf, O.; Walz, K. (1945): Fortschritte und Forschung im Bauwesen Reihe B, Heft 3, S. 62
14. Helmuth, R.A. (1972): Investigations of the Low Temperature Dynamic Mechanical Response of Hardened Cement Paste. Dept. of Civil Engineering, Stanford Univ., Technical Report 154
15. Helmuth, R.A.; Turk, D.H. (1966): Elastic Moduli of Hardened Cement Paste and Tricalcium Silicate Pastes: Effect of Porosity. Special Report 90, Highway Research Board, Washington
16. Hirschwald, J. (1908): Prüfung der natürlichen Bausteine auf ihre Frostbeständigkeit. Wilhelm Ernst, Berlin
17. ISO 5725 (1990): Accuracy (trueness and precision) of measurement methods and results.
18. Litvan, G.G. (1973): Frost Action in Cement Paste. Materials and Structures, 34, p. 293-298
19. Plähn, J., Golz, W. (1982): Verbesserung des Beurteilungsmaßstabes und der Reproduzierbarkeit bei der Frost-Tausalz-Prüfung von Straßenbeton im Labor. Abschlußbericht zum FA 8.058 G 78 G, Hannover

20. Ludwig, M.; Stark, J.: Freeze-deicing salt resistance of concretes containing cement rich in slag. Proc. of this workshop.
21. Plähn, J., Golz, W. (1984): Vergleichsuntersuchungen an vier Prüfverfahren für den Frost-Tausalzwiderstand. Straße und Autobahn 1, S. 14
22. Podvalnyi, A.M. (1976): Phenomenological aspect of concrete durability theory. Materials & Structures **9**, p. 151
23. Powers, T.C. (1945): A Working Hypothesis for Further Studies of Frost Resistance of Concrete. J.ACI Proc., 41, p. 245-272
24. Powers, T.C. (1949): The Air Requirement of Frost Resistant Concrete. Proc. Highway Res. Board, V 29, p. 184-211
25. RILEM Recommendation, RILEM TC 117 FDC: (1996) CDF Test- Test Method for the Freeze-Thaw Resistance of concrete with sodium chloride solution. Materials and Structures Vol. 29, p. 523-528
26. RILEM TC-117 FDC - Freeze thaw and deicing resistance of concrete, report of the 7th Meeting in Trondheim, Norway, August 1994.
27. RILEM TC-117 FDC - Freeze thaw and deicing resistance of concrete, report of the 9th Meeting in Espoo, Finland, August 1996.
28. RILEM TC-117 FDC - Freeze thaw and deicing resistance of concrete, report of the 1st Meeting in Essen, Germany, May 1990.
29. RILEM TC-117 FDC - Freeze thaw and deicing resistance of concrete, report of the 1st Meeting in Essen, Germany 1989
30. RILEM TC-117 FDC - Freeze thaw and deicing resistance of concrete, report of the 2nd Meeting in Brighton, Great Britain, November 1990.
31. RILEM TC-117 FDC - Freeze thaw and deicing resistance of concrete, report of the 3rd Meeting in Lund, Sweden, June 1991.
32. RILEM TC-117 FDC - Freeze thaw and deicing resistance of concrete, report of the 4th Meeting in Scetauroute, Lyon, France, February 1992.
33. RILEM TC-117 FDC - Freeze thaw and deicing resistance of concrete, report of the 5th Meeting in Dübendorf, Switzerland, October 1992.
34. RILEM TC-117 FDC - Freeze thaw and deicing resistance of concrete, report of the 6th Meeting in Quebec, Canada, August 1993.
35. RILEM TC-117 FDC - Freeze thaw and deicing resistance of concrete, report of the 8th Meeting in Sapporo, Japan, August 1995.
36. RILEM TC117 FDC (1995) Test Method for the Freeze-Thaw Resistance of Concrete. Slab Test and cube test, Materials & Structures 28, 366-371.
37. RILEM TC117 FDC (1995) Test Method for the Freeze-Thaw Resistance of Concrete. Slab Test and cube test. Draft of recommendation. Materials & Structures **28**, 366-371.
38. RILEM TC 117 FDC: (1995) Draft Recommendation for test method for the freeze-thaw resistance of concrete - Test with water (CF) or with sodium chloride solution (CDF) Materials & Structures **28**, 175-182
39. Rösli, A.; Harnik, A.B. (1979): Zur Frost-Tausalzbeständigkeit von Beton. Schweizer Ingenieur und Architekt **46** , p. 1
40. Setzer, M. J. (1977): Einfluß des Wassergehaltes auf die Eigenschaften des erhärteten Betons. Schriftenreihe DAStb. Heft 280, 43-117
41. Setzer, M. J.; Auberg, R. (1995): Freeze thaw and deicing resistance of concrete testing by CDF method -Resistance limit and evaluation of precision. Materials and Structures Vol. **28**, p. 16-31
42. Setzer, M.J., Wittmann, F.H. (1973): Modified Method to Calculate Pore Size Distribution Using Sorption Data. Proc. Intern. Symp. on Pore Structure & Properties of Materials, RILEM/IUPAC Prag, Vol.IV, C-69

43. Setzer, M.J. (1983): Das Gefrieren des Wassers in porösen Werkstoffen. Intern. Koll. Werkstoffwissenschaften und Bausanierung, Esslingen, Sept. 1983, ersch. in Sonderheft Bautenschutz und Bausanierung, S.36-40
44. Setzer, M.J. (1995): Freeze thaw and deicing resistance combined with water and chloride uptake. Proc. Intern. Conference on Mass-Energy Transfer and Deterioration. Paris
45. Setzer, M.J. (1976): New Approach to Describe Frost Action in Hardened Cement Paste. Proc. Conf. on Hydraulic Cement Pastes: Their Structure and Properties, Sheffield p. 312-325
46. Setzer, M.J.; Auberg, R. (1998): Reliable testing of frost resistance of concrete against frost attack with CF test. Proc. 2nd CONSEC Conference, Tromsø 1998
47. Snyder, J. (1965): Protective Coatings to prevent deterioration of concrete by deicing chemicals. Nat. Highway Res. Prgrm. Rep. 16, Highway Res. Board Nat. Acad. Sci., Nat Res. Council, Washington
48. Stockhausen, N., Setzer, M.J. (1980): Anomalien der thermischen Ausdehnung und Gefriervorgänge in Zementstein. Tonindustrie Zeitung **104/2**, p. 83-88

Effect of finishing, forming and curing on de-icer salt scaling resistance of concretes

R.D. HOOTON and A. BOYD
Department of Civil Engineering,
University of Toronto, Toronto, Ontario, Canada

Abstract

A series of concretes with 0, 25, 35 and 50% mass replacement of portland cement by ground granulated blastfurnace slag were tested for de-icer salt scaling resistance using a modified ASTM C672 test where scaling mass loss was measured. In one series of tests, finishing times were varied, while in another series of tests, curing periods were varied. In addition, inclined, formed surfaces, similar to those of highway barrier walls, were tested. The results indicate that scaling performance is highly sensitive to these variables. Field trials were also undertaken which show that the laboratory test is too severe.
Keywords: De-icer salt scaling; slag; fly ash

1 Introduction

In the Canadian CSA A23.1 standard, pavement, sidewalk and curb concrete which is to be exposed to freezing in the presence of de-icer salts is required to have a maximum w/cm = 0.45, a minimum 28 day specified strength of 32 MPa and contain 5 to 8% air (for 20 mm aggregate). Bridge decks require 35 MPa strength and maximum w/cm = 0.40. Due to poor performance in the Ontario-modified ASTM C672 de-icer salt scaling laboratory test, the Ontario Ministry of Transport currently limits slag replacement levels to 25%, fly ash to 10%, or a combination not exceeding 25%. In spite of this, field experience shows that this laboratory tests are unnecessarily restrictive and that slag contents of at least 35%, and fly ash contents of 20% (Class F ash) are acceptable [1].

Frost Resistance of Concrete, edited by M.J. Setzer and R. Auberg. Published in 1997 by E & FN Spon, 2–6 Boundary Row, London SE1 8HN, UK. ISBN: 0 419 22900 0.

2 Experimental

2.1 Materials
CSA Type 10 normal portland cements (OPC) were used. For laboratory tests a 6.1% C_3A, low-alkali (0.31% $Na_2O_{eq.}$) cement was used, while field trials used an 8.0% C_3A, 0.7% alkali cement. The separately ground granulated blastfurnace slag was from Hamilton, Ontario with a density of 2920 kg/m^3 and Blaine fineness of 432 m^2/kg. Fly ash ASTM class C from Wisconsin, USA with a density of 2480 kg/m^3 and CaO = 35.6%. A 20 mm crushed dolostone with a density of 2850 kg/m^3 and 0.37% absorption was used. 35% of the combined aggregates was a glacial sand with a fineness modulus of 2.58, density of 2680 kg/m^3 and absorption of 0.86%.

MicroAir air entraining admixture and a Type A light sulphonate based water reducer were used in all mixtures.

2.2 Concrete Proportions
All concretes were batched at a water to cementitious material ratio (w/cm) = 0.45 according to Ontario Ministry of Transport (MTO) de-icer salt scaling test requirements (OPSS LS-412). Target air content were 6.5 ± 1.5% with slumps of 60-100 mm. Cementitious materials contents were 375 kg/m^3. Concretes were mixed in an Eirich, 150 ℓ flat pan mixer.

2.3 Concrete Specimens and Curing
For each strength test, two 100 x 200 mm cylinders were cast. For each salt scaling test, two slabs 250 x 350 x 75 mm were cast, with a raised test surface of 200 x 300 mm. An outer 25 mm formed rim was used to bond 25 mm square styrofoam beams, using silicone sealant, to form a dike for salt ponding.

Cylinders were moist cured at 23°C for 28 day strength tests. Slabs were finished and cured by a variety of methods as will be detailed later, in order to study these effects on salt scaling performance.

2.4 Scaling Resistance Test
A modified form of ASTM C672 was used which has been standardized in Ontario by the MTO and Ontario Provincial Standard (OPSS) LS-412. In this test, after moist curing the slabs for 14 days, the surface are allowed to dry until 28 days of age. Then a 3% NaCl (0.5 M) solution is ponded on the trowelled surface and the slabs are subjected to 50 cycles of freezing and thawing (one cycle per day from 23°C to -22°C) with the freezing period being 16 h). At the end of each 5 cycle period, the loose surface scale is washed off and collected in a filter paper, dried, and weighed. The cumulative mass loss is measured. The MTO limit is 0.8 kg/m^2 after 50 cycles. A typical cycle, based on an embedded thermocouple is shown in Figure 1.

3 Finishing Time and Type Experiments

The OPSS LS-412 procedure specified only two passes with a wood float immediately after strike off. A light bright texture is then applied 45-60 min later. In field practice, magnesium (Mg) floats are used but reported to be detrimental with respect

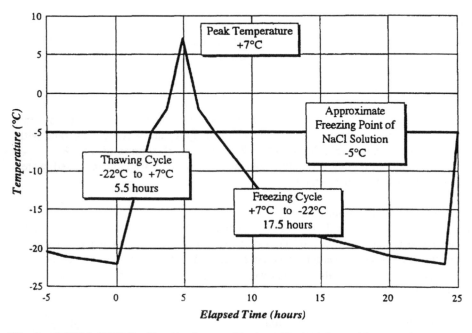

Fig. 1. ASTM C672 Scaling Resistance Testing: Typical freeze/thaw cycle

to salt scaling due to sealing of the surface. As well, in field practice flatwork is not finished until the bleed water is gone - and the time for this to occur will often be extended with slag or fly ash concretes.

A 40% slag by mass replacement concrete mixture was cast and finished by Procedures 1 to 5. The experiment was repeated with a 25% slag mixture, except that Procedure 4 was replaced by Procedure 6.

1. MTO standard test method, brushed immediately after bleed water disappeared (80 min. after the standard wood float, mimicking the standard MTO laboratory practice).
2. Mag. float and brushed 10 minutes after the initial wood float (i.e. before any significant bleeding had occurred).
3. Mag. float and brushed during the later stages of bleeding (60 min. after the standard wood float).
4. Mag. float and brushed immediately after the bleed water disappeared (80 min. after initial wood float).
5. Mag. float and brushed 30 minutes after the bleed water disappeared (110 min. after the initial wood float).
6. Wood float and brushed immediately after the bleed water disappeared (80 min.).

3.1 Results

The concrete properties are shown in Table 1 and the 50 cycle are cumulative scaling losses are given in Table 2 and illustrated in Figures 2 and 3.

Table 1. Finishing Time and Type: Mixing Results

Slag Content (%)	Air Content (%)	Slump (mm)	Plastic Density (kg/m³)	28-Day Strength (MPa)
40	6.6	85	2380	49.2
25	6.6	70	2370	47.0

Table 2. Finishing Time and Type: Scaling Resistance Results

Mix ID	Average Cumulative Mass Loss (kg/m³)					
	MTO Standard	Mag. Float @ 10 min.	Mag. Float @ 60 min.	Mag. Float @ 80 min.	Mag. Float @ 110 min.	Wood Float @ 80 min.
40% Slag	1.42	1.14	1.51	1.48	1.56	-
25% Slag	1.15	0.67	1.27	1.89	-	1.82

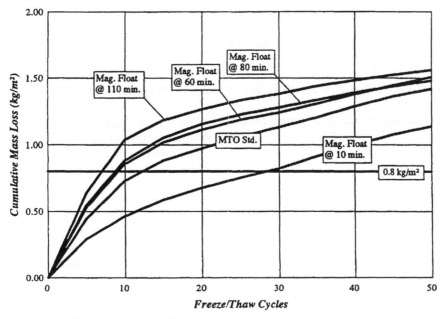

Fig. 2. Finishing Time & Type: Scaling Resistance Results - 40% Slag

It is interesting to note that the standard MTO finishing resulted in a scaling loss in excess of 0.8 kg/m³ for the "MTO approved", 25% slag concrete. In fact, the only test to pass the 0.8 kg/m³ limit was where the surface was pre-maturely finished with a magnesium float. However, it should be noted that in the field trials described later, associated laboratory tests resulted in much lower scaling losses (< 0.60 kg/m² for 25% slag - MTO standard finishing); but a higher C_3A, higher alkali portland cement was used.

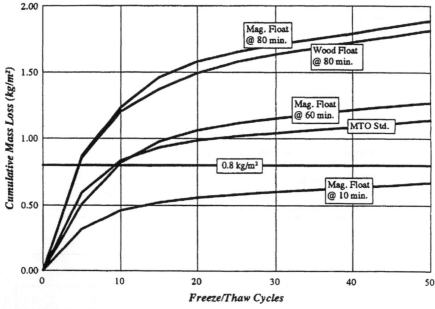

Fig. 3. Finishing Time & Type: Scaling Resistance Results - 25% Slag

4 Formed Surface/Hot Weather/Short Curing Period Experiments

Many scaling problems in Ontario have been observed on highway medium barrier walls, typically formed with steel. Several experiments were undertaken where the test surface of the slab was inclined to 56° from the horizontal to match typical barrier wall designs and formed with a 6 mm thick steel plate. Three typical MTO approved curing regimes were used.

The steel formed surfaces were either kept at 25°C or heated with infrared lamps to 50°C prior to casting (and maintained at 50°C for 8 h to typify hot summer conditions with radiant heat). Five temperature/curing regimes were studied. The short curing period (Regimes I and II are typical of MTO practice) is mainly responsible for the poor performance as can be seen by the dramatic impact of standard 14 day moist curing (Curing Regime VI). After the specified curing, concretes were exposed to laboratory air until 28 days of age.

Curing Regime	Form Temperature (°C)	Moulds stripped at		Moist curing period after stripping (days)
		1 day	4 days	
I	25	X		0
II	25		X	0
III	50	X		0
IV	50		X	0
V	50	X		3
VI	23	X		13

4.1 Results

Fresh concrete properties and strengths are shown in Table 3 for the seven concrete mixtures used. Scaling results are shown in Table 4. It can be observed that the 50°C form surface temperatures had a dramatic, adverse effect on sealing resistance of the portland cement concrete. All of the fly ash and slag mixtures exceeded the 0.8 kg/m^3 limit, even though hardened air void analysis of selected concretes showed air void distribution to be satisfactory (Table 5) even when polished surfaces were taken 1 to 2 mm below the cast faces.

Table 3. Hot Weather Effects: Mixing Results (Average of 2 Batches)

Cementitious Material Content (%)			Air Content (%)	Slump (mm)	Plastic Density (kg/m^3)	28-Day Strength (MPa)	Initial Set (h:min.)
OPC	Slag	Fly Ash					
100	-	-	6.5	100	2381	35.6	4:50
75	25	-	6.9	105	2329	38.2	4:57
75	15	10	7.2	90	2339	38.8	4:58
70	30	-	7.3	75	2343	37.2	4:53
65	35	-	6.4	100	2339	33.4	5:01
60	40	-	6.6	85	2358	32.6	4:55
65	25	10	7.5	100	2318	34.3	5:31
75*	25	-	6.3	100	2380	37.8	-
65*	35	-	6.2	95	2349	37.2	-

*Mixtures Cast for Curing Regime VI.

Table 4. Hot Weather Effects: Scaling Resistance Results

	Average Cumulative Mass Loss (kg/m^2) by Curing Regime					
	I	II	III	IV	V	VI
100% OPC	0.47	0.30	1.60	2.33	3.41	-
25% Slag	4.12	4.23	4.36	4.02	2.81	0.10
15% Slag+10% Fly Ash	4.66	2.40	4.56	1.82	1.33	-
30% Slag	3.06	2.62	4.11	4.14	3.67	-
35% Slag	1.45	2.39	2.56	2.82	2.99	0.43
40% Slag	3.02	2.99	3.71	3.77	2.98	-
25% Slag+10% Fly Ash	2.31	3.94	3.34	4.08	2.96	-

Table 5. Hot Weather Effects: Air Void Analysis Results

Parameter	B1 TI Surface	B1 TIII Surface	B1 TIII 30 mm	B2 TI Surface	B2 TI 30 mm
Air Content (%)	13.8	13.7	6.0	10.9	6.5
Spacing Factor (mm)	0.22	0.19	0.17	0.18	0.21
Specific Surface (mm^{-1})	14.1	18.5	25.6	20.2	20.7
Paste Content (%)	42.1	47.7	25.9	42.5	29.5

5 Effect of Air Content and Slag Fineness

After observing the poor deicer salt scaling performance of inclined, formed concrete surfaces with normal (short) field curing, it was decided to repeat some (25% and 35% slag mixtures) tests with both 14 day moist curing and increased air contents. As well, the same slag ground to higher fineness was tested to see if a faster rate of slag hydration might improve performance. Several other slag sources in North America are normally ground to Blaine fineness in excess of 500 m²/kg.

5.1 Results

Concrete properties and scaling results are detailed in Table 6. Air contents of 6.0, 7.5 and 9.0% were controlled to ±0.3% and concrete temperatures were 20.0 to 21.5°C.

Firstly, by all of the scaling results the 14 day moist curing period resulted in a dramatic improvement in scaling resistance. The results for all of the 25% slag mixtures are very low and differences are insignificant. For the 35% slag mixtures, either increasing air content from 6.2% to 7.5% (both within the CSA range of 5 to 8%) or increasing slag fineness resulted in an improvement. Increasing both air and slag fineness reduced scaling losses to those found for the 25% slag mixtures.

Table 6. Effect of Air Content and Slag Fineness Effects on Scaling Resistance

Slag Content (%)	Slag Fineness (m²/kg)	Air Content (%)	Slump (mm)	Plastic Density (kg/m³)	28-Day Strength (MPa)	Scaling Mass Loss (kg/m³)
25	432	6.3	100	2360	37.8	0.10
		7.6	110	2304	33.1	0.16
		9.2	115	2267	27.8	0.19
	499	6.2	100	2350	38.7	0.09
		7.6	100	2299	33.0	0.11
		9.1	120	2251	29.2	0.15
35	432	6.2	95	2349	37.2	0.43
		7.6	110	2311	30.8	0.25
		9.1	115	2254	26.4	0.24
	499	6.2	105	2342	40.4	0.26
		7.5	115	2296	35.8	0.19
		9.0	135	2247	28.1	0.16

Table 7. Field Trials: Mixing Results - June 1994

Mix ID	Cementitious Material Content (%)			Air Content (%)	Slump (mm)	Plastic Density (kg/m³)	Batch Temperature (°C)
	OPC	Slag	Fly Ash				
L1	50	50	-	7.8	90	2310	22
L2	65	35	-	7.2	85	2330	22
L3	75	25	-	5.8	80	2358	24
L4	65	25	10	7.0	80	2332	22
L5	85	-	15	5.8	60	2371	25
L6	100	-	-	7.5	80	2329	25

6 Field Trials

Because of discouraging laboratory results, a field trial was undertaken in conjunction with Lafarge Canada Inc. and the MTO in June 1994. Six concrete mixtures were placed at w/cm = 0.42 in a series of exterior pavement slabs, each 1.8 x 7.3 m x 0.2 mm thick. These formed an approach slab to a weigh scale for heavy slag tanker trucks. Cementitious components of the mixtures and fresh concrete properties are given in Table 7. The portland cement had a higher C_3A (8.8%) and alkali content (0.7%) than the one used in the laboratory tests. Areas of the slabs were either cured with a curing compound or covered in wet burlap and plastic for 4 days. Some sections were also finished prematurely, before bleeding was complete. All finishing was done with a magnesium float. Time of set and strength results are shown in Table 8. Comparison laboratory slabs were cast, finished and cured with the pavement slabs.

Surface tensile strength were measured on laboratory slab specimens. 50 mm diameter, roughened steel disks were epoxied to surfaces; then pulled off in tension using a Lok-Test device. Results are shown in Table 8. The lowest surface strengths were found with the 50% slag mix and this concrete was the only one that has exhibited scaling in the field slabs after 3 winters.

After 28 days, one pair of slabs cured according to SL412, was tested for scaling resistance at each of three laboratories and the results are given in Table 9. The other slabs were buried on site, with their cast faces exposed to the environment for 4 months, then removed (Oct. 1994) for scaling tests at the MTO laboratory as shown

Table 8. Field Trials

Mix Composition	Initial Set	Final Set	7 Day Strength (MPa)	28 Day Strength (MPa)	127 Day Strength (MPa)	Surface Pull Off Strength (MPa)
50% Slag	5:58	7:59	24.1	35.4	44.1	0.77
35% Slag	5:07	6:14	31.2	41.2	50.2	1.22
25% Slag	4:54	5:59	34.7	44.5	57.0	1.50
25% Slag+10% Fly Ash	4:44	6:03	33.5	43.3	49.9	1.54
15% Fly Ash	4:05	4:53	34.8	42.5	54.1	2.04
100% OPC	4:19	5:30	28.4	37.7	42.3	1.72

Table 9. Field Trials: Scaling Resistance Results - Laboratory Specimens Tested at 28 Days

Mix Composition	Average Cumulative Mass Loss (kg/m^2)		
	Lab #1	Lab #2	Lab #3
50% Slag	1.60	2.02	1.44
35% Slag	0.50	1.24	-
25% Slag	0.60	0.52	-
25% Slag + 10% Fly Ash	1.40	1.55	1.40
15% Fly Ash	0.36	1.24	-
100% OPC	0.14	0.13	-

Table 10. Field Trials: Scaling Resistance Results - Specimens Exposed 4 Months

Mix Composition	Average Cumulative Mass Loss (kg/m^2)			
	Burlap Early Finish	Burlap Normal Finish	Curing Compound Early Finish	Curing Compound Normal Finish
50% Slag	1.28	0.81	1.12	0.68
35% Slag	0.15	0.08	0.08	0.12
25% Slag	0.12	0.05	0.08	0.14
25% Slag+10% Fly Ash	0.07	0.44	0.11	0.15
15% Fly Ash	0.41	0.26	0.05	0.07
100% OPC	0.05	0.06	0.07	0.08

in Table 10. From Table 9, the between-lab variations in some cases were quite large and only the 50% slag and the 25% slag + 10% Class C ash failed consistently. The 6 month old slabs were far more resistant to scaling and the only one to exhibit a high loss was the 50% slag concrete with premature finishing.

After 3 winters with frequent salting and exposure to truck traffic, the pavement slabs are all performing well (ie: no scaling) except for the moist cured, 50% slag concrete which has light, localized scaling (this occurred during the first winter). Unfortunately no trend between surface pull off strengths and scaling had been found in the earlier laboratory tests.

7 Other Tests

In addition to scaling tests on slabs, several other tests were performed in an attempt to better understand the scaling performance. Initial rate of absorption tests (sorptivity) were performed at 28 days to observe whether high capillary suction of de-icing salts would be an influence. No trend was found. Samples of mortar near the surface were removed at 28 days for mercury intrusion porosimetry to see if porosity or pore structure might affect scaling performance. Again no trends were found. Surface tensile strengths were measured, as described in the field tests. Except for concretes in the field trials, no trends were found. Due to this lack of correlation and space limitations, none of this data is presented.

8 Conclusions

1. Both the finishing procedure and timing have a significant effect on the scaling resistance of concrete surfaces tested by a modified ASTM C672 test used in Ontario. Simply striking off the surface with a wood float, followed by the application of a brushed finish at the time when the bleed water just disappears from the surface, resulted in the surface being more resistant to scaling than a surface finished using normal industry practice. A typical industry procedure would involve finishing the surface with magnesium float immediately following the completion of bleeding and prior to application of the brush texture.

This poor performance was reversed, however, by shortening the time period elapsed between the initial wood float strike off and the magnesium float finish. In the laboratory, use of a magnesium float before bleeding, resulted in significant improvement in the surface's scaling resistance, contrary to expectations.

2. Placing concrete against inclined steel forms at an elevated temperature of 50°C can be extremely detrimental to its resulting scaling resistance. Care should be used to offset these harmful effects by taking precautions against high temperatures, such as insulating the moulds or wetting the outer surface of the moulds to provide evaporative cooling to decrease surface temperature.

3. The curing regime is, by far, the most important single factor affecting the scaling resistance of concrete surfaces. When dealing with formed faces, demoulding at one day of age followed by a period of curing with a burlap and plastic overlay was found to be superior to simply leaving the specimens in the moulds for the same total time period. Replacing the burlap and plastic overlay with a water drip (after loosening the mould) did improve the scaling resistance further.

   The length of curing is even more significant than the method used and should be continued as long as possible. The four day moist cured specimens performed far worse than those cured for 14 days.

4. Increasing either entrained air content or slag fineness improved scaling resistance.

5. The surface pull off strength test, although showing promise in the field trials, was not useful in predicting scaling performance in the laboratory tests. Attempts failed to relate scaling performance to sorptivity and pore structure of the surface.

6. The ASTM C672 test, as modified in the Ontario Provincial Standards (OPSS LS-412) is too severe, relative to field performance for concretes incorporating slag or fly ash. Its reproducibility is also questionable. Other test procedures such as those developed in Europe [2] need to be explored. Other influences, such as surface carbonation [3] and differences in portland cement need to be explored, to understand the poorer scaling performance of 50% slag mixtures.

9 Acknowledgements

The funding provided by the Ontario Ministry of Transportation, Lafarge Canada Inc., and the Ontario Centre for Materials Research is gratefully acknowledged.

10 References

1. Thomas, M.D.A. (1997) *Laboratory and Field Studies of Salt Scaling in Fly Ash Concrete*. Proceedings, RILEM International Workshop on Freezing Resistance of Concrete, with or without Deicing Chemicals, Essen.
2. Rilem, (1995) *Draft Recommendations for Test Methods for the Freeze-Thaw Resistance of Concrete, Tests with Water (CF) or With Sodium Chloride (CDF)*. Materials and Structures, No. 28 pp. 175-182.
3. Stark, J. and Ludwig, H.M. (1996) *Freeze-Thaw and Freeze-Deicing Salt Resistance of Concretes Rich in Granulated Blast Furnace Slag*. ACI Materials Journal, Vol. 94, pp. 47-55.

Influence of preconditioning on scaling resistance for different types of test surfaces

P. UTGENANNT and P.-E. PETERSSON
Swedish National Testing and Research Institute, Borås, Sweden

Abstract

When testing the salt scaling resistance of concrete, different test methods can be used. The difference in preconditioning climate, test surfaces used, the time the test specimens are subjected to the climate etc., give rise to difficulties when comparing results from different methods. In order to accurately be able to interpret and translate the results from these methods, it is necessary to learn more about the causes of the differences in results. In this investigation the influence of the preconditioning climate on the scaling resistance for various types of test surfaces is examined. The freeze/thaw test used is the Swedish standard SS 13 72 44, more commonly known as the "slab test" or the "Borås method". Two concrete qualities, both with OPC as binder, were tested. Three types of test surfaces were investigated, surface cut 7 respectively 21 days after casting and cast surface. Two preconditioning climates were examined, 65% RH/20°C and 50% RH/20°C.

It was found that the scaling resistance is highly dependent on the preconditioning climate, irrespective of the test surface. For concrete with poor scaling resistance, the difference in scaling due to the preconditioning climate is initially large, but after 56 freeze/thaw cycles it is almost negligible. For concrete with good scaling resistance, however, the initial large difference remains throughout the test.

When the test surface was ground to a depth of about 0.6 mm just before resaturation and freeze/thaw testing, the scaling was almost independent of the preconditioning climate. These results indicate that a thin, frost resistant skin is produced on the surface of the test specimens, probably due to carbonation. The resistance of the skin seems to be strongly dependent on the preconditioning climate.

The results of this investigation are only valid for concrete produced with the type of cement used, i.e. OPC. Other cements, such as cement with blast furnace slag may show other results.

Keywords: preconditioning climate; test surface; freeze/thaw resistance; carbonation.

Frost Resistance of Concrete, edited by M.J. Setzer and R. Auberg. Published in 1997 by E & FN Spon, 2–6 Boundary Row, London SE1 8HN, UK. ISBN: 0 419 22900 0.

1 Introduction

When testing the freeze/thaw resistance of concrete according to the Swedish standard SS 13 72 44 [1, 3], "the Borås method", a cut concrete surface is tested, although other test surfaces can be used as well. In other test methods, such as the German CDF-test [2] and the German "Cube Test" [3], cast surfaces are tested. In order to be able to evaluate results from different test methods, the differences in test procedure and pre-conditioning of test specimens have to be systematically studied. In this investigation the influence of the preconditioning climate and different test surfaces on the freeze/thaw resistance is examined.

To be able to improve the precision of the test methods, it is important to learn more about the mechanisms leading to freeze/thaw damage. As concluded in [4], carbonation seems to play an important role in the initial scaling. The carbonation of a concrete surface is strongly dependent on the conditioning climate and to some extent also on the structure of the surface. Consequently it is important to understand the influence of preconditioning and of the structure of the test surface on the freeze/thaw resistance.

2 Preconditioning climates and test surfaces

Two preconditioning climates and three types of test surfaces were investigated. The climates were 65% RH/20°C and 50% RH/20°C respectively, the wind velocity was below 0.1 m/s in both cases. In table 1 the types of test surfaces studied are presented.

Table 1. Test surfaces studied

Type of test surface
Surface cut 7 days after casting, subjected to the preconditioning climate for 21 days
Surface cut 21 days after casting, subjected to the preconditioning climate for 7 days
Surface cast against a plastic coated wooden board, subjected to the preconditioning climate for 21 days

To investigate if the influence of different climates on the frost resistance is dependent on the quality of a thin layer of the test surface, tests were carried out after grinding a layer approximately 0.6 mm thick from the test surface immediately before resaturation. These results were compared with results from unground specimens.

3 Materials

Two concrete mixes were used in the investigation, quality A with poor predicted freeze/thaw resistance and quality B with good predicted freeze/thaw resistance. The compositions are given in table 2. The aggregate used is natural gravel with particle size 0-8 mm and crushed gneiss with particle size 8-16 mm. Both are classified as being frost resistant. The plasticizer, Melcrete, is naphthalene based and the air entraining agent, L 16, is a tall-oil derivative.

One single cement quality, Ordinary Portland Cement, Degerhamn std [5], was used for all mixes. Test results for concrete with other cement types such as, for example, blast furnace slag cement would probably lead to other results.

Table 2. Composition of the two concrete qualities used

Quality	Cement type[1]	Cement (kg/m³)	Aggregate (kg/m³) 0-8	8-16	W/C	Plasti-cizer	Air entr. agent	Air (%)	Slump (mm)
A	OPC	420	927	890	0,40	yes	no	1,2	95
B	OPC	380	911	841	0,50	no	yes	2,9	80

[1] OPC=Ordinary Portland Cement (Degerhamn std [5] is a low alkali, sulphur resistant cement)

4 Specimens and curing

For each concrete quality 8 cubes (150 mm) and 8 slabs (150·150·50 mm) were produced. The cubes were cast in metal moulds. The slabs were cast in cube moulds with a 50 mm thick separating layer dividing the mould into two parts, each 50 mm thick. The separating layer was plastic coated wooden board. The surfaces cast against the board were those to be tested.

Cubes and slabs were demoulded 24 hours after casting. They were then immediately placed in a water bath, at 20±2°C, where they remained until the concrete was 7 days old. After water curing, the cubes were placed in the two different climates, the same quantity of cubes in each climate. Half of the cubes were cut into specimens, (150·150·50 mm) 7 days after casting, i.e. immediately after water curing. The remaining cubes were cut into specimens 21 days after casting. In table 3, the different curing conditions and test surfaces are listed and numbered.

In this report, the numbers in table 3 will be found combined with concrete qualities "A" and "B" in table 2, e.g. "A1" means specimens with concrete mixture "A", with a test surface cut 21 days after casting and stored in a climate room with 65% RH and 20°C. For each number in table 3, three specimens were tested for each concrete quality.

Table 3. Curing climates and test surfaces investigated

Number	Test surface	Climate
1	Cut 21 days after casting	65% RH/20°C
2	Cut 21 days after casting	50% RH/20°C
3	Cut 7 days after casting	65% RH/20°C
4	Cut 7 days after casting	50% RH/20°C
5	Cast surface	65% RH/20°C
6	Cast surface	50% RH/20°C

To determine the conditions in the climate rooms, the evaporation from a free water surface was measured. Table 4 presents an average value for the evaporation measured.

Table 4.   Evaporation from a free water surface in the two climate rooms

Climate	Evaporation ($g/m^2 \cdot h$)
65% RH/20°C	33±2
50% RH/20°C	45±2

When the concrete was 26 days old, a rubber sheet was glued to all surfaces of the specimens except the test surface.   The edge of the rubber sheet reached 20±1 mm above the test surface. Figure 1 shows the test setup.   After the rubber was applied, the specimens were returned to the climate rooms.

After 28 days a layer of tap water (approx. 3 mm deep) was poured onto the surface of the three specimens in each series.   They were then returned to the climate rooms and the resaturation continued for 72±2 hours.   Three specimens from each series "A1-A6" and "B1-B6" were then freeze/thaw tested according to SS 13 72 44.

In order to study the influence of the concrete skin on frost resistance, complementary tests were performed.   Four specimens were produced and cured identically for each test series.   Directly before resaturation, an 0.6±0.2 mm thick layer was ground from the test surface on two of the four specimens.   Then all four specimens were identically resaturated and tested.   The results from ground and unground specimens were then compared.

These complementary tests were only carried out on quality B´, which is identical to quality B in table 2.   Only the two types of test surfaces *cut 21 days after casting* and *cast surface* were included.   For both types of test surfaces, the specimens were conditioned at 65% RH/20°C and 50% RH/20°C.   This means that the combinations preconditioning/test surface 1, 2, 5 and 6 according to table 3 were studied.

5 Test procedure

The freeze/thaw test was performed in accordance with SS 13 72 44.   After the resaturation period, i.e. when the concrete was 31 days old, all surfaces of the specimen except the test surface were thermally insulated with 20 mm thick polystyrene plastic as shown in figure 1.   The tap water used for resaturation was removed and a 3% NaCl solution was poured onto the test surface.   By applying a flat polyethylene sheet, as shown in figure 1, the solution was prevented from evaporating.   The specimens were then placed in the freezers and subjected to repeated freezing and thawing in accordance with the freeze/thaw cycle shown in figure 2.   The temperature in the NaCl solution was continuously measured during the test.   The freezer is programmed so that the temperature falls within the limits in figure 2.   A typical temperature cycle is also shown in the figure.

After 7, 14, 21 (not required by the standard), 28, 42 and 56 cycles the scaled material was collected and dried at 105°C until the material was completely dry, after which the weight of scaled material was measured. The results are given as the amount of scaled material per unit area.

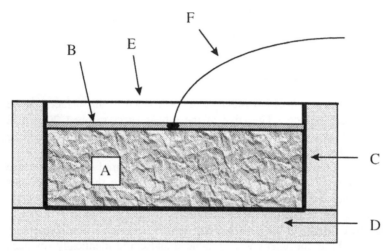

Figure 1. Test setup used for the freeze/thaw test. A- concrete specimen, B- 3% NaCl solution, C- rubber sheet, D- thermal insulation, E- polyethylene sheet, F- temperature measuring device.

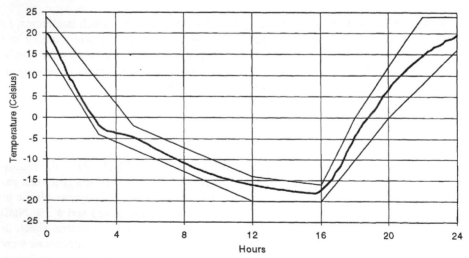

Figure 2. Temperature cycle used, in accordance with SS 13 72 44.

6 Results

The results from the freeze thaw tests are summarized in figures 3-7. The results are presented as the sum of scaled material in kg/m^2 as function of the number of freeze/thaw cycles.

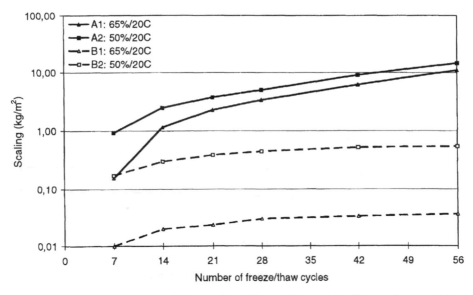

Figure 3. Scaling as function of the number of freeze/thaw cycles for specimens with the test surface cut 21 days after casting.

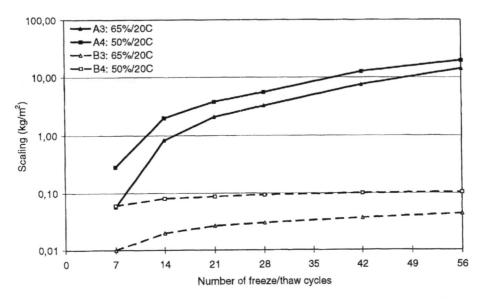

Figure 4. Scaling as function of the number of freeze/thaw cycles for specimens with the test surface cut 7 days after casting.

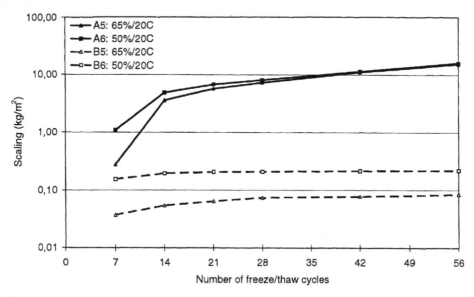

Figure 5. Scaling as function of the number of freeze/thaw cycles for specimens with cast test surface.

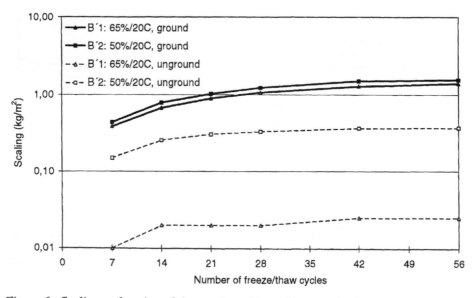

Figure 6. Scaling as function of the number of freeze/thaw cycles for specimens with the test surface cut 21 days after casting. Ground and unground test surfaces.

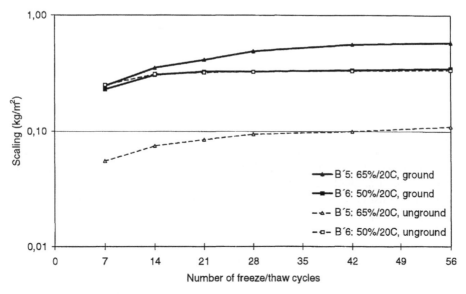

Figure 7. Scaling as function of the number of freeze/thaw cycles for specimens with cast test surface. Ground and unground test surfaces.

7 Discussion

Figure 3 shows the scaling for concrete cut 21 days after casting. From this figure it can be seen that for concrete with poor scaling resistance, the initial difference in scaling is strongly dependent on the preconditioning climate. Specimens conditioned at 50% RH, A2, have after 7 freeze/thaw cycles nearly six times greater scaling than specimens conditioned at 65% RH, A1. During the test, the difference in scaling gets gradually smaller, and after 56 cycles the difference is almost insignificant.

For concrete with good scaling resistance, the difference in scaling due to the different preconditioning climates after 7 cycles is much larger for specimens conditioned at 50% RH, B2, than for specimens conditioned at 65% RH, B1. This difference in scaling is, however, almost constant during the test, and still after 56 cycles, the scaling is much higher for B2 than for B1.

These observations show that specimens preconditioned at 65% RH have a better scaling resistance than specimens conditioned at 50% RH. For concrete with poor scaling resistance this difference can only be seen at the beginning of the test, at the end the scaling is approximately the same regardless of the preconditioning climate. This implies that a thin resistant skin is produced on the test surface and the resistance of this skin seems to be dependent on the climate in which the specimens have been stored. On specimens with poor scaling resistance, the resistance of the skin is too low to withstand scaling during the entire freeze/thaw test. After 7 cycles, the skin is still influencing the results but already after 14 cycles the skin seems to be entirely scaled off, which leads to the rate of scaling becoming less and less dependent on the preconditioning climate. On the other hand, on specimens with good scaling resistance, the resistance of the skin is high enough to influence the result throughout the test. The skin is only broken locally and the initial difference in scaling for concrete conditioned in different climates is maintained during the test.

For specimens with other test surfaces, cut 7 days after casting, figure 4, and cast surface, figure 5, the tendency described above is exactly the same. For concrete with poor scaling resistance, the initial large difference resulting from the conditioning climate decreases while the test is running, and for concrete with good scaling resistance, the initial difference resulting from the conditioning climate remains almost constant during the test.

It can be observed that specimens with cast surfaces have approximately the same amount of scaling as specimens cut 21 days after casting. This is a bit surprising as the surface cast against mould mainly consists of cement paste while only about 30% of the cut surfaces consists of cement paste. Therefore, the scaling ought to be about three times higher for specimens with a cast test surface compared to a cut surface, at least if all other parameters are kept constant.

The explanation is that the duration of preconditioning is very important. The preconditioning period is 21 days for cast test surfaces while it is only 7 days for the cut surfaces. The effect of such a difference in curing conditions is illustrated in figure 8, where scaling results for different curing conditions, 7 or 21 days, for cut surfaces are presented. As can be seen, a longer preconditioning period leads to lower scaling in the interval 0.1-1.0 kg/m^2, when all other parameters are kept constant. Below 0.1 kg/m^2 the skin remains unbroken and the scaling will therefore be independent of the preconditioning time. Above about 1 kg/m^2 the skin is totally broken and the rate of scaling will thus be independent of the preconditioning time. The improved scaling resistance may be due to a higher degree of carbonation in the surface skin when the preconditioning period is prolonged.

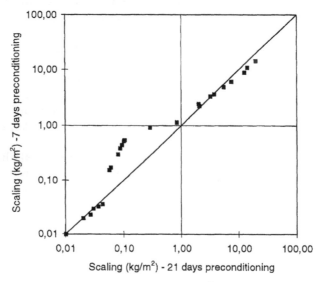

Figure 8. Scaling after the same number of freeze/thaw cycles for specimens cut 7 and 21 days after casting, i.e. preconditioned for 21 and 7 days respectively.

In order to study the importance of the concrete skin, additional tests were performed. A number of specimens were produced and treated according to the procedures described above. Directly before resaturation a layer of the test surface, approximately 0.6 mm thick, was removed by grinding some of the specimens. The other specimens were left unground. After this, all the specimens were identically treated and tested.

The results are presented in figures 6 and 7. The results show that there is a strong difference between the results for unground specimens, when preconditioned in different climates. This difference, however, more or less disappears when a thin layer of the test surface is removed. This investigation can not explain why the concrete skin is so important for the frost resistance. The carbonation of the skin, however, probably plays an important role.

These results are relevant for the type of concrete and binder used in this investigation. Other types of binder would probably lead to other results, see e.g. [6] and [7].

8 Conclusions

From this investigation, the following conclusions can be drawn:
1. The results from freeze/thaw tests are dependent on the preconditioning climate. Specimens conditioned at 50% RH/20°C have higher scaling, in freeze/thaw tests, than specimens conditioned at 65% RH/20°C.
2. For concrete with low freeze/thaw resistance the difference in scaling dependent on the preconditioning climate is large in the beginning of the freeze/thaw test, after 56 cycles the difference is insignificant. For concrete with good freeze/thaw resistance the initial difference in scaling is maintained during the test and is after 56 cycles almost as large as after 7 cycles.
3. The scaling resistance is increased with prolonged preconditioning time at least for concrete with OPC cement.
4. During preconditioning a thin carbonated skin is produced on the surface which strongly affects the scaling resistance. The properties of the skin seem to be highly dependent on the preconditioning climate.

9 References

1. Swedish Standard, SS 13 72 44, Edition 3 (1995), Concrete testing - Hardened concrete - Scaling at freezing.
2. RILEM Recommendation (1996), RILEM TC 117-FDC, Test method for the freeze-thaw resistance of concrete with sodium chloride solution (CDF), *Materials and Structures*, Vol. 29, No. 193. pp. 523-528.
3. RILEM TC 117-FDC (1995), Test method for the freeze-thaw resistance of concrete. Slab Test and Cube Test, *Materials and Structures*, Vol. 28, No. 180. pp. 366-371.
4. Setzer, M. J. (1997) Report, RILEM TC 117-FDC: Freeze-thaw and deicing resistance of concrete. *Materials and Structures*, Supplement March 1997.

5. Malmström, K. (1990) Cementsortens inverkan på betongs frostbeständighet, Rapport 1990:07, Swedish Testing and Research Institute.
6. Matala, S. (1995) Effects of carbonation on the pore structure of granulated blast furnace slag, Report 6, Helsinki University of Technology, Espoo.
7. Stark, J., Ludwig, H-M. (1997) Freeze-Thaw and Freeze-Deicing Salt Resistance of Concretes Containing Cement Rich in Granulated Blast Furnace Slag, *ACI Materials Journal*, Vol. 94, No. 1. pp. 47-55.

An experimental study on frost resistance of concrete considering drying effects

Y. HAMA and E. KAMADA
Hokkaido University, Sapporo, Japan
C.G. Han
Chong-ju University, Chong-ju, Korea

Abstract
Frost deterioration of concrete is influenced by freezing and curing conditions, such as number of freezing-and-thawing cycles, lowest freezing temperature, a period of freezing and drying during summer. In this paper, the frost resistance of ordinary-strength concrete and high-strength concrete which had different air content was investigated considering drying effects and the number of freezing-and-thawing cycle on actual weathering condition. It could be concluded that the effect of air content was significant from the results of the accelerative tests according to ASTM C666 procedure A and B. The difference of frost resistance with air content was shown clearly by long term freezing-and-thawing cycles. The frost resistance of concrete was improved significantly by drying before freezing. However, the improvement of frost resistance on high-strength concrete was not so large as on ordinary-strength concrete.
Keywords: Frost resistance, air content, curing, drying, high-strength concrete

1 Introduction

For standard tests evaluating frost resistance of concrete, freezing-and-thawing tests based on the method designated in ASTM C666 procedure A and B, "Resistance of Concrete to Rapid Freezing and Thawing", are widely used in Japan and many other countries. Though the results of these standard tests are not related to the effects of actual weathering condition. It is expressed in ASTM C666 that neither procedure is intended to provide a quantitative measure of the length of service life. Since every condition of testing procedure is defined as accurately as possible in these standard tests, useful results on the frost resistance of concrete can be obtained to be some comparative data, but by the same token, there is no actual cue to understanding the performance of concrete exposed to actual weathering condition.

Frost Resistance of Concrete, edited by M.J. Setzer and R. Auberg. Published in 1997 by E & FN Spon, 2–6 Boundary Row, London SE1 8HN, UK. ISBN: 0 419 22900 0.

Frost deterioration of concrete is influenced by freezing and curing conditions, such as number of freezing-and-thawing cycles, lowest freezing temperature, a period of freezing and drying during summer. Formerly, we proposed an empirical equation which expressed the effects of environmental conditions as the number of equivalent cycles to ASTM C666 procedure A [1] . Applying this equation to the actual weathering condition of some cities in Hokkaido of Japan, the number of freezing and thawing cycles in one year were about 10 to 50 cycles as equivalent to ASTM C666 procedure A. Furthermore, it is considered that frost deterioration would be restored by drying during summer.

The aim of this investigation is to compare the frost resistance of ordinary-strength concrete and high-strength concrete which have different air content and were cured on different conditions considering drying effects during summer and the number of freezing-and-thawing cycles on actual weathering condition.

2 Outline of experiment

2.1 Materials and mix proportions
The materials used were ordinary portland cement (specific gravity 3.15), tap water, Ohigawa river sand (specific gravity 2.62, percentage of water absorption 1.12%, fineness modulus 2.9) and Chichibu crushed sand stone (specific gravity 2.90, percentage of water absorption 0.54%, fineness modulus 6.62). As chemical admixtures, air-entraining and water-reducing agent, air-entraining agent, air-reducing agent and air-entraining and high-range water-reducing agent (superplasticizer) were used.

Mix proportions and properties of concrete are shown in Table 1 and Table 2. As

Table 1. Mix proportions and properties of concrete (series Ⅰ)

Kind of concrete	45A	45B	45C	45D	65A	65B	65C	65D
W/C (%)	45	45	45	45	65	65	65	65
Air content as target(%)	4.5	4.5	1.0	1.0	4.5	4.5	1.0	1.0
s/a(%)	41.7	40.9	40.0	40.0	46.4	46.9	47.0	47.0
Water (kg/m³)	171	205	223	223	164	195	210	210
Cement (kg/m³)	380	456	496	496	252	300	323	323
Fine aggregate (kg/m³)	726	650	639	639	865	817	836	836
Course aggregate (kg/m³)	1045	967	988	988	1029	952	968	968
AE water-reducing agent (kg/m³)	0.948	0.228	—	—	0.633	0.150	—	—
Air-entraining agent (kg/m³)	0.019	0.032	—	0.001*	0.013	0.018	—	0.001*
Slump (cm)	9.0	18.5	18.0	19.0	8.5	20.0	19.0	19.0
Air content (%)	4.4	4.5	0.9	0.8	4.9	4.0	1.5	1.1
Compressive strength** (MPa)	52.2	44.4	45.3	44.4	33.1	25.3	33.7	30.5

* air-reducing agent   ** curing in water for 28 days

Table 2. Mix proportions and properties of concrete (series Ⅱ)

Kind of concrete	35-1	35-2	35-3.5	35-5	55-2	55-5
W/C (%)	35	35	35	35	55	55
Air content as target (%)	1	2	3.5	5	2	5
s/a(%)	46.6	45.8	44.5	43.2	49.8	47.6
Water (kg/m³)	172	172	172	172	172	172
Cement (kg/m³)	155	155	155	155	99	99
Fine aggregate (kg/m³)	812	786	747	707	925	846
Course aggregate (kg/m³)	1024	1024	1024	1024	1024	1024
Superplasticizer (kg/m³)	4.920	4.920	4.920	4.920	1.565	1.565
Air-entraining agent (kg/m³)	0.020*	0	0.020	0.039	0	0.002
Slump (cm)	22.0	20.4	20.2	21.5	18.9	19.3
Air content (%)	1.2	2.2	2.4	5.5	2.2	6.0
Compressive strength** (MPa)	65.9	65.2	64.8	58.9	35.8	27.6

* air-reducing agent   ** curing in water for 28 days

shown in Table 1, eight kinds of concrete were used in series Ⅰ. The water cement ratios were 45% and 65%. The target air contents were $1 \pm 0.5\%$ and $4.5 \pm 0.5\%$. The slump of 45A and 65A was 8cm, and the others were 18cm. Air-reducing agent was used on 45D and 65D. As shown in Table 2, six kinds of concrete were used in series Ⅱ. The water cement ratios were 35% and 55%. The target air contents were varied from $1 \pm 0.5\%$ to $5 \pm 0.5\%$. The slump was 18cm.

2.2 Experimental method

Experiments were carried out basically according to ASTM C666, changing the curing conditions before testing and the process of freezing-and-thawing cycles. The specimens used in the freezing-and-thawing tests were $75 \times 75 \times 400$mm prisms.

Table 3 shows the conditions of freezing-and-thawing tests. The specimens of series Ⅰ which cured on two different conditions for 4 weeks were subjected to 300 cycles freezing-and-thawing action by ASTM C666 procedure A and B. In series Ⅱ, three different curing conditions before freezing-and-thawing test, ASTM C666 procedure A, were used. 1000 cycles were performed for the conditions of Ⅱ-A and Ⅱ-B. On the condition of Ⅱ-C, the process of drying after every 30 freezing-and-thawing cycles was repeated until 300 freezing-and-thawing cycles. The specimens which subjected to drying process were immersing in water for 2 days before freezing-and-thawing test.

The performance of the test specimens was measured primarily by the change in their resonant frequency, which is proportional to the dynamic modulus of elasticity. The resonant frequency was measured prior to test and at selected intervals during test period and is reported as the relative dynamic modulus which is the ratio of the square of the frequency at a given cycle to the square of the initial frequency before freezing-and-thawing cycling begins. Weight change and length change of the specimens were also measured.

Table 3.   Conditions of freezing-and-thawing tests

Series		Curing condition before freezing-and-thawing test	Condition and number of freezing-and-thawing cycles
I	I A	4 weeks in water at 20℃	ASTM C666 procedure A continuously 300 cycles
	I B	dried at 20℃ for 2 weeks after 2 weeks cured in water	ASTM C666 procedure A continuously 300 cycles
	I C	4 weeks in water at 20℃	ASTM C666 procedure B continuously 300 cycles
	I D	dried at 20℃ for 2 weeks after 2 weeks cured in water	ASTM C666 procedure B continuously 300 cycles
II	II A	2 weeks in water at 20℃	ASTM C666 procedure A continuously 1000 cycles
	II B	dried at 20℃ for 1 weeks after 2 weeks cured in water	ASTM C666 procedure A continuously 1000 cycles
	II C	dried at 20℃ for 2 weeks after 2 weeks cured in water	ASTM C666 procedure A Repeat drying for 7 days and immersing in water for 2 days after every 30 freeze/thaw cycles until 300 freeze/thaw cycles

3 Results and discussion

Frost resistance is evaluated by durability factor which is calculated as follows:

$$DF = PN/M \qquad (1)$$

where,   DF : durability factor of the test specimen

P : relative dynamic modulus of elasticity at N cycles, %

N : number of cycles at which P reaches 60% or the specified number of cycles at which the exposure is to be terminated, whichever is less

M : the specified number of cycles at which the exposure is to be terminated

Figure 1 is a comparison of the results of durability factor by curing condition before freezing-and-thawing test which is ASTM C666 procedure A in series I. On the condition I A, the specimens of non-AE concrete (45C, 45D, 65C and 65D) were deteriorated only after a few freezing-and-thawing cycles and the durability factors were less than 15. Though, those specimens were became durable by dry curing before freezing-and-thawing test on the condition I B. This tendency was significant when the water cement ratio was high.

Figure 2 is a comparison of the results of durability factor by curing condition before freezing-and-thawing test which is ASTM C666 procedure B. The results of procedure B show a same tendency to that of procedure A. The specimens except 45C, 45D, 65C and 65D which were non-AE and not dried before freezing-and-thawing test were not deteriorated at all. Comparing the results of procedure A and B, procedure B is appropriate to define the difference of frost resistance between susceptible concrete specimens.

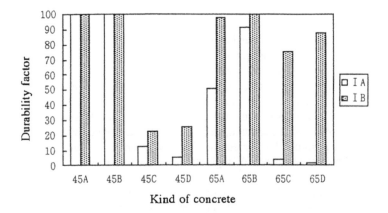

Figure 1. Comparison of durability factors by curing condition before freeze/thaw test
(ASTM C666 procedure A - series I)

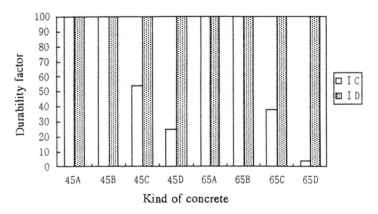

Figure 2. Comparison of durability factors by curing condition before freeze/thaw test
(ASTM C666 procedure B - series I)

From these results, it is clear that the effects of the air content and drying before freezing
are significant to frost resistance of concrete. However, on 5 years exposure test using
these specimens deterioration in cold regions was not observed even in specimens
deteriorated only after a few freezing-and-thawing cycles on these accelerative tests.
This fact suggests that the condition of ASTM C666 is very severe to actual weathering
conditions.

In series II, the effect of dying during summer was investigated on ordinary-
strength concrete and high-strength concrete which have different air content. The
number of freezing-and-thawing cycles in one year on natural weathering condition was
estimated 30 cycles as equivalent to ASTM C666 procedure A and the specimens on the
condition II C were dried after every 30 cycles.

Figure 3 shows the results at 300 cycles of freezing-and-thawing tests in series II.

High-strength concrete was durable to frost damage, however air-reduced high-strength concrete were deteriorated within 300 cycles on the condition Ⅱ A and Ⅱ B. Besides, all kinds of concrete were not deteriorated on the condition Ⅱ C which was assumed actual weathering conditions. Improvement of frost resistance by drying effects on high-strength concrete was not so large as on ordinary-strength concrete. The weight change of specimens by drying and immersing before freezing-and-thawing tests is given in Table 4. The weight change of high-strength concrete was smaller than that of ordinary-strength concrete. The capillary of high-strength concrete is so dense that the capillary water could not be released. In addition, the amount of water which was almost same to that by drying was absorbed by immersing after drying on high-strength concrete. For these reasons, the frost resistance of high-strength concrete was not so improved by drying.

Figure 4 is the results of durability factor at 1000 cycles on the conditions of Ⅱ A and Ⅱ B. Difference of frost resistance with air content was shown clearly by long term freezing-and-thawing cycle.

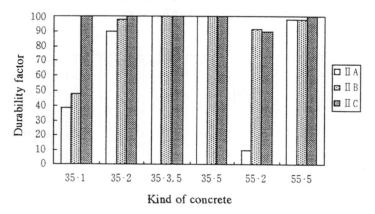

Figure 3. Comparison of durability factors at 300 cycles by testing condition (series Ⅱ)

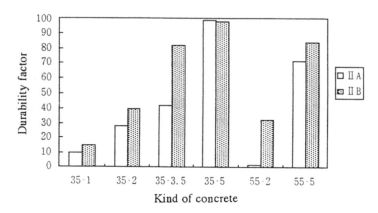

Figure 4. Comparison of durability factors at 1000 cycles (series Ⅱ)

Table 4.   Weight change of specimens by drying and immersing before freeze/thaw test

	35-1	35-2	35-3.5	35-5	55-2	55-5
Weight change by drying : D(g)	29.0	32.3	35.8	39.8	63.8	80.8
Weight change by immersing : I (g)	27.0	30.3	31.3	34.3	43.8	60.5
Percentage of I∕D (%)	93.1	93.8	87.4	86.2	68.7	74.9

4 Conclusions

From the results of this investigation, following conclusions can be drawn.
1. The effects of the air content and drying before freezing action were significant to the results of ASTM C666 procedures.
2. As a characteristic of testing procedures, procedure A evaluates the frost resistance of concrete more severely than procedure B. Procedure B is appropriate to define the difference of frost resistance between susceptible concrete specimens.
3. Difference of frost resistance with air content was shown clearly by long term freezing- and-thawing cycle.
4. Improvement of frost resistance by drying on high strength concrete was not so large as on ordinary concrete.

5 References

1. Y.Hama, E.Kamada, M.Tabata and Y.Koh. (1993) Study on relationship between environmental conditions and standard method of freezing-and-thawing test (ASTM C666 A) on frost resistance of concrete. Durability of Building Materials and Components 6, E & FN Spon, London, pp.1162-1170
2. V.Sturrup, R.Hooton, P.Mukherjee and T.Carmichael.(1987) Evaluation and Prediction of concrete durability - Ontario Hydro's Experience. ACI SP-100, pp.1121-1154

A mechanism of frost damage of concrete under supercooling

O. KATSURA
Hokkaido Prefectural Housing and Urban Research Institute,
Sapporo, Japan
E. KAMADA
Faculty of Engineering, Hokkaido University, Sapporo, Japan

Abstract
It is the purpose of this paper to describe a frost damage mechanism of concrete numerically. A mechanism considering freezing of water under supercooling is proposed. Freezing of capillary water is the origin of frost damage. It depends on supercooling and depression of melting point. When water under supercooling freezes, ice crystal grows very rapidly and the volume increases about 9%. It causes a rapid flow through the capillary pore toward the air void. Consequently viscosity of unfrozen water produces high hydraulic pressure. The relation between hydraulic pressure, flux, the air void system and the pore structure is described by Poiseulle's equation on viscous fluid. When hydraulic pressure goes beyond the tensile strength of hardened cement paste, a micro-crack must occur. Subsequent growth of ice crystals clearly increases the plastic deformation. The cumulative plastic deformation is observed as a phenomenon of frost damage. The relation between the durability factor (ASTM C 666 A) and plastic deformation is verified experimentally.
Keywords: Air void system, Growth of ice crystals, Frost damage mechanism, Pore structure, Super cooling

1 Introduction

Frost damage is a big problem on the durability of concrete in cold regions. The frost damage mechanism of concrete was explained as hydraulic pressure theory by T. C. Powers [1]. The relation between freezing of capillary water, pore structure and hydraulic pressure is not explained enough numerically today.

Freezing of capillary water is the origin of frost damage, and is studied by many

Frost Resistance of Concrete, edited by M.J. Setzer and R. Auberg. Published in 1997 by E & FN Spon, 2–6 Boundary Row, London SE1 8HN, UK. ISBN: 0 419 22900 0.

researchers [2] [3] [4]. It is well known that freezing of capillary water depends on both supercooling and depression of melting point in freezing cycle and depression of melting point depends on the pore radius. Some studies have shown that ice crystal grows very rapidly when water freezes under supercooling [5]. Rapid growth of ice crystals makes fast capillary flow then hydraulic pressure is generated depending on the pore structure. That relation is described by Poiseulle's equation. The hydraulic pressure must cause frost damage of concrete. The relation between hydraulic pressure, tensile strength, the pore structure, freezing of water and temperature have to be clear to establish the mechanism of frost damage numerically.

In this paper the mechanism of frost damage of concrete in a saturated condition is described numerically.

2 Freezing of capillary water

2. 1 Estimation of frozen water
The electric method to estimate frozen water content of cementicious material was proposed by the authors [4]. In this paper that method is modified a little.

Electrical conductivity is sorted to electronic conduction and electrolyte conduction. In the case of electrolyte conduction, the higher temperature is, the lower is resistivity. Such phenomenon was observed on cementicious material too. Capillary water includes some kinds of electrolytes. Therefore it is seemed that the conductivity of cementicious material is according to the mechanism of electrolyte conduction.

2.1.1 Conductivity of solution
Conductivity of electrolyte solution is a product of molar conductivity and concentration. Using mobility, conductivity is described in eq. 1.

$$K = \Lambda \cdot C = C \mid z e \mid u = C \mid z e \mid u_0 \exp.(-U / kT) \tag{1}$$

where $z e$: electric charge u : mobility $= u_0 \exp.(-U / kT)$
          U : mobility energy k : Boltzmann constant
          T : absolute temperature

Mobility energy U in eq. 1 is regarded as Gibbs energy then it is also the chemical potential of the solution. Every electrolyte has individual chemical potential. Therefore mobility energy U of the solution included electrolytes is described as the sum of each mobility energy as eq. 2.

$$U = \sum (\mu_i + RT\ln(\gamma_i \cdot c_i)) \tag{2}$$

where γ_i : activity coefficient c_i : concentration

2.1.2 Conductivity of hardened cement paste
Hardened cement paste has many pores in which water is adsorbed and is condensed under equilibrium with relative humidity. Therefore the chemical potential of capillary water is described as the sum of the electrochemical potential and chemical potential of the water. Activity coefficient γ_i in eq. 2 and 3 is described as eq. 4.

$$U = \sum (\mu_i + RT\ln(\gamma_i \cdot c_i)) + \mu_w + RT \ln (p/p_0) \tag{3}$$
where $\quad p/p_0$: relative pressure

$$\ln(\gamma_\pm) = - z_i^2 e^2 b/(8\pi \varepsilon_0 \varepsilon kT) \tag{4}$$
where ε_0 : dielectric constant of vacuum $\quad \varepsilon$: dielectric constant of solvent
$\quad 1/b$: Debye length $\quad b=(e^2\sum(c_i z_i^2)/(\varepsilon_0 \varepsilon kT))^{1/2}$

Conductivity of capillary water can be described as a function of concentration, temperature and relative pressure by substituting mobility energy and activity coefficient into eq. 1. Equation 1 is rearranged and expressed as eq. 5 simply.

$$\ln K = a_1 \ln C + a_2/T + a_3(C/T^3)^{0.5} + a_4 \ln(p/p_0) \tag{5}$$
where $\quad K$: conductivity of capillary water $\quad C$: concentration
$\quad T$: absolute temperature $\quad p/p_0$: relative pressure

2.1.3 Water content
An experiment was carried out. The specimen is hardened cement paste cured in water for 91 days. The water cement ratio is 0.3. After curing, the specimen are dried under 5 controlled humidities in the range of 0.33 and 0.95. Electric resistance and temperature was measured in a chamber in the range of -70 and 20 °C. The slope of temperature was controlled 1.7 °C/hr. Isothermal adsorption at 20 °C was carried out too.

At 10 °C the relation between conductivity and relative pressure in the logarithmic scale is linear as shown in fig. 1. At every relative pressure the relation between the logarithm of conductivity and reciprocal of temperature is linear as shown in fig. 2. The concentration of each electrolyte in capillary water must be known to apply eq. 5. The first and third term of eq. 5 includes concentration. The effect of the concentration must be observed in the relation between conductivity and the reciprocal of absolute temperature. In this experiment it can be considered that the amount of electrolytes in capillary water is constant and the concentration changes according to the water content. The slope of RH 33% in fig. 2 is merely bigger than one of the others therefore it is assumed that the concentration is constant to make the analysis simple. Eq.5 is rewritten approximately to eq.6.

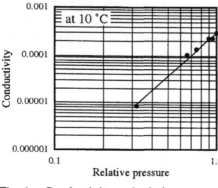

Fig. 1. Conductivity and relative pressure

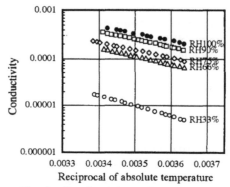

Fig. 2. Conductivity and temperature

$$\ln K = a_1 + a_2 / T + a_3 \ln(p/p_0) \qquad (6)$$

Constants in eq. 6 are decided by multiple regression experimentally. Relative pressure that is equivalent to one at 20 °C can be calculated from the absolute temperature and conductivity. An accurate relation between calculated relative pressure and controlled one is shown in fig. 3. Water content is estimated using the result of isothermal adsorption shown in fig. 4.

2.1.4 Frozen water content

The conductivity of ice is very low and dielectric constant of ice is very high compared with that of water. Consequently ice is regarded as an insulator. It is assumed that freezing of capillary water is equivalent to drying in regard to the state of the liquid phase. A phase transition from water to ice makes electric resistance high as shown in fig. 5. If water did not freeze, electric resistance will depend on temperature only. The relation between resistivity and temperature is obtained above 0 °C. It is considered that the distance of known relative humidity and one calculated using eq. 6 is caused by freezing. The relation between calculated relative pressure and temperature is shown in fig. 6. Consequently frozen water content is estimated from the isothermal adsorption curve shown in fig. 4.

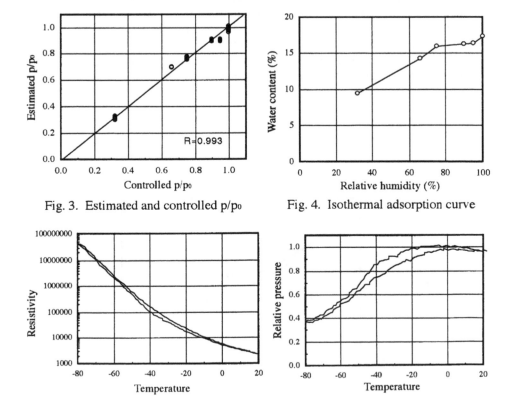

Fig. 3. Estimated and controlled p/p₀

Fig. 4. Isothermal adsorption curve

Fig. 5. Resistivity and temperature

Fig. 6. Estimated relative pressure

2. 2 Freezing under supercooling

An experiment on mortar was carried out. 18 freezing and thawing cycles with various minimum temperatures in the range of -3 and -70 °C were repeated and the electric resistance and temperature were measured. Isothermal adsorption at 20 °C was carried out too. Some results are shown in fig.7.

The obvious hysteresis loops are observed in the freezing and thawing cycles. In the freezing cycle, the freezing point is lowered by both supercooling and depression of melting point. In the thawing cycle the melting point is lowered by depression of melting point only. Consequently supercooling decreases the frozen water content and the hysteresis loop is observed. The minimum temperature at which supercooling occurs is regarded at about 220 °K. The freezing and thawing curves are corresponding below this temperature. The curve in the thawing cycle describes the distribution of the depression of melting point. At the same temperature melting water (the slope of the curve) increases as the minimum temperature lowers. The relation between the freezing rate of water and the degree of supercooling must be estimated by analyzing melting curves empirically. It is also well-known that the smaller the pore size is, the lower the degree of depression of melting point becomes [3][4]. The freezing rate of water can be estimated by the degree of supercooling and the pore size distribution.

The frozen water content in each thawing cycle is calculated by the method mentioned above. The curve that minimum temperature is below -50 °C has no water under supercooling. The rate of frozen water content to the frozen water content without supercooling is calculated at each melting point. The relation between the rate of frozen water and minimum temperature at each melting point is shown in fig. 8. The lower minimum temperature is, the higher the freezing rate.

The relative degree of supercooling is calculated by eq. 7.

$$R_s = (T_m - T_f) / (T_m - 223) \tag{7}$$

where R_s : relative degree of supercooling

           T_m : melting point (°K) T_f : minimum temperature (°K)

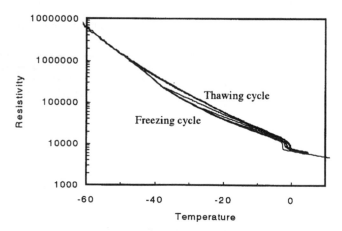

Fig. 7. Resistivity and temperature

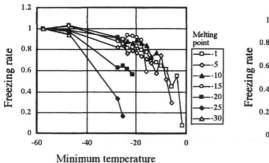

Fig. 8. Freezing rate and minimum
    temperature

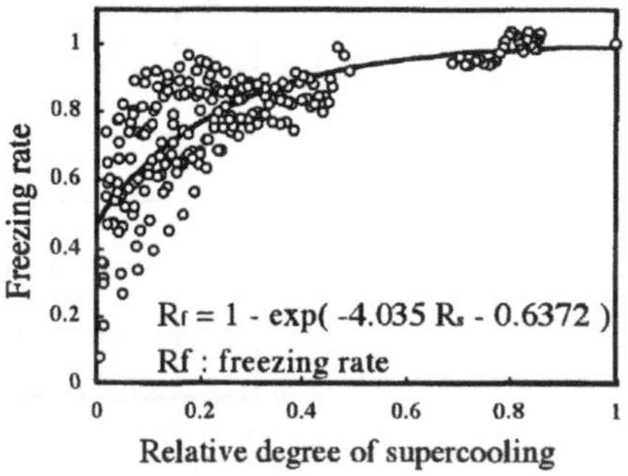

Fig. 9. Freezing rate and relative degree
    of supercooling

The relation between the freezing rate and the relative degree of supercooling is shown in fig. 9. A bold line shows the approximate curve calculated by the least squares method as follows.

$$R_f = 1 - \exp(-4.035\, R_s - 0.6372)\tag{8}$$

The freezing of water under supercooling has not been theoretically clear. It must be considered that eq. 8 describes the approximate freezing rate of water under supercooling empirically.

The relation between the depression of melting point and the pore radius was shown by some researchers. Therefore the freezing rate of water under supercooling is able to be estimated from temperature and pore structure. Furthermore it is considered that the effect of adsorbed water must be taken into account.

3 Mechanism of frost damage considering supercooling

It is well-known that the lower the durability factor is, the larger the residual strain of damaged concrete becomes. We have one result on the dilation of mortar which the water cement ratio is 0.38 as shown in fig. 10. The dilation is measured above -60 °C using a thermal mechanical analyzer (TMA). Strain increases rapidly at -5 °C. It is regarded that water under supercooling freezes rapidly at this temperature. After that strain increases with lowering of temperature above -50 °C. The minimum temperature that supercooling occur is about -50 °C mentioned above. Both temperatures are corresponding. Rapid freezing of capillary water may be a driving force of movement of unfrozen water toward air void. The mechanism of frost damage taking into account the supercooling is proposed as follows.

3. 1 Rapid growth of ice crystal under supercooling
Velocity of growth of ice crystal under supercooling is described by Hillig [5].

$$V = 0.16\, T_{sc}^{1.7}\tag{9}$$

where V : velocity of ice crystal growth (cm/s)   Tsc : degree of supercooling

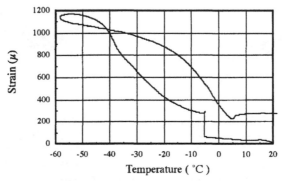

Fig. 10. Dilation of mortar specimen

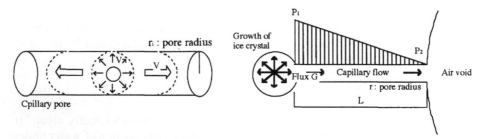

Fig. 11. Simple cylindrical model   Fig. 12. Concept of hydraulic pressure

When the degree of supercooling is 10 °C the velocity V gains about 8 cm/s. It is considered that growth of ice crystals under supercooling is extremely rapid compared with one under thermal equilibrium. The growth of ice under thermal equilibrium must go with the movement of heat quantity. The isotherm in concrete does not move so rapidly and actual building material will never be exposed in such a severe condition. Rapid growth of ice crystal makes rapid increase of the volume and rapid movement of unfrozen water as like as explosion in a small capillary pore.

3. 2 Generation of hydraulic pressure
It is assumed that the capillary pore is simply cylindrical as shown in fig. 11. Nucleation occurs at one point in the cylindrical pore then an ice crystal grows spherically until it reaches surface of pore. Subsequently an ice crystal grows along the cylindrical pore. As water changes phases from liquid to solid, volume increases about 9%. It makes capillary flow through the pore. The maximum flux is caused by spherical growth of an ice crystal when it reaches the wall. The maximum flux G is described by eq. 10.

$$G = 0.0826 \cdot 4 \pi \, r_i^2 \, V \qquad\qquad (10)$$
where G : flux (cm³ / s) r_i : pore radius (cm)
V : velocity of ice crystal growth (cm/s)

The simple model is shown in fig. 12. The relation between flux (G), pressure (P) , pore radius (r) and pore length (L) is described by Poiseulle's equation. Pressure

(P_2) at the surface of the air void is assumed to be 0.

$$G = (n \pi r^4) P_1 / 8 L \eta \tag{11}$$
where η : coefficient of viscosity (N· s · m^{-2})
n : number of cylindrical pore with radius r and length L.

The pressure P_1 is estimated by the pore radius, the degree of supercooling, length and radius of the pore in which unfrozen water moves toward the air void. Equation 11 describes the permeability of hardened cement paste too. Luping proposed the relation between permeability and pore structure measured by mercury penetration method [6] . Exactly the number of cylindrical pores with radius r is estimated by the three-dimensional cylindrical pores model. Application of such model must be studied.

Pores in which water does not freeze function differently from those in which unfrozen water moves, even at the same temperature and pore radius. In the application of the pore structure model, permeability must be calculated from the pore volume in which water does not freeze.

3. 3 Destruction and plastic deformation

It is assumed that the stress of the pore wall is approximately estimated by eq. 12. This describes the relation between pressure and stress on a thick cylindrical pipe. The model of a thick cylindrical pipe is shown in fig. 13. If r_2 is much bigger than r_1, eq. 12 is rewritten to eq.13.

$$\sigma_\phi = P (r_2^2 + r_1^2)/(r_2^2 - r_1^2) \tag{12}$$
where σ_ϕ : stress (N/mm^2) r_2 : outer radius r_1 : inner radius

$$\sigma_\phi = P \tag{13}$$

When σ_ϕ goes beyond the tensile strength of hardened cement paste (σ_t), micro-crack must occur on the pore wall, subsequently stress decreases rapidly and its width increases as water continues to freeze. The cumulative plastic deformation must be observed as the deterioration by frost damage. The plastic deformation of one freezing point is estimated as the increase of frozen water volume after destruction of the pore wall. The rate of water that causes deformation is described in eq. 14. The total volume of deformation V_d is obtained by summing each product of the rate R_d and the pore volume with a radius r_i.

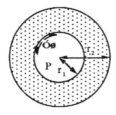

Fig. 13. Thick cylindrical pipe model

$R_d = (\pi r_i^3 - \pi r_c^3) / \pi r_i^3 = 1 - (\sigma_t / P)^{1.5}$ (14)

where r_c : radius of ice crystal when pressure goes beyond tensile strength

$V_d = \sum R_{di} \cdot V_{pi}$ (15)

3. 4 Experimental verification

The result of the freezing and thawing test (ASTM C 666 A) that had been carried out was used to verify the mechanism [7] . The water cement ratio of concrete is in the range 0.37 and 0.55. The air content is in the range 2% and 5%. The pore size distribution is measured by mercury penetration method. The air void system is obtained by linear traverse method (ASTM C 457).

The calculation is carried out on the 5 assumptions as follows.
1. Higuchi's equation describes the relation between depression of melting point and the pore radius. Higuchi's curve is almost same as Fagerlund's curve.

   $\Delta T / T0 = 2 \ \sigma \ M/(r \ \rho \ \Delta H_{SL})$
   where ΔT : depression of melting point σ : surface tension of water
            M : molecular weight of water   r : pore radius
            ρ : density of water ΔH_{SL} : heat of fusion

2. The tensile strength of concrete is approximately used as the tensile strength of the hardened cement paste.
3. Water under supercooling intermittently freezes at every one degree of super-cooling.
4. At each radius, the number of cylindrical pores leading to air void is in proportion to the number of all pores with length (L) (spacing factor).
5. The total volume of deformation is in proportion to the logit of durability factor ($\ln(D_f/(1 - D_f))$).

The pore radius that is equivalent to the every one degree of melting point is calculated by Higuchi's equation. The following procedure is done at every melting point and every minimum freezing temperature.

The freezing rate is calculated using eq. 8. The amount of freezing water is calculated as a product of the pore volume and the freezing rate. The flux is estimated from the pore radius and the degree of supercooling by eq. 9, 10. The pressure and the total volume of deformation are calculated by eq. 11,14,15 using the assumed value of n. Finally the relation between the total volume of deformation and the durability factor is analyzed by nonlinear least square method.

The results of this analysis are shown in fig. 14 and fig. 15. The accurate relation between the durability factor and the total volume of deformation is shown in fig. 14. The one relation between the estimated durability factor and the minimum freezing temperature is shown in fig. 15. The water cement ratio is 0.55. The lower the air content is and the lower the minimum temperature is, the lower is the estimated durability factor.

The assumptions in this paper must be verified in further studies. In particular, the movement of capillary water depending on the pore structure must be verified.

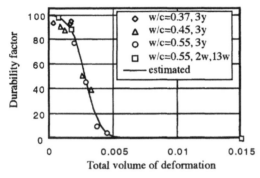

Fig. 14. Durability factor and total volume of deformation

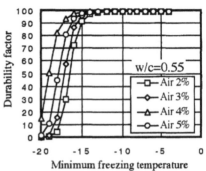

Fig. 15. Durability factor and minimum freezing temperature

4 Conclusion

The numerical mechanism of frost damage under the conditions of rapid freezing of water is proposed in this paper. Rapid freezing of water under supercooling causes rapid flow of unfrozen water then high hydraulic pressure occurs.

The frost durability of concrete is estimated from the pore structure, the tensile strength, the air void system and the temperature. The relation between movement of capillary water and pore structure must be investigated furthermore.

5 Reference

1. Powers, T.C. (1945) A Working Hypothesis for Further Studies of Frost Resistance of Concrete, Journal of American Concrete Institute, Vol. 16, No. 4
2. Helmuth, R.A. (1960) Capillary Size Restriction on Ice Formation in Hardened Portland Cement Paste, Proc. of 4th International Symposium on the Chemistry of Cement
3. Sellevold, E.J. , Bager, D.H. (1980) Low Temperature Calorimetry as Pore structure Probe, 7th International Congress on the Chemistry of Cement, Vol. 4
4. Katsura, O., Yoshino, T.,Kamada, E. (1996) Estimation of Frozen Water in Cementicious Material and Dependance of Ice Formation on Pore Structure, Concrete Research and Technology, Vol. 7, No. 1,
5. Fletcher, N.H. (1970) The Chemical Physics of Ice, Cambridge University Press
6. Luping, T., Nilsson, L.O.(1992) A Study of the Quantitative Relationship Between Permeability and Pore Size Distribution of Hardened Cement Pastes, Cement and Concrete Research, Vol. 22, pp. 541-550
7. Katsura, O., Yoshino, T. (1996) Effect of Coarse Aggregate on Frost Damage of Concrete, Research Report No. 64, Hokkaido Prefectural Cold Region Housing and Urban Research Institute

Length changes of concrete specimen during frost deicing salt resistance test

J. KAUFMANN and W. STUDER
Swiss Federal Laboratories for Materials Testing and Research,
Duebendorf, Switzerland

1. Introduction

Since grit as means of ice control on roads has been replaced by deicing salts and snow and ice are removed totally from our roads during winter, concrete structures not only are damaged by frost but also frost deicing salt attack. Severe damage with very high repair costs is the consequence. It is well known that in laboratory tests (Boraas, CDF-Test, etc.) scaling is 1-2 orders of magnitude higher when as frost medium a 3% salt solution instead of water is applied. Different authors report that damage is maximal at a pessimum salt concentration of about 3 mass-%. Meanwhile great progress was made in avoiding damage in situ by use of adequate mix designs (e.g. air pore entrainment) but this is more the consequence of experience than that of knowledge of deterioration mechanisms.

The crucial influence of deicing agents is still not fully explained. Neither are other observations like the influence of minimum temperature [1], freezing rate and for example the fact, that organic frost mediums which contract upon freezing can cause similar damage. One method to analyse the phase transition of water within concrete when freezing, is to measure the macroscopic consequences such as length changes. Most authors [2-5] who studied the dilatation of concrete worked with samples that were preconditioned to certain degree of water content, mostly water saturated or vacuum saturated. Very little research has been carried out studying volume or length changes of concrete with frost medium layer and especially the length changes perpendicular to such a layer.

Aim of our investigation was to characterise the outermost surface layer during freeze-thaw action with simultaneous application of a frost medium layer.

Frost Resistance of Concrete, edited by M.J. Setzer and R. Auberg. Published in 1997 by E & FN Spon, 2–6 Boundary Row, London SE1 8HN, UK. ISBN: 0 419 22900 0.

2. Apparatus and methods

2.1 Measurement parallel to frost medium
To study the length changes of concrete during frost thaw cycles, two inductive extensionmeters (HBM W1T3) were fixed on an invar frame with little thermal contraction ($\alpha_T \approx 1\times10^{-6}$ K^{-1}). On both lateral faces of the test specimen (150x150x10 mm^3) a steel plate ($\varnothing = 10$ mm, h = 3 mm) was fixed just below the tested surface. An aluminium tape, additionally fixed with glue, allowed the application of the frost medium layer (3 mm) to the top surface. The dilatation was measured beetween the centers of the steel plates parallel to the frost medium layer (Fig. 1). To prevent frost medium loss by vaporisation a thin polyethylene foil was covering the whole apparatus.

In the experiment one concrete side was mounted to the invar frame and the corresponding extensionmeter on this side was used for control purpose only.

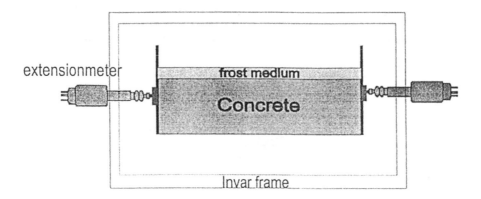

Fig. 1 Test arrangement of dilatation measurement parallel to frost medium layer

The whole apparatus was then put into a climate chamber and submitted to frost cycles. Cooling and heating rates were kept constant at 10°C/h. These rates were equal for all tests covered in this paper. The temperature in the frost medium layer and the specimen temperature (at the centre of the specimen) were continuously monitored.

Through measurement of a test steel bar with well known thermal contraction the deformation of the apparatus itself was defined and allowed to calculate the real deformation of the concrete specimens.

2.2 Measurement perpendicular to frost medium
As the scaled material in frost thaw resistance test often consists of flat plates, the measurement of volume changes perpendicular to the frost medium is of particular interest. For this purpose a special test arrangement was chosen (see Fig. 2).

First, special tapered concrete specimen were cut. From cores (∅ = 50 mm; h = 50 mm) a ring of 7.5 mm of concrete to a depth of 10 mm was removed. After that, an aluminium tape was fixed (glued) around the remaining part to allow the later application of a frost medium layer. Then an invar cylinder (∅ = 3 mm; h = 6 mm) was fixed upon the concrete surface by means of a glue (Araldit). An invar frame, which was pasted onto the lower concrete surface, permitted an exact positioning of an extensionmeter (HBM W1T3) over the invar cylinder.

By means of this apparatus the dilatation of the topmost 10 mm of concrete with simultaneous application of a frost medium layer (3 mm) could be measured. The apparatus was constructed so that the invar cylinder (the measurement position) could be placed at almost any point of the tested surface.

To find the adequate measurement position the concrete specimen were studied first by means of X-ray computer tomography. This allowed a non-destructive analysis of aggregate distribution and therefore the determination of the suitable measurement position.

To calibrate the thermal deformation of the apparatus itself, the latter was fixed on a steel cylinder. The deformation of the concrete specimen only was derived by using the correction obtained in the calibration experiment.

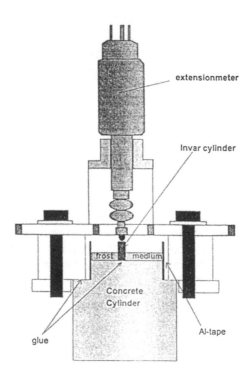

Fig. 2 Test arrangement of dilatation measurement perpendicular to frost medium layer

2.3 Concrete mix design

In a first step only two concrete mixes were studied in detail. Some experiments have been carried out on other mixes not mentioned in the table below. The concrete mixes with maximum grain size 32 mm were designed so that two different resistances to frost thaw action resulted.

	Mix 1	Mix 2	
Sand 0/4 mm	640	722	kg/m^3
Gravel 4/32 mm	1360	1178	kg/m^3
Cement CEM I 42.5	300	300	kg/m^3
Water	150	189	kg/m^3
W/C-ratio	0.5	0.63	
Air pore content	1.6	0.8	vol-%
Bulk density	2447	2452	kg/m^3
Frost resistance	"mean"	"low"	

3. Results and discussion

3.1 Measurement perpendicular to frost medium

We consider specimen with saturation conditions like those in frost deicing salt resistance tests.

First, the results of dilatation measurement perpendicular to the frost medium are presented. Specimen were stored during one week in water before the frost medium was applied and the test was carried out.

The result for the first eight frost cycles of a specimen of mix 2 with 3% salt solution as frost medium are presented. The expansion is shown as a function of time in Fig. 3 and in Fig. 4 as a function of frost medium temperature.

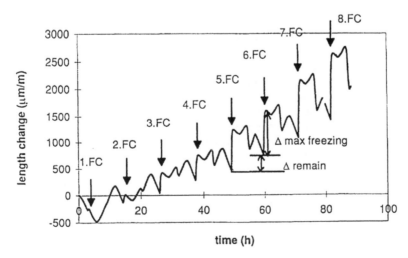

Fig. 3 Expansion of mix 2 perpendicular to frost medium (3% salt solution) vs. time

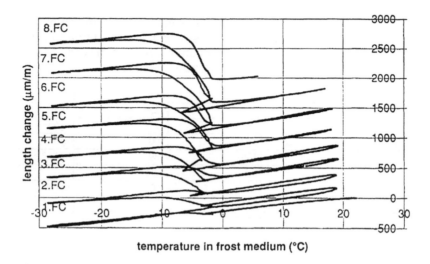

temperature in frost medium (°C)

Fig. 4   Expansion of mix 2 perpendicular to frost medium (3% salt solution) vs. temperature in frost medium

Results from temperature measurement within the concrete specimen at different distances from the frost medium indicated that the concrete temperature in the outermost layers is almost identical to the temperature in the frost medium [6].

After a thermal contraction phase above freezing point (≈ -3°C), with some supercooling, a rapid expansion occurs. The onset of this rapid expansion is indicated for each frost cycle in Fig. 3. Significant expansion continues until about -12°C. Taking the thermal contraction of the specimen into account (e.g. thermal contraction of ice α_T= 55x10^{-6} K^{-1}) the expansion continues steadily with decreasing temperature. With increasing number of frost cycles this rapid expansion becomes larger and larger.

In Fig. 4 it is shown clearly that the rapid expansion does not occur at a fixed temperature but that the phase transition is continuous. The water in finer pores freezes later according to pore radius freezing point depression.

In Fig. 5 the expansion measured at freezing (difference between freezing peak maximum and length at freezing point) as well as the remaining deformation after each frost cycle are shown.

While the remaining deformation increased moderately, the spontaneous expansion at freezing increased almost linearly with increasing number of frost cycles.

The measurement of other specimen (e.g. saturated specimen without frost medium application) showed that often the freezing peak did not appear until a certain number of frost cycles was reached. Without a freezing peak, no remaining deformation was detected. This indicates, that before damage occurs the specimen must be saturated to a certain degree of saturation, i.e. a critical degree of saturation [7].

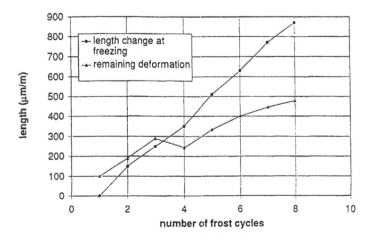

Fig. 5 Expansion at freezing and remaining deformation of mix 2 perpendicular to 3% salt solution

3.3 Measurement parallel to frost medium

Earlier [8] we described length change measurement parallel to the frost medium on specimens with dimension 200x150x50 mm³.

To verify the results obtained with the perpendicular method, additional tests with parallel measurements were carried out with specimen 150x150x10 mm³. The following results were obtained on such specimen. All specimen were first stored for one week under water and then the frost medium layer was applied. Then they were submitted to frost cycles and the expansion parallel to the frost medium was measured. The result of a specimen of concrete mix 1 with frost medium water is presented in Fig. 6.

The expansion curves look quite similar to the expansion measured perpendicular to the frost medium. The freezing peaks however, are much smaller which can be explained by the higher frost resistance due to the lower water cement ratio and the difference in frost medium.

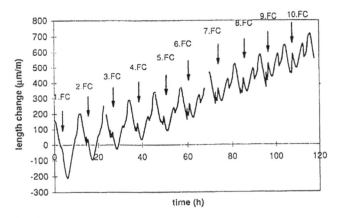

Fig. 6 Expansion parallel to frost medium of concrete mix 1 with frost medium water

For further analysis of the damaging process the situation at freezing point has to be analysed in detail (Fig. 7). First the temperature behaviour is described. At the beginning, the temperature is lowered linearly to the onset of freezing of the frost medium. This occurs below 0°C as the water is supercooled. When freezing begins, the temperature rises immediately to the freezing point of the frost medium (water: 0°C) due to the exothermic ice formation (crystallisation heat). The temperature remains at 0°C until all frost medium is frozen and then drops rapidly. Now we discuss the expansion curve. First we observe the thermal contraction phase above freezing. When freezing starts (see temperature curve) very fast expansion occurs (ΔL_{sp}). Similar expansion is found on wet specimens (water storage) without frost medium application [3/4].

A part of this expansion is due to thermal expansion ΔL_{th} because of the sudden temperature rise ($\Delta L_{th} \approx 35$ µm/m). The additional expansion can be explained as the result of ice formation within the specimen and of hydraulic pressure [9]. The latter explanation would agree with the relaxation that was detected only a few minutes later. It has to be mentioned that ice formation in the supercooled frost medium layer probably starts at the concrete surface because of the presence of nucleation centres. As the rate of ice formation for the first ice layer is very fast because of the supercooling [10] the formation of hydraulic pressure is possible. The fast expansion (ΔL_{sp}) does not occur when there is no supercooling present.

The strong correlation of ΔL_{sp} to the degree of supercooling is demonstrated in Fig. 8. The spontaneous expansion increases with increasing supercooling where as the remaining expansion after each cycle does not significantly depend on the degree of supercooling.

The increasing number of frost cycle is influencing this value too, probably because damage occurs and therefore the total porosity increases.

This phase was also found in the measurements perpendicular to the frost medium.

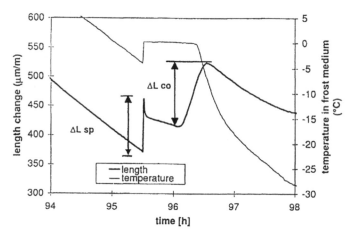

Fig. 7 Detail of expansion at freezing point (mix 1; 9.frost cycle; frost medium water)

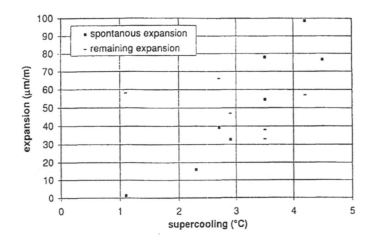

Fig. 8   Spontaneous expansion ΔL_{sp} and remaining expansion ΔL_{remain} vs. degree of supercooling

After the first expansion phase, a phase of only moderate length changes follows. The temperature is constant (0°C) and no ice formation within the specimen seems to occur. When the ice formation in the frost medium is almost terminated and the temperature therefore begins to lower softly, a second continuous expansion (ΔL_{co}) can be detected. The length increases continuously with falling temperature to reach a maximum at $T_{max} \approx -8\ °C$. The continous expansion can be explained with a continuous freezing of water in smaller and smaller pores according to the pore radius freezing point depression.

As the volume of water increases 9% on the phase transition a hydrostatic pressure may be built up continuously. The amount of this second expansion increases almost linearly with the increasing number of frost cycles (analogue Fig. 5).

As ΔL_{co} is bigger than ($\Delta L_{sp} - \Delta L_{th}$), at least for moderate supercooling and big number of freeze cycles, the damage of specimen due to the continuous expansion is more severe. Locally, the dynamic pressure may be of greater importance, but the second process, the continuous expansion, becomes even more dominant with increasing number of frost cycles. In cases where no continuous expansion occurs no remaining expansion (damage) at the end of a completed frost cycle was detected.

It seems that less importance can be attributed to direct damage through water transport within the specimen (ice lens effect) and the large thermal expansion coefficient of ice ($\alpha_T \approx 55\ 10^{-6}\ K^{-1}$) when temperature increases again [3]. Nevertheless this process may be responsible for the increase of ΔL_{co} as a bigger degree of saturation and a refilling of damaged zones with water results. In the presence of deicing salt the refilling may be increased by osmotic effects.

The effect of a salt solution as frost medium is shown in Fig. 9.

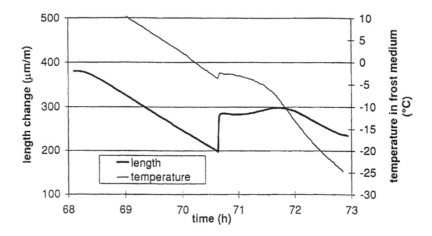

Fig. 9 Detail of expansion at freezing point (mix 1; 7. frost cycle; frost medium 3% salt solution)

Principally the same statements as above can be repeated. But there are certain characteristic differences to be discussed.

First, the freezing temperature of the salt solution is lower and the temperature at freezing point is not as constant as before but decreases continuously. This may be an indication for a non homogenous concentration of deicing salt within the frost medium (some segregation upon freezing is possible).

The spontaneous expansion ΔL_{sp} seems to be larger than with frost medium water. In fact the spontaneous expansion in this case is the sum of the spontaneous expansion and a fraction of the continuous expansion ΔL_{co}. As the freezing point is lower, certain portion of pore water may freeze on the same moment when freezing starts.

The first peak therefore gets more important as both mechanisms occur simultaneously. Nevertheless the hydraulic pressure seems not to be too important as the relaxation is small. The continuous expansion is therefore much reduced, but still visible.

4. Conclusions

As illustrated, the measurement of the length changes of concrete during frost cycles is an effective tool to study phase transformation of water and to analyse damage mechanism of concrete due to frost action. The application of a frost medium seems to be essential for useful result as well as the monitoring of subsequent frost cycles. It is shown, that the expansion of concrete perpendicular and parallel to the frost medium layer is comparable.

The freezing process can be devided up into two different phases, a spontaneous expansion at freezing point and a continuous expansion below freezing point. This can be explained by partial freezing of macropores and a subsequent build up of hydraulic pressure. The spontaneous expansions is strongly dependant on the degree of supercooling. Therefore the importance of supercooling in situ has to be analysed.

The second phase, the continuous expansion, starts after all of the frost medium is frozen and the temperature is lowered further. This expansion does not occur at a fixed temperature but increases continuously with falling temperature. This is the consequence of the pore size freezing point depression in pores with continuous size distribution. The continuous expansion peak increases with increasing number of frost cycles. This can be explained by a refilling process in the damaged zones with frost medium or an increase of the degree of saturation in subsequent frost cycles. The transport of water may be caused by the lower vapour pressure over ice (ice lens effect). It is more effective with a salt solution as frost medium (osmotic effects).

References

1. Studer, W. (1993) *Internal comparative tests on frost deicing salt resistance,* Int. workshop on the resistance of concrete to scaling due to freezing in the presence of deicing salts CRIB Québec, pp. 175-187.
2. Litvan, G.G. (1975) *Phase transition of adsorbates-Part 4 Effect of deicing agents on the freezing of cement paste,* Journal of the American Ceramic Society, Vol. 58 No.1-2, pp. 26-30.
3. Grübl, P. (1981) *Über die Rolle des Eises im Gefüge zementgebundener Baustoffe,* Beton 2, pp. 54-58.
4. Erbavdar, S. (1987) *Eisbildung , Volumendilatation und Wassertransport im Gefüge von Beton bei Temperaturen bis -60 °C,* Dissertation TU München.
5. Powers T.C. and Helmuth R.A. (1953) *Theory of volume changes in hardened portland cement paste during freezing,* proceedings of the highway research board, Vol. 32, pp. 285ff
6. Studer, W. (1994) *Temperaturmessungen bei verschiedenen Frost-Tausalzbeständigkeitsprüfungen,* EMPA test report No 4117/3.
7. Fagerlund G. (1975) *Significance of the critical degrees of saturation at freezing of porous and brittle materials,* ACI Publication SP-47 pp. 13-65.
8. Kaufmann J. and Studer W. (1995) *Scaling process on specimen surfaces in freezing and thawing test,* proceedings of the international Conference on Concrete under severe conditions, Consec'95, Sapporo, pp85-94
9. Powers T.C. (1945) *A working hypothesis for further studies of frost resistance of concrete,* Journal of the american concrete institute, Vol. 16, N° 4,
   pp.245-272
10. Meier U.G. (1978) *Die Entstehung und Auswirkung der Unterkühlung von Porenwasser in Zementstein und Beton,* Material und Technik Nr.1 S.133

Moisture absorption from salt solutions in cement mortar discs during freezing

S. LINDMARK
Lund Institute of Technology,
Division of Building Materials, Lund, Sweden

Abstract
Moisture uptake at constant sub-freezing temperatures was studied on three different mortars (w/c 0.40, 0.55 and 0.65). The specimens were frozen by submerging them in pre-cooled salt solutions. The main purpose was to test a previously presented hypothesis on the mechanism of salt frost scaling of porous materials, a hypothesis which predicts weight increases due to water absorption caused by the growthof ice crystals formed in the pores. Effects of temperature level, NaCl-concentration (both in the pore solution and in the outer solution), pre-drying, rate of cooling and air-entrainment were studied. Only results from the main test, moisture uptake at different temperature levels, are given in this paper. Large weight increases were observed, a fact which supports the hypothesis.
Keywords: Absorption, frost, moisture, mortar, salt

1 Introduction

In a previous paper [1], a first draft of a hypothesis on the mechanism of saltfrost scaling of concrete and other porous, brittle building materials has been presented. The key idea of the hypothesis is that ice crystals which form inside the specimen are able to feed from the remaining pore solution just as was shown by Powers and Helmuth [2]. They explained this ice lens growth would come to a stop when the pressure difference between ice and remaining pore solution was large enough. In specimens with a low enough degree of saturation, such a pressure difference may be established mainly as a reduced pressure in the pore solution. In the present hypothesis, it is assumed that close to the specimen surface, this reduced pressure cannot be established, as new liquid will be sucked in from the outer salt solution. Thus the only way to establish equilibrium is by applying a positive pressure on the ice. At normal freezing temperatures, the pressure needed will be too large for normal building materials to withstand and thus the surface will be destroyed.

Frost Resistance of Concrete, edited by M.J. Setzer and R. Auberg. Published in 1997 by E & FN Spon, 2–6 Boundary Row, London SE1 8HN, UK. ISBN: 0 419 22900 0.

If this mechanism is active, there has to be a net flow of moisture into the specimen during freezing. Thus, if a test on moisture uptake during freezing shows that there is no weight increase, the hypothesis is either wrong or has to be further developed. Possibly, a water flow into the specimen could be caused by some other phenomenon as well, and this is briefly discussed in the Discussion-section of this paper.

2 Method

The idea was to study moisture uptake under temperature conditions that were as stable as possible. Circular, approximately 5mm thick discs of diameter 100 mm were immersed in pre-cooled saltsolutions. In this way, a rapid cooling was secured: The time for cooling the specimens to the set temperature was about 1 hour. About once an hour the discs were taken up, free water was wiped off with a moist sponge and the specimens were weighed. Each specimen was kept out of the salt solution for about 30 seconds during the weighing procedure. it was chosen not to measure weight changes by weighing under water, as specimen volume changes might cause an error in such a measurement.

The entire test program comprised studies on the effects of the following parameters:
- Water cement ratio (0.40, 0.55, 0.65)
- Air entrainment
- Effect of a mild pre-drying (two days at ca 65% RH, CO_2-reduced atmosphere)
- Temperature level (-4°, -10°, -16°C)
- Salt concentration in the pore solution. (6.6, 14.4, 19.7% by weight NaCl)
- Salt concentration in the outer solution (6.6, 14.4, 19.7% by weight NaCl) and
- Effects of low salt concentration in the pore solution (1, 3 and 5%)

Only results from the main study, effect of temperature level measured on air-entrained specimens, are presented in this paper.

The major part of the tests were done on air entrained specimens, as it was expected that scaling would be severe enough to give misleading results in the case of non-air-entrained specimens.

Three specimens of each material was run at each temperature with matched salt solution concentrations. The term "temperature matched" is used to denote that the freezing point of the salt solution is equal to the temperature at which the test is run. The choice of these high concentrations is motivated by the fact that at any temperature below the freezing temperature of a salt solution used in a standard test, the remaining salt solution will have a concentration which is determined by its phase diagram. Thus, by using these concentrations, all of the solution remains liquid ; that is, there is no ice at the specimen surface which might hinder any moisture exchange. It also means that the specimens are readily accessible for weighing. In the present study, temperature levels of -4°, -10° and -16°C were chosen, and so the salt concentrations chosen were 6.6, 14.4 and 19.7% b.w., respectively.

In those cases where scaling occured, scaled-off material was collected, dried and weighed. The weights of the specimen torsos were then compensated by adding the weight of the scaled off material. Thereby, it was assumed that the scaled-off material, which was collected after the test had been terminated, was scaled off at a constant rate after scalings had first been observed. (This observation was done during the measurements and taken to the protocol). Using this procedure means that the calculated weight increases are at an absolute lower limit and so there is no risk of overestimating the moisture uptake.

3 Materials

Cement mortars were cast in cylinder forms of diameter 100 mm and length 200 mm. The cylinders were demoulded the day after casting and the cylinders were immediately stored in lime water. One week after casting, discs of thickness approximately 4.9±0.4 mm were sawn and stored in lime water and different salt solutions. Testing was begun 12 weeks after casting, and except for a few completing tests, all of the testing was done in 16 days.

The cement mortar mixes are given in table 1. Properties of the hardened materials are given in table 2.

Table 1. Mortar mixes.

	W/C:	0,40	0,65	0,40 Air Ent.	0,55 Air Entr.	0,65 Air Entr.
Water	kg/m³	290	290	290	290	290
Cement	kg/m³	725	446.2	725	527.3	446.2
Sand (0-3mm)	kg/m³	1271.6	1506.2	1271.6	1437.9	1506.2
AEA/C,	% b.w.			0.005	0.005	0.005
Plasticizer/C	% b.w.	0.23	0.23	0.23	0.23	0.23
Sand moisture content	% b.w.	8.79	6.43	7.80	6.95	5.69
Fresh Air Cont.	%	3	----	13	11.5	13.5

The plasticizer and the air-entraining agent were of melamin resin type and talloil type respectively. The dosage figures given above refer to the active substances, not the bulk weight. Maximum aggregate size was 4mm. Air content of the non-air-entarined 0.65-mortar was not noted. The mixes given above are not corrected for actual air content, but simply give the mixture proportions.

Table 2. Properties of the hardened materials, mean values

Property	0,40	0,40 - A.E.	0,55 - A.E.	0,65	0,65 - A.E.
Dry density [kg/m³]	2089.7	1864.9	1851.4	2032.1	1796.4
Porosity [%]	21.7	30.5	30.9	24.0	33.1
S, natural [%]	98.7	76.3	75.1	96.8	70.6
Air, natural [%]	0.3	7.2	7.7	0.76	9.7

S = Degree of saturation. "Natural" denotes values obtained after pre-storage

4 Results from series 1: Moisture uptake in lime water stored specimens submerged in temperature-matched salt solutions

In the main series, three specimens of each material were run at each temperature with a matched salt concentration. As a reference, air-entrained specimens of water-cement ratios of 0.40 and 0.65 with and without salt in the pore solution were run in pure water and strong salt solution respectively (figure 1). The results from tests on dried and remoistened specimens at low temperatures are given in figure 2a-c, while those on never dried specimens are given in figure 3a-c.

Generally, all specimens tested at low temperature show large weight increases during the first hour, i.e. during the time when temperature drops from +20° to the set temperature. This is probably a temperature effect and is further commented on below.

Spreadings are small; In a plot of each and every single value, no single result ever deviates so much from its series' mean value plot that it overlaps with the other results. The only exception to this is the results for w/c 0,40AE tested at -10° and -16°, in which case some overlaps occur (as is seen in figure 1a, the mean values for these two series are very close to each other).

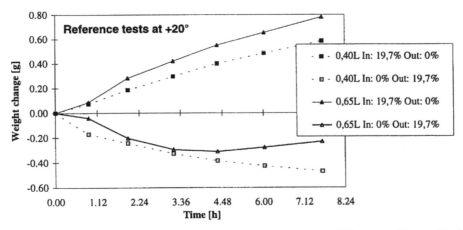

Figure 1: Weight changes at +20°C due to salt concentration differences. Never dried specimens. Each dot = mean value for three specimens.

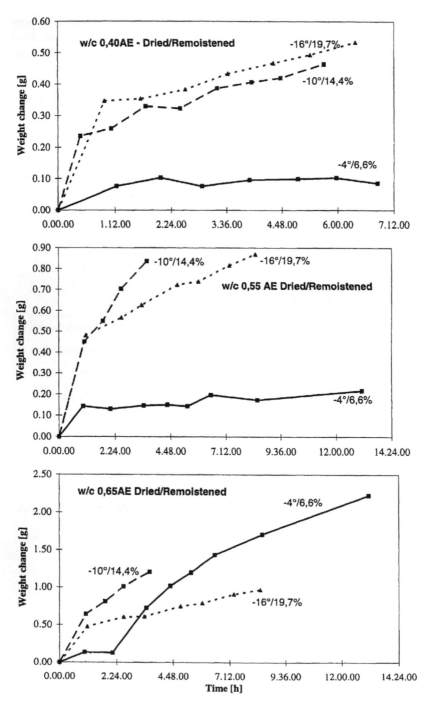

Figure 2: Weight changes for dried and remoistened specimens. Note scales!

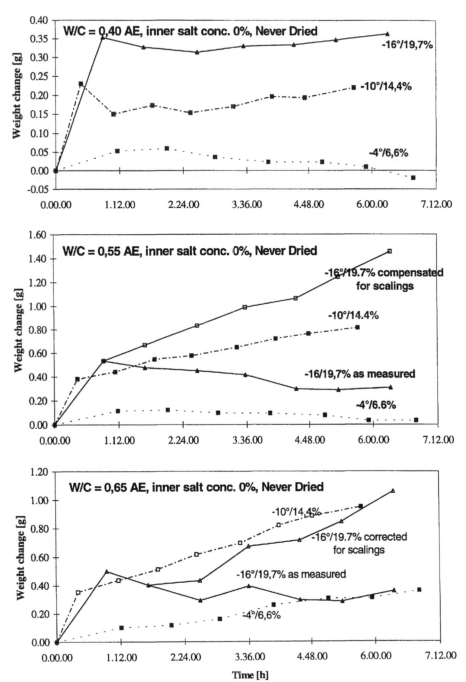

Figure 3: Weight changes during freezing. Never dried specimens. Corrections for scaling is done by adding the dry weight of the scaled-off material to the specimen weight, assuming scaling took place at a constant rate after it was first observed. All points are mean values of three specimens. Note scales!

4.1 Reference tests -Figure 1

The reference tests at +20°C were carried out in order to check the order of size and qualitative behaviour of moisture changes caused by pure salt effects. Mortars stored in 19,7% salt solution were tested in pure water and specimens stored in lime water were tested in a 19,7% salt solution. The results are given in figure 2. The first result to observe is that the large change in weight during the first hour, which is seen in figures 2 and 3, does not occur in these tests. This supports the idea that such a weight change is caused by the temperature drop itself. Secondly, specimens stored in water and tested in salt solution loose weight during the test and vice versa. This is perfectly in accordance with what was expected. Interestingly, the 0,65-mortar tested in a 19,7% solution first looses weight and then, after 4 hours, start gaining weight again. (All three specimens behave in this way; it is not due to one "misbehaving " specimen.) At present ,no explanation for this can be given.

Finally, it is seen that specimens tested in a strong salt concentration loose weight at this temperature. From figures 2-3 it is seen that when the same tests are run at sub-freezing temperatures, the specimen weights generally increase. This indicates that moisture absorption due to ice lens growth dominates the moisture exchanges caused by an outer salt concentration.

4.2 Dried and remoistened specimens - Figure 2

Firstly, none of these specimens were damaged, probably because the air pore systems were empty enough to accommodate all the moisture taken up during freezing. It is clearly seen from figure 2, though, that moisture uptake had not stopped when the tests were interrupted - probably, the weights would have kept increasing at the same rates until damages occurred.

Clearly, final weight changes are higher the higher the water-cement ratio. For w/c 0,40, the absorption rate is equally high at -10° and -16°. For w/c 0,55, it is highest at -10°, and for w/c 0,65 it is about equally high at -4° and at -10°. The 0,65 mortar has a strange initial behaviour when tested at -16°: No large weight change accompanying the temperature drop is registrated. Apart from this last fact, the results are in agreement with what may be expected from the hypothesis: The 0,40 mortar contains less freezable water at high temperatures, and also, its permeability is lower than that of the other materials. Thus, it is not until very low temperatures are applied that moisture uptake becomes large in this high-quality material. The w/c 0,65 mortar on the contrary, contains a lot of freezable water already at high temperatures, and so there are many ice crystals working together to absorb moisture from the outer solution. At very low temperatures though, so much ice may have formed that the remaining permeability is too low to allow moisture absorption rates that are as high as at high and intermediate temperatures. Finally, the 0,55 mortar is between these two extremes.

4.3 Never dried specimens - Figure 3

Still, weight changes are smallest for w/c 0.40 (also during the first hour), independent of temperature level. The explanation is presumably the same as that given above. At -4° these specimens are even loosing weight. This may be because so little ice, if any, has formed in the pore system that the drying effect of the outer salt solution dominates the moisture transport processes.

As for never dried specimens of w/c 0,55 and 0,65, weight changes are smaller for 0.55-mortars at -4°/6.6%, while they seem to be independent of material quality at -10°/14.4%. The scaling compensated results for 0.55 and 0.65 mortars show that at -16°/19.7% the 0.55-mortar increases more in weight than does the 0.65-mortar.

Generally, the rate of moisture absorption is lower in these specimens than in those that were dried and remoistened before testing. This may be explained by the increased amount of freezable water and increased permeability which follows on the first drying.

5   Discussion

5.1 Other reasons for weight changes to occur
Generally, on a temperature decrease, the weight of the immersed specimens will change due to many factors, some of which will be briefly commented on in the following. (This is to check that the measured weight increases are not caused by pure temperature- or salt-effects.)

Air contracts approximately according to the gas law. If the air was enclosed in a sealed, rigid container, there would be a pressure reduction. In the present case, the result will be a suction of water into the specimen and the weight change is calculated:

$$\Delta m = \Delta V_a \cdot \rho_{w,f} = V_{a,i} \cdot \left(1 - \frac{T_f}{T_i}\right) \cdot \rho_{w,f} \tag{1}$$

where Δm is the mass change, ΔV_a is the change of air volume, $\rho_{w,f}$ is the density of water at the final temperature, $V_{a,i}$ is the initial air volume, T_i is the initial temperature and T_f the final one [K]. For the air-entrained specimens in this study, which had maximum air contents of about 9,7%, i.e. 3.88 cm³ (w/c 0,65, normal specimen volume 40cm³), the maximum weight increase is obtained in the tests run at -16°C, and it is calculated to be approximately 0.47 g. This figure is in good agreement with the values shown in figure 2c at 1 hour.

Further, the specimen itself will contract, whereby the pore volume is reduced, and the density of water will change with temperature, and so, depending on which temperatures are chosen, the change in specific volume of the pore solution will cause either an increase or a decrease in moisture content. The following expression may be used for calculating the entire effect of both these effects *and* the effect of air contraction described above (the derivation is omitted to save space):

$$\Delta m = m_{w,f} - m_{w,i} = \rho_{w,f} \cdot V_i \cdot P \cdot \left\{ \left(1 + \alpha\left(T_f - T_i\right)\right)^3 - \left(1 - S_i\right) \cdot \frac{T_f}{T_i} \right\} - V_i \cdot P \cdot S_i \cdot \rho_{w,i}$$

(2)

Here, α denotes the coefficient of thermal contraction, S the degree of saturation and the other symbols and indices are as above (i for initial, i.e. start values, f for final values). For mortars of w/c 0.65 (air entrained), the average degree of saturation was 0.706. The densities of water at +20 and -16°C are 998.2 and 995.7 kg/m³ respectively. The coefficient of thermal contraction is about $11*10^{-6}$ for concrete and a similar value may be expected for mortars of this kind. With these values the equation gives $\Delta m = 0.44$ g. Obviously, the air contraction effect is dominant, the other effects actually act to reduce the specimen weight.

5.2 Rate of moisture transport - Dependence of viscosity on temperature

According to Darcy's law, the rate of water transport into the specimens is dependent on the driving force, the intrinsic permeability and the viscosity of the flowing pore solution. The latter is strongly dependent on temperature, as is seen in table 3:

Table 3. Viscosity of water. From [3]

Temperature, °C	Viscosity, centipoise
+20	1.002
±0	1.798
-16	3 *

* (extrapolated value)

From this table it is clear that the temperature-dependence of viscosity has to be taken into concern when trying to estimate either the driving force or the intrinsic permeability.

5.3 Final remarks

As was seen above, the absorption rate reaches its maximum at different temperature levels for different water-cement ratios. The explanation given in the results section, admittedly, is speculative, but still it is in line with what was foreseen in the first presentation of the hypothesis; There will be an optimum pore size distribution (with respect to damages), which depends on the way the material is tested. In this case, it seems as though mortars of w/c 0,55 and 0,65 are damaged equally fast at -10°, while the latter will be damaged faster at higher temperatures and the former at lower temperatures.

Generally, all specimens show large increases in weight during the first hour, i.e. during the period when temperature drops from +20° to the set temperature. It was assumed above that this is caused by the contraction of air inside the specimen, which causes a suction of the outer solution into the specimen. A numerical calculation, which cannot be given here, shows that for samples of the thicknesses and qualities used in this study, the contracted air volume is replaced by water within about 15 minutes after a step change of temperature from initial to final value. Thus it is primarily the rate of temperature decrease which governs this process. In this test, the temperature reduction was completed in one hour, and so water uptake due to air contraction will be completed in about that time as well. There fore, the large weight increases during the first hour should not be classified as caused by the growth of ice crystals, but rather as a pure temperature effect.

6   Conclusions

The primary purpose of this study was to check that moisture absorption during freezing exists. It was clearly seen that the specimens undergo weight increases (except for specimens of low water-cement ratio tested at relatively high temperatures). After checking for order of size and rate of weight changes caused by other plausible phenomena, it is concluded that the test does not reject the hypothesis put forward.

---

1   Lindmark, S: "A hypothesis on the mechanism of surface scaling due to combined salt frost attack", Report TVBM-3072, Div. of Building materials, Lund Institute of Technology, 1996.
2   Powers, T.C., Helmuth, R.A.: "Theory of volume changes in hardened portland cement pastes during freezing", Highway Research Board, Proceedings 32/1953
3   Handbook of Chemistry and Physics, 70th Ed., CRC Press Inc, Florida 1989-1990

Influence of water uptake during freezing and thawing

R. AUBERG and M.J. SETZER
IBPM – Institute of Building Physics and Materials Science,
University of Essen, Essen, Germany

Abstract
The frost resistance of concrete is guaranteed when a sufficient air void system is available. Practical experiences show that in Central Europe concretes without air entrainment can also have a high durability and serviceability. This practical experiences are confirmed by testing frost resistance with CF test. The frost resistance essentially depends on the degree of saturation and porosity. During the first freeze thaw cycles the water absorption increases until a critical degree of saturation is reached. If this critical value is surpassed the water uptake decreases and the internal damage increases. The investigations of water absorption during freeze thaw test confirms that reduction of capillary pore and coarse pore volume can increase the frost resistance. In case of no application of air entrainment in concrete the performance concept is necessary to test the frost resistance.
Keywords: Capillary suction, CF test, internal damage, frost resistance, water absorption

1 Introduction

Freezing of free water leads to pressure due to volume expansion. Since water in gel pores generally is unfrozen above -20°C a hydraulic pressure increases. The freezing point depression of the gel pore water results from interaction mechanisms of specific surfaces. These also lead to tension in the gel pore system and hysteresis effects between freezing and thawing. Due to the different coefficient of thermal expansion of water, solid matrix and ice the hysteresis leads to additional pressure. Finally, during ice formation different soluble salts of the pore liquid are crystallized or increase the concentration of unfrozen water. The arising osmotic pressure causes on one hand transport mechanism, e.g. capillary suction and diffusion and on the other hand tensile tension in the matrix and micro cracks. Therefore, the pore system,

Frost Resistance of Concrete, edited by M.J. Setzer and R. Auberg. Published in 1997 by E & FN Spon, 2–6 Boundary Row, London SE1 8HN, UK. ISBN: 0 419 22900 0.

especially the proportion of gel and capillary pores, is decisive. Air entrapped pores are effective as expansion space and make the structure more flexible. This prevents high local tensions [1, this conference].

If this resulting tension exceed the critical value of strength of microsystem the structure damaged and micro cracks develop. Due to the damage of the concrete structure disintegration of the matrix arises, which leads inevitable to irreversible volume expansion, reduced modulus of elasticity and strength.

Besides these physical effects changing of the chemical situation during the freeze thaw attack can not be neglected. Depending on temperature clinker phase transition can take place. The reason is displacement in thermodynamic equilibrium of the pore solution. Lower temperature leads to higher solubility of calcium hydroxide and formation of secondary ettringite, e.g. due to phase transition of the available monosulfate. Changing temperatures accelerate this process and can intensify the frost damage [2].

2 Principles of the Measuring Procedure -"CF Test"

Following the RILEM recommendation CDF test [3] defines precisely the boundary conditions. The principles of CDF test are also used for CF test [4]. The characteristic arrangement of the test can be seen in figure 1. By the exact definition of design, storage (after demoulding 6 days in water and 21 days pre-drying at 20°C/65% rel. hum.), capillary suction of solution and temperature cycle and by the simple handling, observation errors are minimised. The aim of the test method is high selectivity and very good comparability, i.e. precision [5].

Following significant test parameters were changed in CF test [4] with respect to CDF:
- Demineralized water is used as test liquid.
- The duration of the test is 56 freeze thaw cycles.
- Besides the surface scaling the internal deterioration is determined.

2.1 Running the test

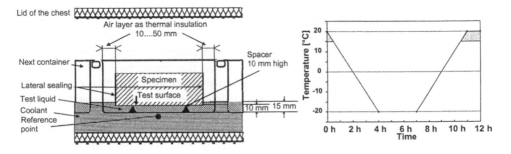

Fig. 1.   Principle arrangement I of the CDF/CF test during the freeze thaw cycles. During capillary suction the specimen is stored in the test container with a cover at 20 °C. For removing loosely adhering scaled particles the container is immersed in an ultrasonic bath

2.1.1 Capillary suction at room temperature

During the capillary suction and during the freeze-thaw cycles, the specimens are stored with the test surface inverted in stainless steel containers. They are placed on 10 mm high spacers. The level of the test liquid is kept at 15 mm so that the specimens dip 5 mm into the liquid. By this kind of capillary suction, a degree of saturation is reached, as in practice, in a very well defined manner.

2.1.2 Freeze thaw cycles

It is of essential importance to monitor the temperature control during the freeze-thaw cycles (ftc). A liquid coolant is used for the heat transfer in the CDF/CF chest. Thus a precise actual temperature curve of ± 0.5 K can be achieved.

2.1.3 Measuring scaled material and ultrasonic transit time

CF test usually lasts 56 freeze thaw cycles, i.e. the duration is 28 days of freeze thaw cycles plus 7 days of capillary suction. In order to reduce errors caused by laboratory operators, the loosely adhering scaled material is removed from the test surface by an ultrasonic bath. The scaled off material is filtered and dried at 105°C. Every four or six freeze-thaw cycles the scaled off material and the ultrasonic transit time of the specimens are measured. The detailed test description of CF test is given in the annex. and in [4].

3 Test program

The concrete series were tested according to CF test. For each series with 3 specimens each the following measurements were performed:

- weight of scaled material
- weight of the specimens (in consideration of scaled material)
- ultrasonic transit time parallel to the test surface (see figure 2)
- length change of the specimens in a height of 35 mm from test surface
- porosity with mercury intrusion porosimetry (MIP).

Due to the measurement of length change the specimen size is 250 x 100 x 75 mm. After water curing until the concrete age of 7 days in CF test the specimens are stored for 21 days in climate chamber (20°C/65% rel. hum.). This storage makes a pre-drying of the material possible. At the age of 28 days the specimens are re-saturated for 7 days by capillary suction in test solution before starting the freeze thaw test. During the freeze thaw cycles the specimens take up water continuously during the thawing period. This water uptake and the influence of frost damage were investigated in this project.

Therefore, concrete series with one type of cement, CEM III/A 32,5 (BFSC with 55% slag content), one type of aggregate and nearly the same workability were manufactured with different w/c ratio. Due to these conditions the measured effects can be correlated to water-cement ratio and porosity. Additionally, one concrete series with a CEM I 32,5 R (OPC) and w/c ratio 0,6 was tested.

The following table shows the design and properties of fresh and hardened concrete.

Table 1: Design and properties of fresh and hardened concrete

	Series H1	Series H2	Series H3	Series H4	Series P1
cement type		CEM III/A 32,5			CEM I 32,5 R
w/c ratio	0.7	0.6	0.5	0.4	0.6
cement content [kg/m³]	270	320	320	360	320
density [kg/dm³]	2.22	2.34	2.36	2.35	2.37
flow table [cm]	38	43	40	44	41
air content in fresh concrete [%]	0.4	0.4	0.6	0.8	1.2
compressive strength [MPa]	26	44	54	59	47

CEM III/A 32,5	blast-furnace slag-cement (BFSC) with 55% slag content
CEM I 32,5 R	ordinary portland cement (OPC)

4 Results and Discussion

4.1 Frost resistance measured by CF test

Figure 3 shows the results of CF test. The relative dynamic modulus of elasticity is taken to estimate the degree of internal damage. For CF test the reduction of dynamic modulus of elasticity of concrete with high frost resistance has to be less than 40% after 56 freeze thaw cycles [4]. Therefore, the tested concrete series with w/c ratio > 0.5 failed the test.

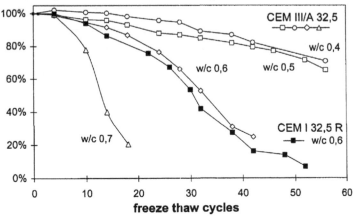

Fig. 3: Relative dynamic modulus of elasticity versus freeze thaw cycles of the 5 tested concrete series (mean value of 3 specimens)

Although the frost resistance of the concrete series is very different the behaviour during the test is comparable. By reduction of dynamic modulus over 30% the internal damage increases rapidly when continuing the test. The duration of these two stages of the test depends on cement type and content, w/c ratio and degree of hydration.

Although the results of CF test are reached by relatively simple and cheap methods they can be correlated to parameters of sophisticated materials research, e.g. mercury porosimetry, permeability and electron microscopy.

4.2 Mercury porosimetry testing and calculation of porosity

The total, gel and capillary porosity can be calculated with consideration of w/c ratio and degree of hydration by using the model of Powers and Brownyard [6, 7]. The problem is the unknown degree of hydration and the amount of chemical bound water.

For calculation of the porosity following degrees of hydration are used:

- CEM III A 32,5   68 % degree of hydration
- CEM I 32,5 R   82 % degree of hydration

These values are based on data of measured chemical boundary water after 28 days standard curing and on totally hydrated specimens [8].

Table 2: Calculated porosity according to Powers model and measured porosity by MIP related to concrete matrix.

	Series H1	Series H2	Series H3	Series H4	Series P1
total porosity					
calculated data	16.4%	16.0%	12.9%	10.8%	14.9%
MIP data (r > 3,8 nm)	13.7%	10.1%	9.5%	5.7%	10.2%
gel porosity					
calculated data	3.5%	4.1%	4.2%	4.8%	4.9%
MIP data (r < 10 nm)	2.5%	1.7%	2.7%	1.9%	1.4%
capillary porosity					
calculated data	12.9%	11.9%	8.6%	6.0%	10.0%
MIP data (r > 10 nm)	10.3%	7.6%	5.9%	3.0%	7.9%

The classification of gel and capillary pores with regard to pore size distribution can be criticised. Nevertheless, the microstructure of hardened cement paste is well described by distinguishing moisture transport and freezing mechanism. According to Powers' model the average width of gel pores is $r = 15$ Å. Setzer [9] extends the pore classification based on classification of IUPAC and according to physical behaviour of pore water and defines a hydraulic radius ($R_H = 2\ r$). Therefore, pores with $R_H \geq 30$ nm belong to capillaries. In addition a remarkable depression of freezing point starts at $R_H < 30$ nm. Using bulk data for the heat fusion and the surface interaction the depression of freezing point can be estimated by the following equation [10]:

$$\Delta T = \frac{-32\ [K \cdot nm]}{R_H\ [nm]} \tag{1}$$

The problems are discussed in [1] in this conference proceedings. Due to this the separation made here is fair.

The correlation of calculated and measured porosity data is given in figure 4. There is a high agreement between measured and calculated data of the amount of capillary pore volume. The measured total porosity is of course smaller, because in this case the minimum measurable pore radius of MIP is 3,8 nm ($\Delta T \approx -20°C$).

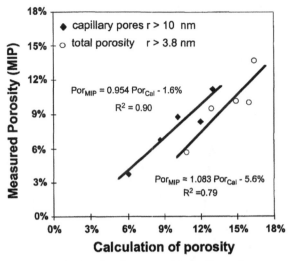

Fig. 4: Correlation between calculated and measured data of porosity

The following figures 5 and 6 describe the pore size distribution of the 5 concrete series measured with MIP. All concretes were tested at the age of 28 days. The series with CEM III/A 32,5 should have nearly the same degree of hydration. Therefore, the effects on pore size distribution results from change of w/c ratio (figure 5). Using the classification of Setzer [9] the main change in porosity takes place in the pore size range of micro-capillaries (30 nm < R_H < 1 µm). The micro-capillaries are of high importance due to moisture transport and freezing behaviour under normal laboratory test conditions (minimum temperature $\leq$ 20°C).

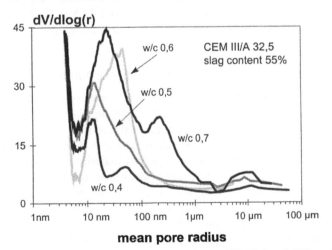

Fig. 5: Pore size distribution of concrete series with CEM III/A 32,5 and different w/c ratio measured with MIP in the age of 28 days

By increasing w/c ratio the pore structure gets coarser. In this case the change in pore size distribution from a w/c ratio of 0,6 to 0,5 is interesting, because the pore volume of micro-capillaries is reduced. This means that the amount of freezable macroscopic water decreases immensely.

Figure 6 shows the difference in pore size distribution between two cement types with the same w/c ratio 0,6.

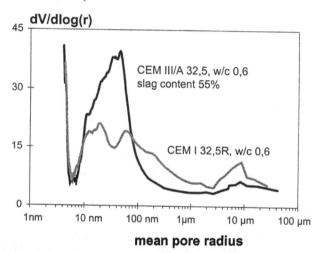

Fig. 6: Pore size distribution of concrete series with w/c ratio 0,6 and different cement types measured with MIP in the age of 28 days

The different cement types imply a different degree of hydration (see above). Although the CEM III/A 32,5 has a lower degree of hydration the pore structure is more dense. In the range of capillary pores the CEM I 32,5 R shows a continuously higher amount of capillary pore volume and represents an intermediate stage to pore size distribution of w/c 0,7 from figure 5.

4.3 Water uptake during capillary suction and freeze thaw test

Figure 7 shows the water uptake during CF test, i.e. capillary suction and freeze thaw cycles. During a single freeze thaw cycle the water uptake is only possible in the thawing phase of 5 hours (see temperature cycle in figure 1).

In CF test the water uptake can be subdivided into 4 stages. Starting with re-saturation the typical water uptake curve of porous material can be measured. After a high suction rate during the first day, corresponding to macro-capillaries, the water uptake flattens rapidly. After 7 days of re-saturation the degree of saturation due to capillary suction is nearly 90%. The amount of water uptake after 1, 3 and 7 days depends on w/c ratio. The maximum increase is from w/c ratio 0.6 to 0.7. This effect is linked to the change of permeability and continuity of capillary porosity. The results are confirmed by the porosity data of chapter 4.2.

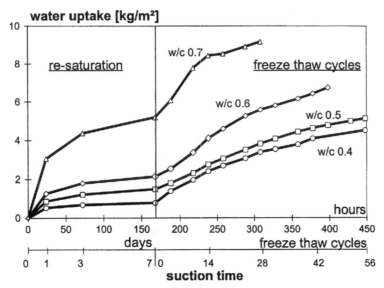

fig. 7: Water uptake during CF test. The suction time is described in hours, in this case the
resaturation time is 7 days (168 hours) and during a single freeze thaw cycle of suction
time of 5 hours is assumed (see figure 1)

During the first freeze thaw cycles the water uptake increases depending on w/c
ratio as well. This „pumping effect" is rapidly reduced after further freeze thaw cycles
and the concrete series show a nick-point in water uptake curve. When the test is
continued the suction rate decreases. This proves the micro ice lense model.

The behaviour during the 4 different stages of water uptake in CF test depends on
the cement type, w/c ratio and porosity.

4.4 Connection between water uptake and dynamic modulus of elasticity and length change

Additional water uptake takes place during the freeze thaw attack due to the contact
with liquid. This effect is described above. By measuring the internal damage with
ultrasonic transit time water uptake and deterioration effects overlap each other. The
effect of water uptake to the transit time is highest during re-saturation. This results
from exchange of air in capillary pores by water with higher sound velocity. During the
first freeze thaw cycle this effect is overlapped by the deterioration of concrete matrix
due to frost attack. When continuing the test the frost damage is principally measured.
The measurement of resonance frequency and/or investigation of damping is more
sensitive due to water uptake than ultrasonic transit time.

The water uptake dependent on reduction of dynamic modulus of elasticity is shown
in figure 8. For all concrete series an increasing water uptake is measured during the
first cycles and is coincided with a small decrease in relative dynamic modulus of
elasticity. After a reduction of dynamic modulus of elasticity above 30% the water
uptake decreases rapidly.

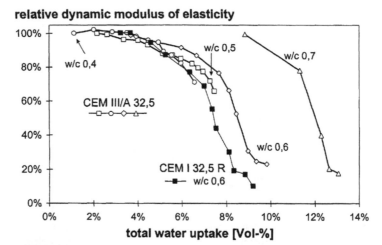

Fig. 8: Relative dynamic modulus of elasticity versus total water uptake during CF test

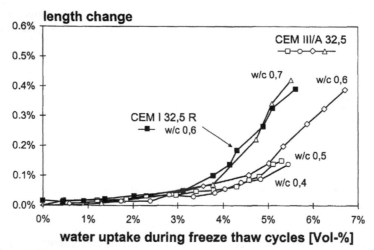

Fig. 9: Relative dynamic modulus of elasticity versus water uptake during freeze thaw cycles
(difference between total water uptake and amount of water uptake after re-saturation)

An equivalent nick point can also be found when considering length change versus water uptake (figure 9). Here, the rate of water uptake is reduced by length changes above 1‰. After this nick point the mean gradient length change to water uptake is nearly constant 0,2. [% / Vol-%]. This means that the additional water uptake can be mainly correlated to volume expansion due to micro cracking.

Both, dynamic modulus of elasticity and length change confirm that water uptake during freeze thaw cycles can not only be explained by filling fine capillary pores and developing micro cracks. Therefore, the amount of water is too high. This situation can be separated into two stages; 1st from starting point of freeze thaw cycle to the nick-point in water uptake as described above and 2nd a further water uptake with much lower gradient by continuing the test.

The initial stage corresponds to filling of fine capillary pores and of developing micro cracks, but mainly air entrapped coarse pores. It is interesting that all series independent on cement type and w/c ratio show nearly the same behaviour during freeze thaw test. The only difference is the duration when the critical nick point is reached. Here, higher w/c ratio and coarser porosity is of course worse.

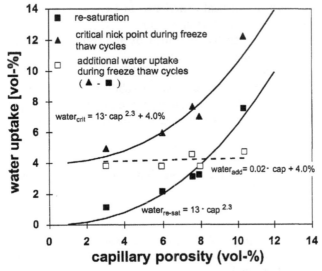

Fig. 10: Water uptake during CF test in comparison to capillary porosity

Figure 10 describes the dependence of water uptake after re-saturation and at the critical nick point in comparison to capillary porosity. The difference in water uptake is nearly 4 vol-% for all kinds of tested concretes. Taking into account that workability and compacting are in the same range coarse pores must have the same amount.

5 Conclusion

Testing of frost resistance is a time depending estimation of the serviceability of concrete structure. Therefore, the actual degree of saturation at the beginning of frost attack and the additional water absorption in the thawing phases are the main parameters to investigate the durability. With CF test the water absorption of concrete before and during freezing thaw cycles can be tested precisely and reproducibly.

In CF test the water uptake can be subdivided in re-saturation due to capillary suction, extensive water absorption during first freeze thaw cycles „pumping effect", due to micro ice lenses effects and decreased water absorption by continuing the test. After re-saturation capillary pores are almost totally filled. Whereas larger air entrapped coarse pores will be saturated during first freezing by moisture transport due to diffusion from unfrozen water in micro-capillaries and meso-gelpores to ice lenses in larger pores. Especially for concrete with lower w/c ratio it leads first to contraction due to shrinkage of fine pores following by expansion after saturation of the larger pores.

The duration of these successive effects are dependent on porosity, e.g. w/c ratio, degree of hydration, cement type and workability. Besides these parameters, the service life time is a function of temperature regime, water absorption and drying periods.

Using Powers model an air entrainment is required to guarantee a high frost resistance. Nevertheless; in practical experience in Central Europe it is common to produce plastic concretes without air entrainment.

This practical experience can be confirmed by the results of CF test. Following requirements must be taken into consideration:

- Reduction of capillary porosity, w/c ratio ≤ 0,50
- Reduction of coarse porosity due to good workability and sufficient treatment.

In case of no application of air entrainment the performance concept is necessary. Therefore , the CF test is a precise tool to measure the frost resistance.

6 Acknowledgement

The basic research projects for the CF test have been supported by the Deutsche Forschungsgemeinschaft (DFG).

7 References

1. Setzer, M.J. (1996) *Testing of Freeze Thaw Resistance: Surface and Internal Deterioration.* In: Frost resistance of concrete, (eds) M.J. Setzer, R. Auberg, RILEM Proceedings, Essen (Germany).
2. Ludwig, H.-M., Stark, J. (1996) *Die Rolle von Phasenumwandlungen im Zementstein beim Frost- und Frost-Tausalz-Angriff auf Beton.* Zement-Kalk-Gips 49 (1996) H.11.
3. RILEM Recommendation, RILEM TC117 FDC (1996) *CDF Test - Test Method for the Freeze-Thaw Resistance of Concrete with Sodium Chloride Solution (CDF),* Materials & Structures 29, 523-528.
4. Auberg, R. (1997) *Phd. Thesis,* University of Essen.
5. Setzer, M.J.; Auberg, R. (1995) Freeze Thaw and Deicing Salt Resistance of Concrete Testing by CDF Method. *Materials & Structures,* Vol 28, pp. 16-31.
6. Power, T.C., Brownyard, T.L. (1948) *Studies of physical properties of hardened portland cement paste,* PCA Bulletin 22.
7. Power, T.C. (1960) *Physical properties of cement paste,* PCA Bulletin 154.
8. Manns, W. (1975) *Über den Wassergehalt von Beton bei höheren Temperaturen.* Beton, H1, 26-30.
9. Setzer, M.J. (1991) Interaction of water with hardened cement paste. In: S. Mindess (eds) Advances in cementious materials. Ceramic Transactions, American Ceramic Soc., Westerville, Vol 16, 415-.
10. Setzer, M.J. (1977) Einfluß des Wassergehalts auf die Eigenschaften des erhärteten Betons. DAfStb Heft 280, Berlin, W. Ernst u. Sohn Verlag.

8.   ANNEX

CF TEST

TEST METHOD FOR THE FREEZE-THAW RESISTANCE OF CONCRETE - TESTS WITH DEMINERALIZED (CF)

1   Determination of surface Scaling

The CF test principles are based on the CDF test description (RILEM Recommendation published in Materials & Structures 29, 1996, 523-528). The following test parameters are changed:

- The duration of the test is **56 freeze thaw cycles**.
- The test liquid is **demineralized water**.
- Besides the amount of scaling the reduction in dynamic modulus of elasticity is measured.

2   Determination of Internal deterioration

2.1   Measurement of sound velocity

2.1.1   Equipment

The ultrasonic transit time is measured with a customary ultrasonic equipment for determination of the transit time of longitudinal waves in porous building materials in accordance to E DIN ISO 8047. The natural frequency of the transducers shall be within the range of 50 to 150 kHz. The diameter of the transducers should not be larger than 35 mm.

2.1.2   Testing

The dynamic modulus of elasticity is determined by measurement of the ultrasonic transit time by direct transmission parallel to the test surface (annex figure 1). The transit path length must be in a distance of 35 mm from the test surface. The coupling point of the transducers must be positioned centric in the lateral edge by rectangular specimens and by cylindrical specimens the transit path have to cross the axe vertical to the test surface. Each specimen must be measured in two directions vertical to another.

Before starting capillary suction the transit path length of the specimens must be measured with an accuracy of ± 0,5 mm. Before starting freeze thaw test the transit time has to be determined with an accuracy of ± 0,1 µs. The marked coupling points of the first measurement must be fixed for all the following determinations.

During the determination of the ultrasonic pulse velocity the temperature of the coupling medium and the specimens shall be 20 ± 5°C. The specimens shall only be without contact to the test liquid for a short time.

Coupling with demineralized water

The coupling medium is demineralized water. Therefore, for determination of the transit time a rectangular container (e.g. Polymethylmethacrylat) is used. The transducers are fixed in the lateral sides of the container in such a way that the axe of the transit path length can be measured in a height of 35 mm from the test surface of the specimen. The container is filled with demineralized water up to 10 mm above the transducers. The upper side of the specimens must be kept dry. The total transit path length of the demineralized water shall be nearly 10 mm (on both lateral sides of the specimens 5 mm). The actual total transit path length of the demineralized water has to be determine with an accuracy of ± 0,5 mm. It must be taken care during the test that no air bubbles tack to the transducers or the lateral sides of the specimens and that the lateral insulation is still fixed.

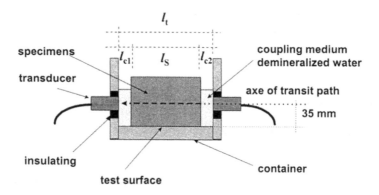

Figure 1.   Principle arrangement of the CF test to measure the ultrasonic transit time

2.1.3 Determination of changing in dynamic modulus of elasticity

The ultrasonic pulse velocity of the material is calculated by the ultrasonic transit time and the path length of the specimens (corrected by the transit time of the preliminary path length of the demineralized water). The change of the dynamic modulus of elasticity after *n* freeze thaw cycles is to be determined:

$$\Delta E_{dyn\ n} = \left(1 - \frac{E_{dyn\ n\ ftc}}{E_{dyn\ cs}}\right) \cdot 100 \text{ in } \%$$ (1)

here
ΔE_{dyn} : reduction of dynamic modulus of elasticity
n : number of freeze thaw cycle
$E_{dyn\ cs}$: dynamic modulus of elasticity after capillary suction (cs)
$E_{dyn\ n\ ftc}$: dynamic modulus of elasticity after n freeze thaw cycles (ftc)

The simplified calculation of the change of the dynamic modulus of elasticity is carried out by neglecting the change in density and geometry of the specimens:

$$\Delta E_{\text{dyn n}} = \left[1 - \left(\frac{t_{\text{t cs}} - t_{\text{c}}}{t_{\text{t n ftc}} - t_{\text{c}}}\right)^2\right] \cdot 100 \text{ in } \% \tag{2}$$

here

ΔE_{dyn} : reduction of dynamic modulus of elasticity

n $\quad$: number of freeze thaw cycle

$t_{\text{t cs}}$ $\quad$: total transit time after capillary suction (cs) [µs]

$t_{\text{t n ftc}}$: total transit time after n freeze thaw cycles (ftc) [µs]

t_{c} $\quad$: transit time of the path length of coupling medium demineralized water [µs]

3 Precision data

Due to the measurement of transit time in a defined height of 35 mm from the test surface the CF test is nearly independent of the geometry of the specimens, e.g. cubes, drilled cores or precast elements can be used. By using demineralized water as coupling medium the determination of transit time can be measured easily with high precision. The first investigation of precision according to ISO 5725 [1] confirms the very good results of the CDF test. Following precision data were found and have to be fixed in future investigations [2][3]:

Table 1: Internal damage: Precision according to ISO 5725.

concrete	undamaged	slightly damaged	damaged
		internal damage	
relative dynamic modulus of elasticity	100% - 90%	90% - 60%	60% - 0%
standard deviation of operators	< 3%	< 4%	< 5%
total standard deviation of			
repeatability	< 5%	< 9%	< 12%
between laboratory	< 2%	< 6%	< 8%
reproducibility	< 6%	< 11%	< 14%

Table 2: Scaling: Precision according to ISO 5725.

	repeatability	reproducibility	between laboratory
functional relation	$s_r = 0{,}70 \text{ m}^{0{,}78}$	-*	-*
m = 1000 g/m²	170 g/m²	226 g/m²*	150 g/m²*
m = 2000 g/m²	280 g/m²	-*	-*

* Due to the present precision data it is not useful to calculate a functional relation.

4 Literature

1. ISO 5725. (1990) *Accuracy (trueness and precision) of measurement methods and results*.
2. Auberg, R. (1997) Phd. Thesis, University of Essen.
3. Setzer, M.J., Auberg, R. (1998) *Reliable testing of resistance of concrete against frost attack with CF test. Assessment and Precision of CF test*. In: Conference proceedings of CONSEC 98, (eds.) O.E. Gjørv, et al., Tromso, Norway, 1998.

Scaling and internal cracking in wet freeze/thaw testing

S. JACOBSEN
The Norwegian Building Research Institute, Oslo, Norway

Abstract
Wet freeze/thaw testing means that the concrete is in continuous contact with liquid during freeze/thaw. This is by far the most common way of frost testing concrete. The behaviour of concrete in two types of wet freeze/thaw tests: slow deicer salt tests and rapid freeze/thaw tests was investigated. 12 concretes varying from normal to high strength qualities were invesigated: **I** - slabs with 3 % NaCl exposed to slow cycles (SS 13 72 44) **II** - beams in water exposed to rapid cycles (ASTM C666 procedure A) **III** - slabs with varying cooling rates to "bridge the gap" between slow and rapid cycles. In all tests measurements were made of surface scaling, internal cracking and liquid uptake during freeze/thaw. The results showed that damage often is related to a "pumping effect", which is an accelerated absorption caused by wet freeze/thaw. Also evaporable water and calorimetric ice formation increased due to wet freeze/thaw. Frost/salt scaling was amplified by reduced cooling rate whereas internal cracking was amplified by increased cooling rate. This observation corresponds to less internal cracking in the slow slab test compared to the rapid beam test. Clearly different mechanisms of deterioration are mobilized in slow and fast cycles. One wet freeze/thaw test method may rank resistance against wet freeze/thaw of a series of concretes, but it cannot predict service life of structures exposed to different frost and moisture conditions.
Keywords: durability, freeze/thaw, test methods, deicer salt scaling, cracking, absorption, ice formation, mechanisms, material ranking, service life prediction

1 Introduction

The choice of appropriate test method is important for assessing frost durability of concrete according to the actual exposure conditions. In the Nordic countries the main frost durability problem today is considered to be surface scaling due to the increased use of deicer salts. However, also frost problems connected to freezing without deicing

Frost Resistance of Concrete, edited by M.J. Setzer and R. Auberg. Published in 1997 by E & FN Spon, 2–6 Boundary Row, London SE1 8HN, UK. ISBN: 0 419 22900 0.

salt occur due to cold climate and freezing of highly saturated concrete. In the Nordic countries the slab test with slow cycles and 3 % NaCl solution and measurement of surface scaling is considered the most relevant test (SS 13 72 44 - similar to ASTM C672). In other places such as North-America (USA, Canada) the rapid freeze/thaw test with measurement of loss of dynamic modulus of elasticity is considered the main type of frost test (ASTM C666).

Freezing and thawing in water or salt solution may cause two distinctly different types of deterioration: surface scaling which is increased by deicer salt, and internal cracking which may occur without surface damage.

Air entrainment is the classical technology to ensure concrete frost resistance. With the introduction of modern low porosity, high strength/high performance concrete, it is however possible to omit air entrainment and still maintain frost resistant concrete. In the slab test, concrete without entrained air with sufficiently high compressive strength or low binder porosity, can show very high resistance to damage. However, many such concretes have shown poor performance in rapid freeze/thaw testing in spite of their excellent performance in the slab test.

The present investigation was carried out to study these two types of wet frost tests (scaling and cracking) to improve the understanding of how they work. For this purpose a variety of concretes were tested in slow freeze/thaw tests with 3 % NaCl solution as well as in rapid freeze/thaw tests with water. In addition a third test with variable cycles based on a new German test principle with cooling liquid, the CDF-test, was applied on some of the concretes. The purpose was to bridge the gap between the slow frost/salt and the rapid freeze/thaw test in order to study the different mechanisms of deterioration in the two types of tests. Absorption during freeze/thaw was measured in all tests since degree of saturation is considered the main parameter for the occurrence of frost damage.

2 Test methods

In wet freeze/thaw testing the specimen or test surface is in continuous contact with the liquid during test. There is no possibility for drying during freeze/thaw. This is the most common way of testing concrete frost resistance, though not the most common type of natural frost exposure of concrete structures. In each of the three tests studied herein the following measurements were performed during freeze/thaw cycling:
• surface scaling by weighing scaled-off and dried concrete
• internal cracking by measuring resonance frequency or pulse velocity
• liquid uptake during freeze/thaw
Pulse velocity was measured with a 54 kHz Pundit. Resonance frequency was measured using a Ono Sokki CF 910 frequency analyser and light tapping with a plastic hammer. In addition several types of diagnosis were used to characterize the effect of freeze/thaw on the concrete [1]. Of these, change in evaporable water content and calorimetric ice formation as affected by wet freeze/thaw will be discussed here.

2.1 Slow cycles with 3 % NaCl - Scandinavian slab test (SS 13 72 44)
This test uses slabs that are insulated on the sides and bottom to ensure uni-directional temperature gradient. The test surface is covered with a 3 mm layer of 3 % NaCl solution during freezing and thawing, normally up to 56 or 112 cycles. The cooling

rate is 2,5 - 3 °C/h and minimum temperature is -18 °C. The damage is measured as accumulated surface scaled material (kg/m²). In figure 1 the test set-up and freeze/thaw cycle used are shown. For each concrete four parallel slabs are used.

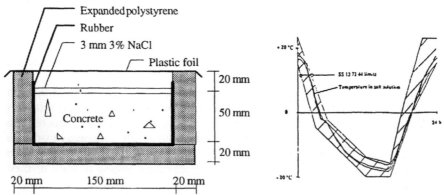

Fig.1. Test set-up and temperature in salt solution for slab test with slow cycles

2.2 Rapid freeze/thaw in water - ASTM C666 procedure A

This test is performed by rapid freezing and thawing of beams submerged. The deterioration is measured as loss of resonance frequency or pulse velocity up to 300 cycles at not more than 35 cycle intervals. Thin-walled open-topped metal boxes were used. Five cycles pr day within the limits of ASTM C666 were used. The cooling rate was 12 °C/h and the minimum temperature was -18 °C. Figure 2 shows test set-up and freeze/thaw cycle. For each concrete three beams were used.

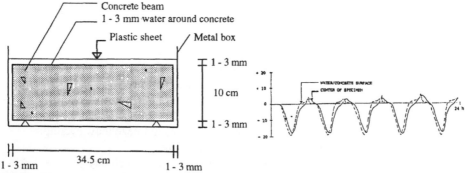

Fig. 2. Test set-up and temperature in centre of beam and water on concrete surface

A Durability Factor (DF) can be calculated according to the test method. $DF = 100 \cdot E_r$ if $E_r \geq 0,60$ at 300 cycles, E_r = relative dynamic E-modulus. If $E_r < 0,60$ at 300 cycles $DF = 0,60 \cdot (N_{E_r = 0,60}/300)$ where $N_{E_r = 0,60}$ = number of cycles where E_r falls below 0,60.

2.3 Variable cycles in 3 % NaCl - CDF test set-up

This test was carried out to bridge the gap between slow and rapid cycles by varying the cooling rate between the cooling rates of the former two tests. To be able to control the freeze/thaw cycle very accurately, a specially designed testing machine was constructed. The test principle is the German CDF-test, which allows a very accurate control of the freeze/thaw cycle by using a cooling liquid. The apparatus built at The Norwegian Building Research Institute, is made with a powerful and well controlled cooling/heating system to produce different types of cycles. Figure 3 shows the test specimen and freeze/thaw cycles used. As can be seen the concrete specimen is upside down in the salt solution in a steel container. The steel container floats in the cooling liquid giving very well defined freeze/thaw cycles.

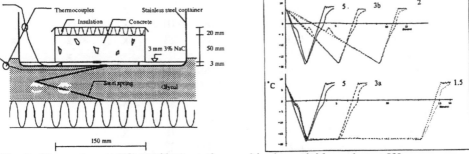

Fig. 3. Test set-up and freeze/thaw cycles used in the variable cycle test [2]

From figure 3 it is seen that the minimum temperature was - 18 °C and the cooling rate varied from 2,8 °C/h to 12 °C/h (constant time at minimum temperature). Figure 3 also shows that a test series with constant cooling rate (12 °C/h) but variable time at minimum temperature was performed. More details about the test can be found in [2].

3 Concretes investigated and ranking in the wet freeze/thaw test

12 different concretes were investigated in the 3 different test set-ups; 11 concretes in the slab and the beam test respectively, and 4 concretes in the variable cycle test. Table 1 shows the main concrete parameters and tests. More details about the concretes are given in [1].

The concretes for the scaling tests with 3 % NaCl were sawn and the surfaces predried for a week at 50 % RH and 20 °C. The sawn test surfaces were then resaturated by capillary suction with pure water for 3 days. After 3 days the water on the surface was exchanged with 3 % NaCl solution and freeze/thaw testing started. For the rapid freeze/thaw test the beams were water cured until start of test.

From table 1 we see that in ASTM C666 procedure A Mix 1, 8 and 9 have good DF > 90, Mix 11 has DF = 79, whereas Mix 2, 3, 4, 5, 7, 10 and 12 have low DF in the range 2 - 23.

In the SS 13 72 44 test Mixes 1, 5, 8, 9, 11 and 12 show very good frost/salt scaling durability (scaling < 0.10 kg/m^2 after 56 cycles). Mixes 2, 3, 4 and 7 have non-acceptable resistance to frost/salt (> 1 kg/m^2 after 56 cycles) and Mix 10 has 0.24 kg/m^2 after 42 cycles. Mix 5, 11 and 12 are therefore examples of concretes with very good durability against frost/salt scaling even though they do not survive internal cracking testing in ASTM C666 procedure A. Clearly the choice of test completely rules the assessment of frost durability for some concretes and may cause unnecessary requirements for air entrainment since the ranking is different in the two tests. To understand more about this behaviour was a main point of the study [1].

Table 1. Concrete mixtures and ranking in the tests performed

No	Mix	Air voids [1]		f_c	m_{56} [2]	DF [3]	Test (cycle)		
	W/(C+S)-S/(C+S)-Air	%	$\overline{L}$ (mm)	MPa	kg/m²		Slab (slow) 3 % NaCl	Slab (variable) 3 % NaCl	Beam (rapid) water
1	049-00A	5,8	0,13	47	0,01	91	x		x
2	035-00-	3,5	0,75	73	1,24	9	x		x
3	040-00-	2,5	0,99	74	4,25	11	x		x
4	040-05-	2,0	0,97	83	3,82	10	x	x	x
5	040-05A	5,3	0,33	73	0,02	23	x	x	x
6	037-00-	1,4	1,07	81				x	
7	035-08-	2,0	1,04	103	2,59	11	x	x	x
8	035-08A	13,2	0,16	65	0,09	94	x		x
9	035-08LWA1	1,7	1,03	75	0,01	99	x		x
10	035-08LWA2	1,9	1,30	72	0,24 [4]	2	x		x
11	030-08-	1,0	1,16	141	0,02	79	x		x
12	030-08LWA3	3,3	0,81	41	0,04	23	x		x

1) ASTM C457 2) Scaling after 56 cycles in SS 13 72 44 (accept. limit = 1 kg/m²)
3): Durability Factor according to ASTM C666 (no damage = 100) 4): 42 cycles (stopped due to severe cracking)

4 Results and discussion

4.1 Effect of wet freeze/thaw on absorption - "pumping effect"
Figure 4 shows the absorption in test specimens in the variable cycle test during 3 days of capillary absorption and during subsequent freeze/thaw cycling. During capillary absorption a typical nick point curve is obtained. However, after start of freeze/thaw the absorption increases again. Apparently, wet freeze/thaw leads to a "pumping effect" that increases the degree of saturation in the specimen.

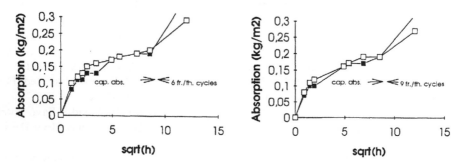

Fig.4. Effect of wet freeze/thaw on absorption (pumping effect) cycle 2 (left) and cycle 3 (right) CDF - test set-up, concrete 040-05 (■) and 040-05A(□)

4.2 Effect of absorption during freeze/thaw on internal cracking
Figures 5, 6 and 7 show how internal cracking can be related to the accelerated absorption in all three tests. From the absorption values we see that the cracking is more severe in the rapid freeze/thaw test than in the slow test in spite of lower absorption per unit exposed surface. Also the CDF test set-up has more cracking than

the slab test, and it was observed that increased cooling rate gave more cracking. Two of the concretes (040-05-/040-05A) were run up to more than 200 cycles in the variable cycle test set-up [2]. For the non-air entrained concrete the deterioration was rather severe for all types of cycles, but for the air entrained concrete there was a clear effect that the cracking was more severe the higher the cooling rate. Therefore internal cracking appears to be ruled by a mechanism that is amplified by fast cooling rate. The scaling was not so clearly related to absorption even though absorption increased steadily during freeze/thaw as shown in figure 4. For some concretes that suffered internal cracking in addition to surface scaling a relation between scaling and absorption could be seen. However for a series of concretes with no internal cracking, the amount of scaling did not correlate to absorption.

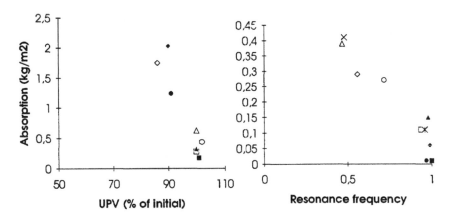

Fig. 5. Absorption vs. cracking in SS 13 72 44 (slow) at 56 cycles

Fig. 6. Absorption vs. cracking in ASTM C666 (rapid cycles) at 70 cycles

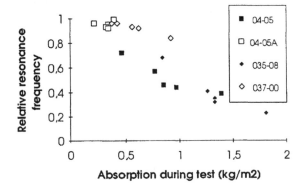

Fig. 7. Absorption vs. cracking in the variable cycle test (CDF set-up) after 28 cycles

4.3 Effect of wet freeze/thaw cycle variations on scaling

Figure 8 shows how scaling was affected by the different cycles in the variable cycle test. Scaling after 28 cycles is plotted versus time in frozen condition (below -3 °C). The results show that for concretes susceptible to scaling the scaling increases more by reducing the cooling rate than by increasing the time at minimum temperature (constant cooling rate). This shows that a time consuming mechanism is governing the deterioration by frost/salt scaling. For the air entrained concrete with very low scaling no clear effect was detected of the cycle variations on scaling.

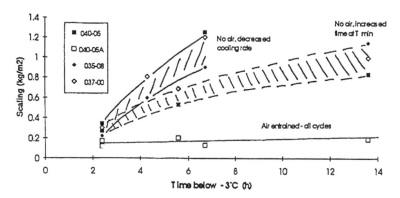

Fig. 8. Time in frozen condition vs. scaling at 28 cycles in the variable cycle test [2]

The scaling is affected by simultaneously occurring internal cracking in a specimen. Concretes that suffered severe internal cracking showed large acceleration of the scaling. On the other hand concretes with very good durability against internal cracking due to a very efficient air void system had significant surface scaling during the 300 cycles of the ASTM C666 test (as much as 0,55 kg/m²). Clearly scaling may occur without cracking. All four possible combinations of cracking and scaling were observed in the tests [1, 3].

4.4 Effect of wet freeze/thaw on evaporable water content and ice formation

Table 2 shows evaporable water content and ice formation in the beams measured at various number of cycles in the ASTM C666 test. Figure 9 shows low temperature calorimeter curves for two concretes at start and after severe damage. The specimens were sawn from and tested in the actual moisture condition of the beams. Evaporable water was measured in slices sawn perpendicular to the axis of the beams whereas ice formation was measured in smaller specimens cut from similar slices. The specimens were moisture-sealed immediately after sampling and prepared for the calorimeter. Details about measurements are given [4].

From table 2 we see that the evaporable water content is increased in concretes both with severe and almost no damage due to the pumping effect. The freezeable water content of the concrete in the temperature range of the ASTM C666 freeze/thaw test is increased on deterioration. An additional calorimeter test at 49 cycles in 035-08- showed more freezeable water at the end than at the centre of a beam, showing that there was a progressive form of deterioration in the beam following the pumping of water. This deterioration started at the concrete surface.

Table 2. Effect of wet freeze/thaw on evaporable water and ice formation (g/g_{dry})

Mix	0 (E_r=1,00)		35 (E_r=0,94)		49 (E_r=0,78)		70 (E_r=0,31/ - / 0,51/0,23)		300 (E_r=0,94)
	W_e	W_{f-20}	W_e	W_{f-20}	W_e	W_{f-20}	W_e	W_{f-20}	W_e
035-08-	,0487	0	,0492	0	,0494	0	,0580	,0030	-
035-08A	,0640	-	-	-	-	-	-	-	,0650
040-00-	,0497	,0010	-	-	-	-	,0541	,0050	-
040-05-	,0500	,0010	-	-	-	-	,0584	,0080	-

Another important feature of the ice formation data is that the freezeable water content to - 20 °C in the 035-08- concrete is zero at start of test (water cured concrete). This implies that if this concrete freezes without possibility of increasing its water content during freeze/thaw exposure, then no frost damage is to be expected unless the freezeability of the water is increased due to ageing or cracking or some other kind of change in the concrete pore structure.

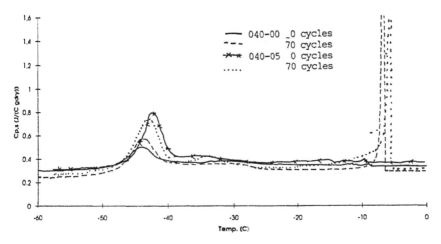

Fig. 9. Ice formation in low temperature calorimetry affected by wet freeze/thaw [4]

From figure 9 it is seen that there is practically no ice formation in undamaged concrete down to -20 °C and a large primary freezing peak in bulk water after deterioration by freeze/thaw (just below 0 °C). There are also changes in ice formation at lower temperatures indicating changes in the porestructure and/or structure of water kept in very small pores. This can also be seen in table 2: the increase of evaporable water is greater than the increase of freezeable water meaning that some of the water absorbed by the pumping effect is kept very tightly in the porestructure of the concrete. This is discussed in further detail in [4].

Other aspects of wet freeze/thaw investigated as part of [1] are: the critical limit for good scaling resistance without air entrainment, the scaling in fresh water in the rapid freeze/thaw test, the volume increase upon frost damage, studies of cracks in microscopy, studies of self-healing of cracked concretes, effect of cracking and healing on chloride transport and the effects of drying and natural weathering on scaling resistance. These results have also been published elsewhere [5-12].

5 Conclusions

Wet freeze/thaw testing (constant contact with liquid during freezing and thawing) gives accelerated absorption ("pumping effect"). Two different types of damage can be observed in such tests: surface scaling (amplified by salt) and internal cracking.

Surface scaling and internal cracking were studied in three different wet freeze/thaw test set-ups with minimum temperature - 18 °C and various cooling rates on a variety of concretes. The purpose was to explore mechanisms controlling scaling and cracking in wet freeze/thaw testing. The results showed that internal cracking correlated to the pumping effect. Surface scaling did not correlate so well to the pumping effect even though the same accelerated absorption occurred. Evaporable water content and ice formation increased during wet freeze/thaw. The pumping apparently gives a progressive form of deterioration from surface and inwards following the absorption.

The cooling rate affects the dominant type of damage in wet freeze/thaw. Frost/salt scaling is controlled by a mechanism that is promoted by slow cooling rate whereas cracking is controlled by a mechanism that is promoted by fast cooling rate.

One wet freeze/thaw test may rank concrete frost resistance, but different wet freeze/thaw tests (slow slab test, rapid beam test) rank concrete frost resistance differently. Service life prediction from wet freeze/thaw test results cannot be performed directly.

6 References

1. Jacobsen S. (1995) Scaling and cracking in unsealed freeze/thaw testing of OPC and silica fume concretes, PhD-thesis 1995:101, The Norwegian Institute of Technology ISBN 82-7114-851-3, 286 p.
2. Jacobsen S. Sæther D. Sellevold E.(1997) Mat. and Struct. V30 Jan/Feb pp.33-42
3. Jacobsen S. and Sellevold E. (1996) 3rd int. symp. HPC/HSC, Paris pp.597-605
4. Jacobsen S. Sellevold E. Matala S. (1996) Cem Conc Res V26 N6 pp. 919-931
5. Jacobsen S. (1993) Proc. Nordic Concrete Res. Meeting, Gothenburg, pp.417-419
6. Jacobsen S. Sellevold E.(1994) Nordic Concrete Res. Publ. 1/94 pp. 26-44
7. Jacobsen S. Sellevold E. (1997) Rilem proc.30 Ed. J.Marchand et al pp.93-105
8. Jacobsen S. Gran H. Bakke J. Sellevold E.(1995)Cem Co Res V25 N8 pp.1775-80
9. Jacobsen S. Marchand J. HornainH.(1995) Cem Co Res pp.V25 N8 pp.1781-1790
10. Jacobsen S. and Sellevold E. (1996) Cem Conc Res V26 N1 pp.55-62
11. Jacobsen S. Marchand J. Boisvert L. (1996) Cem Conc Res V26 N6 pp. 869-881
12. Jacobsen S. Marchand J. Boisvert L. Pigeon M. Sellevold E. (1997) Accepted, Cement Concrete and Aggregates (ASTM)

7 Acknowledgement

The Research Council of Norway, Norcem, NPRA and NC financed the study [1] which was carried out mainly at The Norwegian Building Research Institute in Oslo. Thanks to: Professor Erik J. Sellevold, Norway, who was supervisor, Assistant Professor Jacques Marchand, Canada, who together with Professor Michel Pigeon, was host for a one year stay at Laval University in Québec, Professor Gøran Fagerlund, Sweden, as main opponent on [1], and the other co-authors and contributors.

The infrared thermal image characteristic and injured degree evaluation of freeze-thaw injured concrete

ZHANG XIONG and HAN JI HONG
State Key Laboratory of Concrete Materials Research,
Tongji University, Shanghai, China

Abstract
This paper researched the infrared thermal image characteristic of freeze-thaw injured concrete by using infrared thermography. The results indicated that: the injured area of freeze-thaw injured concrete could be showed remarkably as thermal spot in the infrared image. According to its temperature range and area, the injured degree could be evaluated to provide valuable informations for the building renovation. Otherwise, ultrasonic-pulse testing method was also used to compare and verify the reliability of this infrared thermal test and evaluation.
Keywords: Characteristic, concrete, evaluation, freeze-thaw injury, infrared thermal image.

1 Introduction

Freeze-thaw injury of concrete is the main influence on the useful life of building, how to test and evaluate the injured degree of freeze-thaw injured concrete scientifically is the focus that structural engineers always pay close attention to. Ultrasonic-pulse non-destructive testing method has been widely used in construction test, but it is not suitable for the test of freeze-thaw injured concrete, because it demands strictly high smooth level of test surface however the surface layer of freeze-thaw concrete is injured remarkably. So it is still not sure which non-destructive testing method is

Frost Resistance of Concrete, edited by M.J. Setzer and R. Auberg. Published in 1997 by E & FN Spon, 2–6 Boundary Row, London SE1 8HN, UK. ISBN: 0 419 22900 0.

more suitable and economic for freeze-thaw injury test.

Infrared thermography is a new non-destructive testing method developing recently. It has the virtues of visual, un-contact,large-area scanning and is more sensitive to test surface layer[1] of materials or construction. So it is possibly suitable for the test of freeze-thaw injured concrete. This paper researched the feasibility and reliability by using infrared thermography to test and evaluate the injured degree of freeze-thaw concrete in the laboratory.

2   Fundamentals and methods of experiments

2.1   Fundamental of infrared thermography

Infrared thermography is a specialized technique based on infrared radiation for the test of material surface layer. Infrared radiation is the generality of all materials whose temperature exceeds the absolute zero(0K), but the radiating power and surface temperature of different materials will be   different. So infrared thermography can be used sensitvely   to detect the infrared radiation of materials with smaller thermal conductivity and bigger thermal radiation value such as concrete, brick or stone, etc. So infrared thermography has the adaptability to building test.

If some flaws with different contents and forms exist in different positions of concrete or other construction, thermal conductivity and specific heat or other properties will be changed partly. When heat transmitted,the infrared radiating power will be affected and the temperature distribution will be changed remarkably. So difference in temperature will appear in the thermal image. The injured features of freeze-thaw concrete structure are that loose layer appears gradually from surface to interior, a large number of micro-crack spread inside it ,   more seriously   the surface layer will peel off with or even without finger touching. The present structure of surface layer is essentially made of air and micro-hydration products through weak binding power   and   the   heat transmittion   will   be   blocked   by air .   In   the end the surface temperature of these positions will be raised and   thermal spot will appear in infrared thermal image.The temperature range of thermal spot is conversely corresponding to the injured degree and range of this position,   so structural   evaluation can be made by analyse of infrared thermal image.

Infrared   thermal analyser is   a   detector that   can   be   used in a high scanning speed from far distance without contact with the test surface. So   it is   convenient   and   suitable   for   building   test   in   site.

2.2 Experiment methods

2.2.1 Freeze-thaw concrete specimen preparation

Prepare concrete prismoidal specimen in size of 100×100×400mm. Adopt fast freezing method to make the freeze-thaw test according to the standard GBJ82-85. Within three hundred freeze-thaw cycle, control the stop-timing by testing relative elastic modulus and weight loss.

2.2.2 Infrared thermal testing method

Adopt TH1100 infrared thermal detector made in Japan. Temperature test range: -50 ~ 2000°C, detectable distance: 20cm ~ ∞. Display infrared thermal image in colour screen. Use infrared lamp irradiate in active one-side heating way. Make Data handling and area calculating through internal CPU.

2.2.3 Ultrasonic testing method

Use fast hardening cement paste to plaster the surface of freeze-thaw injured concrete after infrared thermal test. Adopt CTS-25 non-metal ultrasonic-pulse detector to test the ultrasonic-pulse velocity of every position[2].

3 Results analyse and discussion

3.1 Infrared thermal image characteristic analyse of freeze-thaw concrete

Figure.1 showed the infrared thermal image of injured concrete specimen after 220 freeze-thaw cycle. There were three distinct parts in it: temperature distribution of the thermal spot in Part I showed three circular zones with temperature range from $25.2°C$ to $31.0°C$. The highest temperature was at the centre, lower outward and the lowest at the edge. Moreover the transition from centre to edge was very wide. The surface of the corresponding part in concrete specimen indicated that it has been seriously injured and possessed the I degree injury characteristic defined in this paper: serious loose layer existed on the surface, which would peel off with finger touching; the injury depth was deeper than 10mm. In fact , the injury depth tested practically at the centre of Part I has reached 12mm.

Temperature distribution of the thermal spot in Part II showed two circular zones with temperature range from $25.2°C$ at the edge to $28.8°C$ at the centre. The transition from centre to edge was relatively narrower than that of Part I. The surface of the corresponding part in concrete specimen possessed the II degree injury characteristic defined in this paper: more cracks concentrated and less seriously loose existed on the surface; the injury depth was at the range of 5~10mm. The injury depth tested practically at the centre of Part II has reached 6mm.

Part III included all left area of the infrared thermal image, in which the temperature distribution was not distinct with the temperature range only from 23.2°C to 24.2°C. There was no remarkable transition from centre to edge. The surface of the corresponding part in concrete specimen possessed the III degree injury characteristic defined in this paper: only micro-crack existed on the surface, the texture of this part was harder than that of Part I and II and the average injury depth was less than 5mm.

Above analyse showed that I, II and III degree injury of freeze-thaw concrete corresponded successively to different infrared thermal image characteristic. So the injury degree of freeze-thaw concrete could be distinguished and evaluated by the analyse of infrared thermal image characteristic.

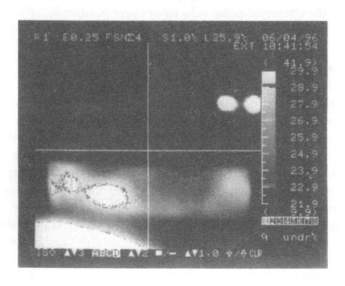

Fig. 1.   Infrared thermal image of freeze-thaw injured concrete

Table 1.   Average ultrasonic verlocity of every injured part in concrete

Part Number	Average Veloccity (km/s)
I	2.03
II	2.40
III	3.55

In order to further verify the reliability of above conclusion, ultrasonic-pulse testing method has also been used in this paper to test the velocity of injured concrete positions corresponding to the three parts in infrared thermal image.   The data were displayed in Table 1 above.

Result in Table 1 showed   that the thermal spot with higher temperature in infrared thermal image was corresponding to lower ultrasonic-pulse velocity, well the latter indicated serious injury or a large number of micro-crack and even loose layer. So the result of ultrasonic-pulse testing method verified the correlation between infrared thermal image and concrete injured position and the reliability of injury degree evaluation on freeze-thaw concrete.

3.2 Infrared thermal comprehensive evaluation of freeze-thaw concrete

Based on above qualitative analyse, this paper tried to research the infrared thermal comprehensive evaluation of freeze-thaw injured concrete. Its purpose was to evaluate the quantity of injured degree of freeze-thaw concrete so as to provide valuable informations for the building renovation.

At first,   image analysis was used to calculate   the area of every thermal spot   in   infrared thermal   image,   then quantity   of injured degree was displayed   according to below formula:

$$P_i = \frac{S_i}{S} \times 100\% \qquad (1)$$

notes: P_i — ratio of some degree injury   i=I, II, III
S_i — calculating area of the thermal spot of some degree injury   i=I, II, III.
S — total area of testing concrete surface

So the total ratio of injury was:

$$P = P_I + P_{II} + P_{III} \qquad (2)$$

The quantity description of injuryed degree of freeze-thaw concrete test in this paper was as follows:

$P = 100\%$, $P_I = 41.0\%$, $P_{II} = 15.0\%$, $P_{III} = 44.0\%$.

As metioned above, structure engineer could then make effective renovating programme from these valuable quantum informations to refrain from any unnecessary economic loss.

4 Conclusions

4.1 The injured position of freeze-thaw concrete specimen was displayed in infrared thermal image as thermal spot with high temperature. More serious the injured degree was , more higher the temperature of thermal spot would be.

4.2 I, II and III injured degree of freeze-thaw concrete possessed different characteristic in infrared thermal image. According to the latter, the injured degree could be evaluated.

4.3 Quantity description of different injured degree of freeze-thaw concrete could be made by the way of calculating first its thermal spot area and then its injury ratio.

5 References

1. Zhangxiong, (1997) Infrared thermography of building, Non- destructive Testing, Vol.19, No.3, China.
2. Li weidu, (1989) Non-destructive Technique of Concrete, Tongji University Publishing House, China.

Testing of freeze-thaw resistance portland cement compositions by low temperature dilatometry

M.A. SANITSKY, V.M. MELNYK and M.Z. LOZA
State University "Lviv Polytechnic", Lviv, Ukraine
I.V. SHICHNENKO
Scientific Research Institute on Building Production, Kyiv, Ukraine

Abstract
Chemistry of cements hardening processes and concrete frost-resistance with a multifunctional complex anti-freeze admixtures at negative temperatures is given. The temperature of the beginning of ice formation and extension strain of quick-freezen mortars and concrete depending on water-cement ratio, grade of cements and amount of anti-freeze admixtures are established by low temperature dilatometry.
Keywords: Concrete, freeze thaw resistance, cement compositions, anti-freeze admixtures, ice formation, dilatometry.

1 Introduction

Methods of concreting without warming up in winter are based on using cements with deicing chemicals. Out of antifreeze admixtures thouse most wide spread are potash, sodium chloride and nitrite. Eutectic temperatures of water solution freezing for NaCl, NaNO$_2$ and K$_2$CO$_3$ make -21,1; -19,6; -36,5^{0}C correspondingly. But operating range of application of these admixtures is limited by corresponding temperatures -10, -15, -25^{0}C. It is connected with insufficiently high rate of concrete hardening under negative temperatures [1]. With the decreasing of temperature by intensive ice formation in concrete considerable ecstension strain can develop bringing about destructive processes. That's why when concreting in the conditions of winter it is necessary to determine the temperature of the begining of ice formation and extension strain depending on water-cement ratio, grade of cements and amount of anti-freeze additive.

Frost Resistance of Concrete, edited by M.J. Setzer and R. Auberg. Published in 1997 by E & FN Spon, 2–6 Boundary Row, London SE1 8HN, UK. ISBN: 0 419 22900 0.

2 Materials and Methods of Investigation

Portland cements ordinary (OPC) and gypsum-free (GFPC) were used. Alkali-containing anti-freeze admixtures in complex with admixtures-plasticizes (lignosulphonates etc) were included in mix of cement compositions. Such cements included active mineral additives, fillers.

A number of physicochemical analysis methods were used for investigation of hydration processes of cement compositions. Physico-mechanical tests of cements and concretes were carried out according to usual procedures. Concrete prism strength and modulus of elasticity of concrete and its frost resistance, imperviousness to water, corrossion resistance during hardening at ambient, low and negative (down to -35^0C) temperatures were defined.

The temperature of the begining of ice formation and cement-sand mortar extension strain was determined by the method of low-temperature dilatometry. To do this special cylinder-shaped form was used, consisting of separate rings allowing fresh mortar to deform freely only in one direction. After its forming the fresh mortar was put in a form fastened to the stand with indicator with scale division 0,001mm. The investigation was carried out in cooling chamber with the rate of cooling 20^0C/hour down to the temperature -40^0C. Temperature in the centre of the sample was determined by means of chromel thermocouple.

To determine the content of ice in concrete colorimetric method was applied. Ice content was determined as the ratio of ice content in concrete formed in the process of freezing to the mass of uncombined water.

3 Results and Discussion

In concrete practice alkali metal salts are widely used as antifreezing admixtures and hardening accelerator of cement. At the same time a number of alkali admixtures (potash) in amounts exceeding 1 mass.% lead to the sharp stiffening of concrete mixes. At the same time other salts (nitrite and sodium chloride) may be used in increased amounts (up to 10 mass %) and they do not exert destructive effect on concrete mixes.

We investigate the reason of different effect nature alkali metals salts on portland cement structureforming processes and concrete resistance at negative and signchangeable temperatures [2]. The results obtained from the time of setting of OPC with alkaly metals salts admixtures show that they can be divided into two groups. The first group consists of sodium salts such as chloride, nitrate, nitrite not affecting strongly on time of setting. The second group consists of sodium carbonate and silicate, and potassium salts. Even small dosage of these compounds results in the sharp acceleration of grout setting and the loss of the system plastic properties.

The destructive role of dihydrate gypsum in the portland cement pastes containing carbonates, silicates, sulphates sodium and potassium is established [3]. Given results affirm the expediency of alkali metals salts of the second grope usage as admixtures to portland cement without gypsum dihydrate. Gypsum-free portland cements which represent a finaly ground portland cement clinker - set retarder - plasticiser - hardening accelerator - H_2O system are used for enchancing the effectiveness of chemical admixtures in portland cement compositions.

This was the aim of investigation of the effect of anti-freeze admixtures on the temperature of the beginning of ice formation and extension strain of quick-frozen cement-sand mortar (composition 1:2) carried out by low temperature dilatometry method.

It was established that at W/C=0.4 extension strains of such mortar on the basis of OPC made 1.82 %, while with the injection of 10 mass.% of anti-freeze additives NaCl, $NaNO_2$ and K_2CO_3 extension strains decrease to 1.7; 1.03 and 0.72%. In case of using GFPC with 10 mass.% admixtures of potash extension strain decreases to 0,55%, mortar mobility increasing. With the decrease of W/C (to 0,35) of the mortar on GFPC preserving the same plasticity as mortar on OPC the extension strain decreases to 0,15%.

Thus, the effectiveness of using potash with low eutectic temperature at its injection into GFPC greatly increases, this fact being of great importance for construction work in the regions with cold climatic conditions. GFPC with admixtures of NaCl and $NaNO_2$ require increased dosages of retarders for obtained set terms of setting. It is caused by the fact that the hydration of aluminian-containing phases with such admixtures have greater solubility in comparizon with hydrosulphoaluminates and hydrocarboaluminates of calcium. But at high dosages of LS strength growth of a binder decelerates sharply, especially under negative temperatures. That's why the use of anti-freeze agents of natrium chloride and nitrite is expedient only with the use of OPC.

Physico-mechanical tests proved that OPC with no more than 4 mass% of potash admixture had very short terms of setting and practically didn't harden under negative temperatures. GFPC with the same amount of potash admixture intensively gains strength under negative temperatures.

A.V.Laghoida [4] found that to provide intensive concrete hardening under negative temperatures the decisive factor is injection of anti-freeze agent of optimum concentration into its composition. This agent preserves liquid phase in concrete in the amount sufficient for the proceeding of hydration processes and rules out considerable volume growth during partial transformation of water into ice. Besides, irreversable destruction of concrete structure at the stage of early freezing must be taken into consideration first of all.

That's why we together with L. G. Shpynova, O. Ya. Shyiko and O. S. Ivanova [5] have studied the influence of potash admixtures on the temperature of liquid phase freezing and extension strain for cement-sand mortar on the base of OPC and GFPC of medium-aluminate clinker (C_3A = 6 mass.%). Maximum amount of potash admixtures made 15%, corresponding to its limiting content in concrete mix on the base of OPC at the temperature of concrete hardening -25^0C.

As it is seen from fig. 1, with the increase of the amount of potash admixtures from 5 to 15% the temperature of the begining of ice formation lowers from -2 to -9^0C for OPC and from -5 to -16^0C for GFPC. With the same amount of potash admixture the temperature of the begining of ice formation for mortars on the base of ordinary and gypsum-free portland cement differs to a great extent. It is significant that temperatures of the begining of ice formation for mortars on the base of OPC and GFPC are the same with potash dosages 10

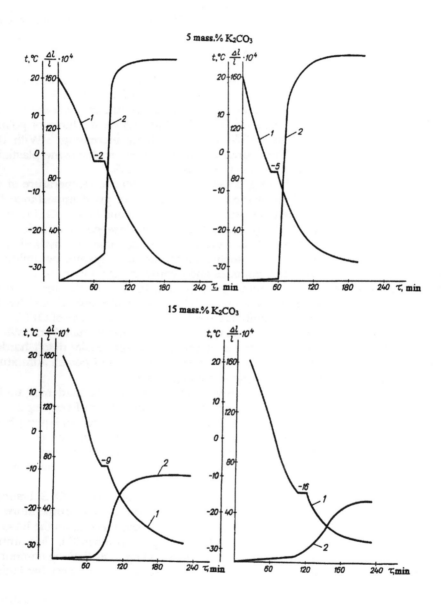

Fig.1. A diagram of system $K_2CO_3 - H_2O$

and 5% correspondingly, as well as with 15 and 10%, that is, the use of GFPC makes it possible to reduce potash expenditure 1,5-2 times.

With the set W/C = 0,4 for GFPC with the admixture of, for example, 15% of potash the concentration of the latter in water solution in the process of setting makes 27%. In the process of its cooling the temperature of the beginning of ice formation corresponds to the similar temperatures (-16^{0}C) of cement-sand mortar where cement is substituted by fine sand and aqueous solution of potash, of 27% concentration. It follows that at the early stage of freezing GFPC has no significant effect on the nature of the process of ice formation. At the same time this relationship cannot be observed for OPC, as the reaction of 5 mass.% of gypsum with 4 mass.% of potash results in the formation of arkanite possessing no anti-freeze effect. It leads to the increase of liquid phase freezing temperature. In case of GFPC the whole amount of injected potash is used for its proper purpose as anti-freeze admixture promoting considerable decrease of liquid phase freezing temperature.

Taking into account that the freezing temperature of potash solution of certain concentration corresponds to the temperature of the beginning of ice formation of liquid phase of GFPC set by potash solution of the same concentration we can determine the amount of ice formed at the liquid phase at the initial period of binder hardening by computation-graphical method with the sufficient degree of accuracy making use of the diagram of the state of system $K_2CO_3 - H_2O$ (fig.2). Thus, with the decrease of temperature from initial (t_i) to design (t_d), the concentration of potash at liquid phase changes along the line of liquidus from initial (A_i) to final (A_f) it corresponding to design temperature. Applying the rule of lewer one can determine the content of ice (I) at the liquid phase of concrete on GFPC portland cement at the same temperature:

$$I = (A_f - A_i)/A_f \cdot 100\%$$

A.V.Lagoida [4] shows that to reach intensive concrete hardening at negative temperatures it is necessary to inject such amount of anti-freeze agent that would cause no more than 70% of initial ice formation. Besides, the material has sufficient amount of liquid phase, providing cement hydration at negative temperatures. It follows that initial ice formation of 70% will be limiting ice formation for concrete with potash admixture. Thus, one can determine the concentration of potash solution (A_i) for the setting of concrete mix on GFPC at given design temperature of hardening (T_d):

$$A_i = (100-i) A_f/(100-0,01A_f)$$

Proceeding from working concentration of the solution one can calculate the amount of anti-freeze agent A_a(%) depending on water - cement ratio:

$$A_a = A_i (W/C)/(1-0,01A_i)$$

To determine the amount of ice experimentally by colorimetric method concrete mix was placed in brass boxes held then at the temperature -25^{0}C for 1, 3, 7 and 28 days. At the same time samples of concrete were formed for determining strength both in normal conditions (R_{28}=30,4 MPa) and at negative

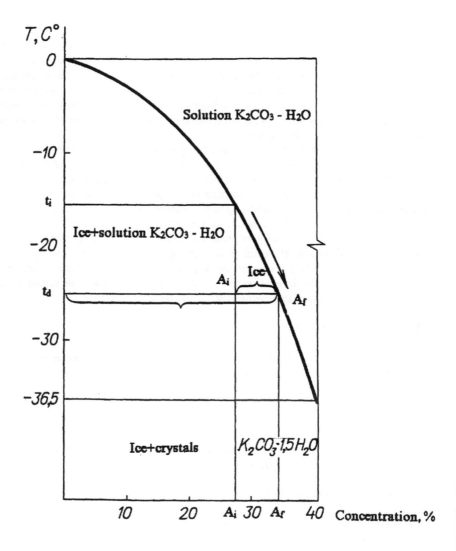

Fig. 2. The temperature of the beginning of ice formation and extension strain of quick-frozen cement-sand mortar (composition 1:2) based OPC and GFPC with LS and potash admixtures.

temperatures. The analysis of the results obtained show that with the injection of the same amount of potash the amount of ice in concrete of composition 1 : 1,8 : 3,5 at W/C=0,47 on the base of GFPC is much less that for concrete on OPC. It results in its more intensive hardening at negative temperatures. It is significant that in 28 days of hardening at -25^0C concrete on OPC with 8mass.% of potash admixture corresponds to concrete on GFPC with 4 mass.% of potash admixture at the same W/C ratio in strength (R_{28}=7.0 MPa) and amount of ice (40%). At the same time, concrete mixes on GFPC are characterized by higher mobility. It testifies to the possibility of decreasing the amount of ice at the expense of extra decrease of W/C. Given data also confirm the conclusion that in concrete mixes on the base of OPC potash in the amount of 4 mass.% is not used for its proper purpose as anti-freeze agent as arkanite $K_2SO_4 \cdot 1,5H_2O$ is formed as the result of its interaction with 5 mass.% of dihydrate gypsum. Arkanite has no anti-freezing effect.

It is significant that for concrete on GFPC with 4 mass.% of potash admixture and for concrete on OPC with 8 mass.% of potash admixture only 23% of concrete grade strength is reached in 28 days, i.e. the amount of liquid phase is insufficient for intensive hardening at -25^0C. With the increase of potash content in concrete on GFPC to 8 mass.% initial amount of ice decreases to 62%, in 28 days - to 10%. It allows to increase twice the strength of concrete at negative temperatures. After further holding in normal conditions the strength of such concrete makes 118% of grade strength.

It must be pointed out that when using anti-freeze potash admixture for concrete on the base of GFPC experimental values are the same as design ones. For concrete on OPC computation method of determining ice content cannot be used as owing to the interaction of gypsum and potash the temperature of initial ice formation is higher than the freezing temperature of aqueous potash solution of given concentration. When we use anti-freeze agents NaCl and $NaNO_2$ not interacting with gypsum computation - graphical method can be used for determining ice content on the basis of diagrams of state of systems NaCl - H_2O and $NaNO_2$ - H_2O.

Concrete on GFPC with LS and potash admixtures basis attains 80-100% of branding strengh in 28 days at temperatures -15^0C and posses the capability of hardening at frost -35^0C. Concrete is characterized by high value of frost resistivity (F300 and greater) and increased water-proofness. Concrete based on the special GFPC compositions can be used in winter concreting without heating, during carrying out of repair works at any season of years and in order to receive the articles with high frost-resistance. When hardening at lower positive temperatures and at frost up to -10^0C the possibility of construction monolithic structures with term of loading which are nearly to summer loading is created.

The processes of ice formation with the use of technical pentaerythritol filtrate (TPF) - secondary product of chemical industry with freezing temperature -17^0C - as antifreeze agent was determined by the method of low-temperature dilatometry too [6]. It is stated that for fresh cement - sand mortar (1 : 2) without admixtures intensive ice formation is observed at temperature - 2^0C, extension strains grows unevenly to 1.82%. With the injection of 4 mass.% of TPF the temperature of the begining of ice formation is -10^0C , at further freezing extension strains begin to graw gradually and making 1,2% at -25^0C.

With the use of complex admixture containing TPF and alkali-containing hardening activators on the base of secondary products of chemical industry, extension strains decrease to 0,96%. It testifies to higher effectiveness of complex admixture on the base of TPF and alkali metal salts in concretes hardening at negative temperatures. Cold resistence of concrete with complex admixtures after 28 days of hardening at -15^0C held in normal conditions to gaining grade strength makes no less than F 150.

Application of the method of low temperature dilatometry allows to extend the field of application of a number of industrial wastes of chemical industry to concrete mixes used for winter monolith concreting without heating. It also solves ecological problem of their utilization.

4 Conclusions

It is shown that the dose of anti-freeze additive is defined first of all by liquid phase freezing temperature. On the grounds of the data of low temperature dilatometry, using structural diagram $NaCl-H_2O$, $NaNO_2-H_2O$, $K_2CO_3-H_2O$ are can define the amount of ice formed at liquid phase by calculating-graphics method. Obtained values of the amount of ice agree with the data on the amount of ice in concrete obtained by calorimetric method. It was established by X-ray method that in the OPC compositions gypsum additive transform potash admixture part (4 mass.%) to arkanite not possesing the antifreezing effect.

Potash used as a highly effective hardening accelerator as well as an anti-frost admixture for GFPC allows us to obtain binders with accelerated strength gain at negative temperatures. The expensive stress of these binders is 5 to 10 time lower at early freezing as compared to OPC.

5 References

1. Mironov, S.A. and Lagojda, A.V. (1975 *) Concretes Hardening at Frost,* Stroyizdat, Moscow.
2. Sanitsky M. A., Sobol H. S. and Shevchuk G. Ya. (1994) High frost durability concrete based rapid-hardening gypsum-free portland cement compositions . *Ibausil 12,* Tagungsbericht - band 2, pp. 232-8.
3. Pashchenko, A. A., Sanitsky, M. A., Shevchuk, G. Ya.(1990) Special features of the portland cement compositions with alkaline metals salts admixtures. *Cement,* No 7, pp. 17-9.
4. Lagojda, A. V. (1984) Winter concreting using antifreezing admixtures. *Concrete and Reinforced Concrete,* No. 9 pp. 23-6.
5. Shpynova, L.G., Sanitsky, M.A., Shyjko O.Ya.,Ivanova O.S. (1988) Gypsum-free portland cement with potash admixture for winter concreting, *Concrete and Reinforced Concrete,* No. 3, pp. 6-7.
6. Sanitsky, M.A., Shichmenko, I.V., Vandalovskaj, L.A. and Jernovoj, S.V. (1993) Admixtures to concrete, hardening at frost. *Building Materials and Constructions,* No 1, p. 11.

The reasons of damping maximums of hardened cement paste (hcp) at extremely low temperature

X. XU and M.J. SETZER
IBPM – Institute of Building Physics and Materials Science,
University of Essen, Essen, Germany

Abstract

The reasons for the different damping maximums of hardened cement paste (hcp) in a temperature range between +20°C and -160°C have been analysed. The bending vibration of a hcp beam has been induced and the eigenfrequency and damping have been measured. Five temperature regions are apparent which correspond to the different damping mechanisms of pore water. Region II from -5°C to -50 °C contains a damping maximum due to the internal friction between the crystals of ice and salt in capillary pores of hcp. In region III between   -30°C and -70°C the damping peak characterizes phase transition of gel pore water. The sharp peak at -90°C in region IV between -60°C and -120°C is mainly due to a mechanical relaxation process of a defect monolayer of water molecules. It can be attributed to the interaction between the pore water and the internal solid surface of hcp.

Keywords: Cement paste, Low temperature, Dynamic elastic modulus, Damping maximum, Internal friction, Pore solutions, Activation energy, Dispersion, Mechanical relaxation

1.   Introduction

The freezing of water in porous systems is determined by the solid-water interaction at the internal surface of hcp which can exceed 100 m²/g. The freezing point is increasingly depressed with decreasing pore size and increasing salt concentration. Several physical and mechanical tests on the interaction between pore ice, the adsorbed water films and the internal surface of the solid have been published [1-10].

Frost Resistance of Concrete, edited by M.J. Setzer and R. Auberg. Published in 1997 by E & FN Spon, 2–6 Boundary Row, London SE1 8HN, UK. ISBN: 0 419 22900 0.

Measuring the dynamic elastic modulus and damping proved to be an efficient test method to study the abnormal freezing of pore water in hcp [1,2,11,12,13]. It was found that the water in hcp exists as adsorbate, prestructured condensate and condensate as well as in the lower temperature range as ice [5]. Below the freezing point of the macroscopic water an equilibrium between water, vapour and ice is possible in a large temperature range from -5°C to -60°C. This is thermodynamically related to the surface interaction at the internal surfaces. In the proximity of the internal surfaces the structure of the adsorbed water films is affected considerably by the surface forces of the solid. The structure, therefore, deviates extremely from that of the macroscopic water. By DSC- (differential scanning calorimetry) measurements the phase

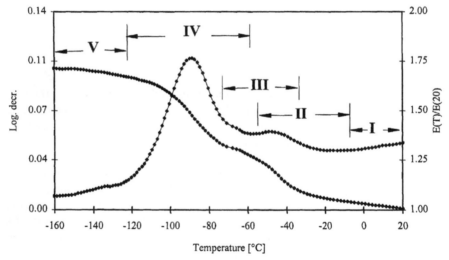

Fig. 1.   Typical temperature dependence of dynamic elastic modulus and damping coefficient. Both the elastic modulus related to the value of room temperature and the logarithmic decrement are plotted as a function of the temperature.

transitions of water at -23.7 °C, -31,0 °C and -39,4 °C could be proved [8]. The freezing of pore water is reflected in the thermal expansion [5,8] of hardened cement paste and of concrete [14,15], as well as in the elastic modulus. In [9] we could even correlate the increase of the elastic modulus quantitatively with the amount of frozen water. Below -70 °C the pore water is completely frozen as we proved by DSC, elastic modulus measurement and thermal expansion. There can be only an unfrozen film at the very great internal surface left which is below 3 molecular layers, probably 1 molecular layer thick. This water is highly structured.

Fig. 1 shows typical temperature dependence of dynamic elastic modulus and damping coefficient of hcp. Five temperature regions are apparent which correspond with the different damping behaviours. The reasons for the damping maximums in different temperature regions have been analysed in this work.

2. Experimental Procedure

2.1 Test Apparatus

The dynamic elastic modulus and the damping have been measured by a test device within a cryogenic chamber (figure 2).

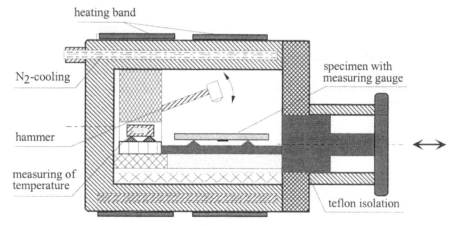

Fig. 2. Measuring chamber. The chamber is additionally thermally isolated

The hardened cement paste beam is placed at its nodal point of vibration on two knife-edged supports which are positioned on a slide-in unit. This unit is placed in the measuring chamber and the chamber is closed tightly. Inside the test chamber the specimens are excited dynamically by a falling hammer which is internally driven. The impact of the hammer is dimensioned in a way that the first basic bending oscillation is generated. Any additional higher oscillation would impair the evaluation. After the impact the hammer is out of contact with the beam, the beam above the support vibrating freely for about 300 msec. Within these 300 msec the damped flexural oscillation is recorded by a strain gauge glued on the bottom of the beam. A second strain gauge is placed on a reference beam in the chamber forming a half Wheatstone bridge with the measuring gauge both at the same temperature. The signal is processed by a high precision Hottinger Baldwin constant current bridge amplifier, it is digitalized and stored by transient recorder. The recording device is only switched on during the recording process to avoid heating up of the specimen.

The measuring chamber is cooled by evaporation of liquid nitrogen. The precision of the temperature is controlled below $\pm\,0.1°C$. The cooling rate was kept constant at $2.0°C/min$.

In order to avoid a possible falsification of the test results by the cooling and heating regime, both cooling and heating were taken into account.

2.2   Specimens

The specimens are made out of ordinary Portland cement CEM I 32,5R and high slag cement CEM III/A 32.5 with the w/c ratios of 0.4 and 0.8. The beams were cut out of the middle of hardened cement paste plates. After curing in saturated Ca(OH)$_2$ solution the specimens were stored in NaCl solutions at 20 °C. The solutions contained up to 3.9 mol Cl/l.

After three months a batch of samples stored in salt solutions was removed and adsorbed at different humidities until equilibrium of weight was reached.

2.3   Determination of the Dynamic Elastic Modulus and Damping

Using the above mentioned test design the dynamic elastic modulus was calculated by the Timoshenko equation [16,17]:

$$E = 0.9464 \frac{\rho l^4 f^2}{h^3} k \qquad (1)$$

where ρ is the density

$$\rho = \frac{m}{hb} \qquad (2)$$

with the width b, height h, the length l and the mass per unit length m of the beam. f is the frequency (eigenfrequency), and k a correction value

$$k = 1 + 6.585(1 + 0.752\mu + 0.8109\mu^2)(\frac{h}{l})^2 - 0.686(\frac{h}{l})^4 \qquad (3)$$

with the Poisson number μ.

In this article the damping was calculated by the logarithmic decrement:

$$\ln(\frac{x_i}{x_{i+1}}) = \delta = 2\pi D \qquad \text{for } D < 1 \qquad (4)$$

where δ is the logarithmic decrement, x the amplitude of two successive oscillations, i and $i+1$, and D the damping coefficient.

3.   Results and Discussion

Freezing of capillary pore water

In Fig. 1 the dynamic elastic modulus and the damping coefficient are plotted as a function of temperature. The elastic modulus increases until -160 °C due to ice formation. However, in measurements by differential scanning calorimetry (DSC) phase transitions could only be observed down –60°C. It is concluded that at -60°C all pore water apart from the water in the last layer is frozen [1,7].

In region II between -5°C and -50°C the damping maximum could only be observed when higher chloride concentrations ($\geq$ 1.6 mol Cl/l) in the pore solution or higher relative humidities of hcp ($\geq$ 85% r.h.) are present. Fig. 3 shows the damping coefficient of hcp as a function of temperature after storage in various NaCl solutions. The damping maximum in this region are shifted to lower temperature when chloride concentration are increased. The shifting of freezing points is correlate with the phase diagram of H_2O-NaCl-System and with the results of DSC-measurement. The damping maximum between -5°C and -50°C is produced by an increased internal transport due to the interaction between the ice and the unfrozen water, the salt and ice crystals as well as the salt crystals and the unfrozen water, because in this case the interface effects between the unfrozen water and the matrix is not dominant.

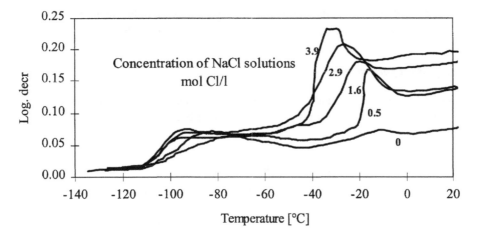

Fig. 3. damping coefficient of hcp as a function of temperature after storage in various NaCl solutions (hcp, w/c=0.8, stored at 97% r.F.)

In region II the damping peaks not only depend on chloride concentration and water content of hcp but also on the cooling rate. It can be seen in Fig. 4 that the cooling rate influences the damping behaviour. There was no damping peak when the cooling rate was very small. The same effect is not observed in region IV between -60°C and -120°C. This result means that both the damping maximum between -5° and -50°C and the damping maximum between -60°C and -120°C are not formed due to same reason. The internal friction between the pore ice and salt crystals by slow cooling is smaller than that by rapid cooling because the crystals size of ice and salts are increased and therefore the interface between the crystals is decreased. For this reason there is no damping peak by smaller cooling.

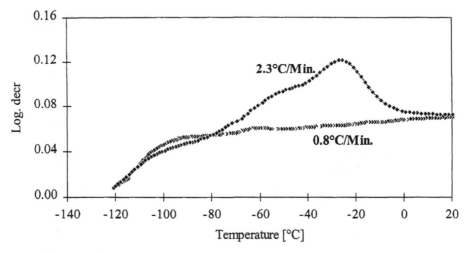

Fig. 4. Influence of cooling rate on the damping maximum between -5°C and 60°C (hcp, w/c=0.8, 3.9 mol Cl/l and stored at 97% r.h.)

Freezing of gel pore water

Region III between -30°C and -70°C contains a low-temperature transition of gel pore water. The detailed description of the damping behaviour in this temperature region is to be found in [1,16].

Mechanical relaxation process of pore ice

A very distinct damping peak is observed in region IV between -60 °C and -120 °C as shown in figure 1. A transition enthalpy was measured [7]. A first order phase transition could be excluded. Either a second order transition or a viscous flow or a relaxation process were discussed. However, for a relaxation process a dispersion has to be proved. This was not possible at an earlier time [12] due to the reduced precision. In any case, the damping maximum characterises the surface interaction between the internal surface of hardened cement paste and pore water or pore ice respectively. Currently, it has not been clarified completely which mechanism generates this damping maximum.

Several authors reported about the damping maximum of hardened cement paste containing pore water in this temperature region. Helmuth [19] assumes that the damping peak is due to a viscoelastic process. Sellevold and Radjy [20] propose a solidification of the adsorbed water film to explain the damping maximum. Zech and Setzer [11,12] showed that the area under the damping maximum is proportional to the size of the solid ice interface.

By testing the dependence of the logarithmic decrement from temperature and frequency we examined whether the mechanical relaxation is the reason for the damping maximum . The activation energy and relaxation time calculated by the Arrhenius equation can be compared reasonably with those of pure water. In the same temperature region where the damping maximum is observed the deviation of the inherent frequency was evaluated to prove the model.

TABLE 1. Measured results of the specimens with w/c ratio = 0.4 and 90% r. h.

Specimen	Damping maximum						Activation energy Q		τ_o	
-No.	T_m (°C)		Frequency (f_m)		δ_{max}·(10⁻³)		kJ/mol		/	
	cool.	heat.	cool.	heat.	cool.	heat.	cool.	heat.	cool.	heat.
CEM I 32,5R										
P-01	-90.9	-86.2	3361	2112	09.48	09.69				
P-02	-89.9	\	3533	\	09.35	09.94				
P-03	-87.8	-82.2	4675	3500	09.51	09.60				
P-04	-87.5	-80.2	5064	4666	09.22	09.16	35.1	33.6	10^{-14}	10^{-14}
P-05	-86.1	-78.7	6438	6510	09.51	09.48				
P-06	-84.6	\	7186	\	10.20	10.40				
P-07	-83.2	-73.8	7863	7747	09.69	08.44				
CEM III/A 32.5										
H-01	-89.9	-80.7	2091	2067	10.22	09.69				
H-02	-84.2	-77.2	2869	2816	10.20	10.20	39.9	35.9	10^{-15}	10^{-14}
H-03	-82.7	-75.7	3917	3896	10.90	10.90				
H-04	-79.9	-71.4	5742	5727	10.20	10.20				

* The subscript *m* signifies the data by damping maximum.

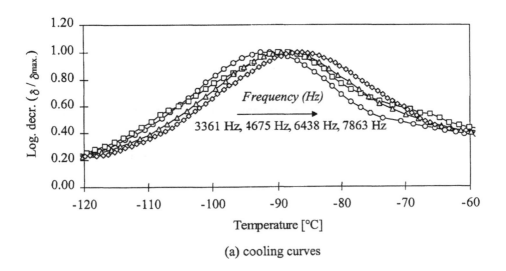

(a) cooling curves

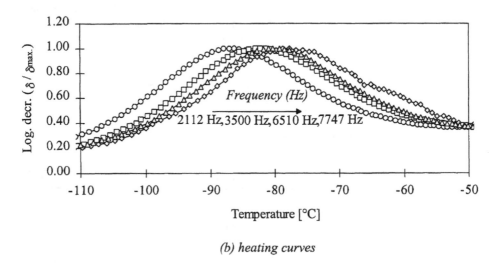

(b) heating curves

Fig. 5.   Temperature dependence of resonance frequency at damping maximum

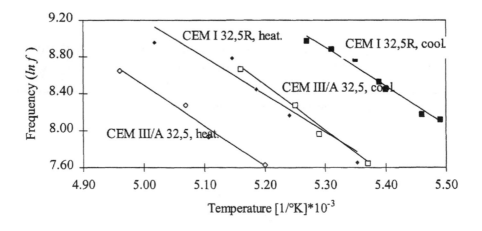

Fig. 6.   Resonant frequency vs. reciprocal of absolute temperature at damping maximum

Table 1 shows the typical dependence of the damping maximum from temperature and the eigenfrequency.

The dependence on temperature of the frequency at the damping maximum during cooling and heating is clearly seen in figure 5. By plotting the temperature at the damping maximum as a function of the logarithm of the frequency, a linear correlation is found (figure 6) according to the Arrhenius equation.

$$f = f_0 \exp(-\frac{Q}{RT}) \tag{5}$$

R being the gas constant and T the absolute temperature. The slope of the line is proportional to the activation energy Q. The numerical results are found in table 1.

From Arrhenius equation and its frequencies f and f_0 the relaxation time τ respectively. τ_0 are evaluated as follows:

$$\tau = \tau_0 \exp(\frac{Q}{RT}) \tag{6}$$

Schiller [21] and Kuroiwa [22] examined the mechanical relaxation of ice crystals with a similar procedure and determined an activation energy of 54.8 kJ/mol. They found a damping maximum of ice crystals at about -30°C. They relate the damping maximum in free ice to two mechanisms:

(a) By the periodic mechanical deformation of the crystal energy differences are generated additionally between the various configurations of hydrogen atoms. This will disturb the dynamic equilibrium between the configurations. After the relaxation time a new equilibrium will be reached.

(b) In the deformed ice lattice the probability of stay of certain lattice defects will be increased on longer oxygen-oxygen bonding than on the shorter ones. The resulting rearrangement of the lattice defects could be connected with the observed relaxation time. In this case, the damping maximum must be proportional to the concentration of the lattice defects.

Helmuth [19] determined the dependence on temperature of the frequency of the damping maximum at water saturated hcp specimens. However, he found a very high activation energy of about 90 kJ/mol. The damping maximum in his calculations was, also, between -80°C to -90°C. Helmuth concludes that a mechanical relaxation as free ice is doubtful, since the height of damping peak in ice crystals is 10 times smaller than in water saturated hardened cement paste although the water content in hcp is only about 10% as a rule.

Overloop and Van Gerven [23] examined the freezing behaviour of adsorbed water on high-area silica gel and pore glass with NMR methods and determined an activation enthalpy of the bound water (the nonfrozen water of the first two or three adsorbed layers) of 31.4 kJ/mol.

To characterise the process we analysed the eigenfrequency. A relaxation process leads not only to a damping maximum but also to a dispersion of the eigenfrequency. The phenomenon is comparable with the resonant dispersion of light [24]. The damping maximum is generated by a relaxation process there should be observed a dispersion of the eigenfrequency at the same temperature. This dispersion superimposes to the frequency change by the modification of the elastic parameters caused by cooling or heating respectively.

Figure 7 shows both the change of the resonant frequency and the damping as function of temperature. The dispersion of the eigenfrequency and the damping maximum coincide in the same temperature range. The dispersion step of the eigenfrequency is, apparently, superimposed to the normal, linear, change of the frequency due to cooling. It confirms that the effect can be attributed to a relaxation phenomenon.

In addition to the dispersion of the eigenfrequency the real part of the dynamic elastic modulus E' in the temperature range of the damping maximum has been investigated. It showed a similar change in the temperature region of the damping maximum.

At the interface between ice and internal hardened cement paste surfaces a strong interaction is generated leading to the unfrozen monomolecular structured water film. Considering the low activation energy and temperature it is unlikely that a mechanical relaxation of the pore water will take place as it is observed at free ice. In free ice this is caused by proton movement and further diffusion due to the defects in the ice lattice. On the other side, it can be assumed that the density of crystal defects increase drastically at the pore ice interface due to the considerable surface interaction with hardened cement paste. We explain the mechanical relaxation at this temperature by an increased movement and diffusion of the protons which are located at these surface defects. The movement has the characteristic relaxation time.

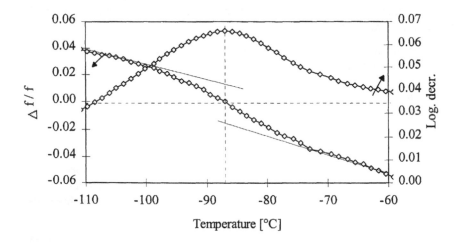

Fig. 7. Dispersion of resonant frequency and damping as a function of temperature

The "interface model" explains, too, why the area under the damping peak is proportional to the ice solid interface as found in [12,13]. since the defects are directly proportional to the area of his solid ice interface. Even for specimens stored at 30 % r.h. a proton movement can be attributed to the water in the adsorbed film even if there is no more ice formed.

4.   Summary

Due to surface interaction different damping mechanisms of hardened cement paste (hcp) at low temperature must be distinguished. Fig. 8 summarizes the results of the present work:

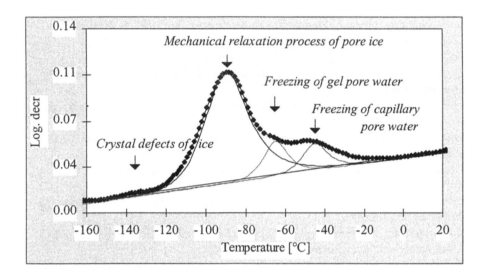

Fig. 8. Damping mechanisms of hardened cement paste (hcp) at low temperature

5.   Acknowledgements

The authors gratefully acknowledge the financial support of the Deutsche Forschungsgemeinschaft (Project Se 336/32)

6.   References

1. Xu, X. *Tieftemperaturverhalten der Porenlösungen in hochporösen Werkstoffen*, Ph.D. Dissertation, Universität GH Essen, 1995
2. Xu, X; Setzer, M.J. *Advanced Cement Based Materials*, Vol.5 # ¾ (1997), S. 69-74.
3. Setzer, M.J. DAfStb **280**, 1977.
4. Bager, D.H.; Sellevold, E.J. *Cement and Concrete Res.*, Part **I**: 16(5), (1986), 709-720, Part **II**: (16(6), 1986), 835-844, Part **III**: 17(1), (1987), S. 1-11.
5. Setzer, M.J. Conf. Advances in Cementitious Materials, (1990: Gaithersburg), Ceramic Transactions Vol. **16**. Westerville, Amer. Cer.Soc. (1991) S.415-439.
6. Stockhausen, N. *Die Dilatation hochporöser Festkorper bei Wasseraufnahme und Eisbildung*, Ph.D. Dissertation, TU München, 1981.

7. Beddoe, R.E.; Setzer, M.J. *Änderung der Zementsteinstruktur durch Chlorideinwirkung, Foschungsberichte* aus dem Fachbereich Bauwesen der Universität GH Essen, 1990.

8. Stockhausen, N.; Dorner, H.; Zech, B.; Setzer, M.J. *Cem. Concr. Res.*Vol. **9**, (1979), S. 783-794.

9. Stockhausen, N.; Setzer, M.J. *Tonindustrie Zeitung,* **104**, (2/1980), S. 83-88.

10. Setzer, M.J. *2nd Intern. Conf. on Durability of Building Materials and Components,* National Bureau of Standards, Gaithersburg, USA, 1981, S. 160.

11. Zech, B.; Setzer, M.J. *Materials and Structures,* Vol. **21**, , 1988, S. 323.

12. Zech, B.; Setzer, M.J. *Materials and Structures,* Vol. **22**, 1989, S. 125.

13. Zech, B. *Zum Gefrierverhalten des Wassers im Beton,* Ph.D. Dissertation, TU München, 1981.

14. Rostasy, F.S.; Schneider, U. and Wiedemann, G. *Cement and Concrete Research,* Vol. **9**, (1979), S. 365-376.

15. Rostasy, F.S.; Wiedemann, G. *Cement and Concrete Research,* Vol. **10**, (1980), S. 565-572.

16. Timoshenko, S. *Vibration problems in engineering,* D,Van Nostrand Co., New York, 1937.

17. Pohl, E. *Zerstörungsfreie Prüf- und Meßmethoden für Beton,* VEB Verlag für Bauwesen, Berlin, 1969.

18. Xu, X.; Setzer, M.J.: Einwirkung wäßriger Salzlösungen auf Zementstein, DFG-Abschlußbericht. Identifikation Se336/26 , 1996.

19. Helmuth, R.A. *Inverstigation of the low temperature dynamic-mechanical response ofhardened cement paste,* Department of Civil Engineering, Stanford University, Techinical Report, No. 154, (1972).

20. Sellevold, E.J.; Radjy, F., J. *Mater. Sci.* **11**, (1976), 1927.

21. Schiller, P. *Zeitschrift für Physik,* Bd.**153**, (1958), S. 1-15.

22. Kuroiwa, D. *Interal friction of H_2O, D_2O and natural glacier ice,* Research Report 131,U.S.Army Materiel Command, Cold Regions Research& Engineering Laboratory, Hanover, New Hampshire, January, (1965).

23. Overloop, K.; and Van Gerven, L. *Journal of Magnetic Resonance,* Series A **101**, (1993), S. 179-187.

24. Gerthsen, C.; Kneser, H.O; Vogel, H.: Physik-Ein Lehrbuch, 16. Auflage, Springer-Verlag, 1992

Subzero temperature investigation of autoclaved concrete with gypsum added

J. ADOLPHS and M.J. SETZER
IBPM – Institute of Building Physics and Materials Science,
University of Essen, Essen, Germany
S. SHIBATA
Sumitomo Metal Mining Company Limited, Yokohama, Japan

Abstract
The dependence of autoclaved concrete with variable amount of gypsum and of different cure times was investigated in the low temperature region down to -120°C. The dynamic elastic modulus and the damping maximum were measured. In addition calorimetric data and pore size distributions with mercury porosimetry and water vapor adsorption isotherms were obtained. The results show a correlation of the change in the pore size distribution with the temperature shifts of the enthalpic and mechanical properties.
Keywords: autoclaved concrete, gypsum, dynamic elastic modulus, damping maximum, calorimetry, phase transitions, low temperature, adsorption

1 Introduction

Environmental pollution is a general problem. To reduce the sulfate emissions in coal power plants the emissions are filtered. Sulfuric acid is transformed to gypsum when calcium carbonate is added. However, the amount of this produced gypsum has increased in such a way that problems have arisen concerning the disposal. Therefore, it is of common interest to use this gypsum as an additive in concrete or mortar. In this contribution the effect of varying quantities of gypsum in autoclaved concrete in the low temperature region is investigated. Autoclaved concrete is used for the production of parts of prefabricated houses.

Frost Resistance of Concrete, edited by M.J. Setzer and R. Auberg. Published in 1997 by E & FN Spon, 2–6 Boundary Row, London SE1 8HN, UK. ISBN: 0 419 22900 0.

2 Experimental

The specimen were produced by the Sumitomo Metal Mining Co., Ltd., Japan. The basic composition of the concrete consists of 65.1% Portland cement and 34.9% SiO_2 for all samples. The ratio CaO to SiO_2 was 0.9. The quantity of added gypsum varied from 0% over 15% to 20% for the whole composition. After 5 minutes mixing the paste was casted for 3.5 until 4.5 hours at a temperature between 40°C and 45 °C. Then aging followed for 24 hours at 40°C. Finally the autoclaved aging started, while the samples were put into a vessel at 180°C and a pressure of 2 MPa either for 24 or 72 hours. Table 1 shows the composition in details.

Table 1. Composition of the autoclaved concrete samples.

	Composition					Autoclaved aging	
Sample	Cement (%)	SiO_2 (%)	Gypsum (%)	$\frac{CaO}{SiO_2}$	w/c	Time (hr.)	Temp. (°C)
1	65.1	34.9	0	0.9	0.50	24	180
2	65.1	34.9	15	0.9	0.50	24	180
3	65.1	34.9	20	0.9	0.50	24	180
4	65.1	34.9	0	0.9	0.37	72	180
5	65.1	34.9	20	0.9	0.44	72	180

Beams (50 mm x 5 mm x 4 mm) were cut from the core of these samples for the elastic modulus experiment. Some broken pieces of a few grams were used for differential scanning calorimetry (DSC), mercury intrusion porosimeter (MIP) and water vapor adsorption. After complete wetting at 100% rel. h. a part of the samples was stored over saturated salt solutions at 90% rel.h. ($BaCl_2$), 75 % rel.h. (NaCl) or dried over an ice trap at 0.5 % rel.h., all at 20°C. For MIP and water vapor adsorption only the dried samples were used.

The dynamic elastic modulus was measured from +20°C down to -120°C with a self constructed apparatus as described by Xu [1]. The cooling rate had been 180 K/h. The calorimetric data were obtained with a DSC 4 from Perkin Elmer in a temperature range from +20°C to -60° at a cooling rate of 120 K/h. For the MIP measurements a Carlo Erba Porosimeter 2000 (max. pressure 2000 MPa) was used.

The water vapor isotherms were measured gravimetrically with a self constructed adsorption apparatus at a temperature of 20%. This high precision apparatus was especially developed for long durance measurements of about ½ year at absolutely stable low pressure conditions. Details are described by one of the authors in [2].

3 Results and discussion

The pore size distribution from the mercury intrusion porosimeter measurements is remarkably changed due to a change in the amount of gypsum (Fig. 1). Increasing the amount of gypsum leads to an increasing amount of the mesopores in the range from 5 nm to 40 nm.

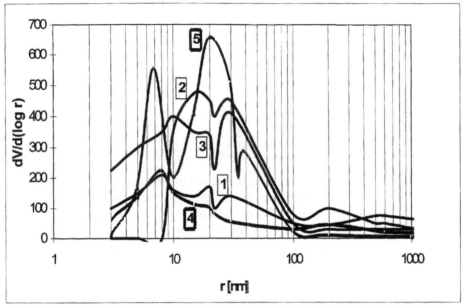

Fig. 1.   MIP pore size distribution. The samples were dried at 0.5 % rel.h. The numbers denote the samples of table 1.

In addition, the pore radii shift from values below 10 nm to values between 30 nm and 40 nm. The microcapillaries (100 nm to 500 nm) diminish completely at 20 % added gypsum.

Considering the microporosity of hydrated cementious materials investigations at subzero temperatures have to take into account a different physical behavior. Since the solid surface and the water interact in molecular ranges surface physics plays an important role. For specimens stored below 100% rel.h. the amount and the size of pores dominate the freezable water content. The pore size itself characterizes phase transitions in two ways. This is firstly the condensation of water vapor at a definite relative vapor pressure below saturation for a given temperature. Secondly the freezing point is decreased [3] [4].

For a first approximation of pore size effected precondensed water the Kelvin equation is taken:

$$p/p_s = \exp(-2\gamma V_{mol}/(RT\,r)) \tag{1}$$

with surface tension $\gamma = 7.23 \cdot 10^{-2}$ Nm^{-1}, molar volume $V_{mol} = 18$ cm^3 mol^{-1}, gas constant $R = 8.314$ Jmol^{-1} K^{-1}, absolute Temperature T (K) and radius r (m). The data of table 2 are recalculated in units of °C , nm and relative humidity (%) for convenience.

Table 2:   Relative humidity dependency of the pore radius and the temperature from the Kelvin equation.

ϑ (°C)	-40	-35	-30	-20	0	20
r (nm)			*rel.h.* (%)			
5	76.4	76.9	77.3	78.1	79.5	80.8
10	87.4	87.7	87.9	88.4	89.2	89.9
20	93.5	93.6	93.8	94.0	94.4	94.8
30	95.6	95.7	95.8	96.0	96.3	96.5
40	96.7	96.8	96.8	97.0	97.2	97.4

The following discussion is restricted to the region of 90% relative humidity. Thus, we expect from figure 1 that all samples independent of the gypsum content have water filled mesopores with pore radii smaller or equal to 10 nm. The calorimetric data show indeed a similar freezing behavior of all samples with freezing temperatures between -35°C and -39°C. The occurring difference of the freezing temperatures according to the damping maxima from the dynamic elastic modulus measurement is well known and described by Xu [1] [5]. The damping maximum in this region depends besides the sample size also on the cooling rate. On the other side the additional amount of gypsum effects drastically the melting temperature which increases from -18°C to about 0°C. This is independent of w/c values or time of aging. The same tendency is observed for the damping maxima.

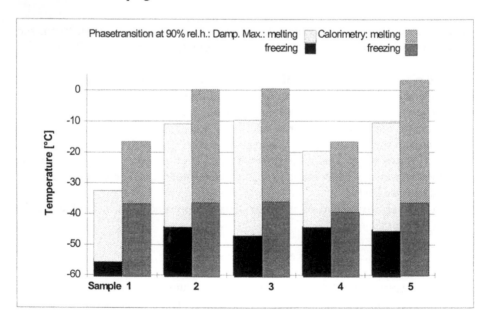

Fig. 2.   Phase transition temperatures from calorimetric data and damping maxima. The samples were stored at 90% rel.h. and the sample numbers denote from table 1. The transition temperatures of freezing and melting are shown.

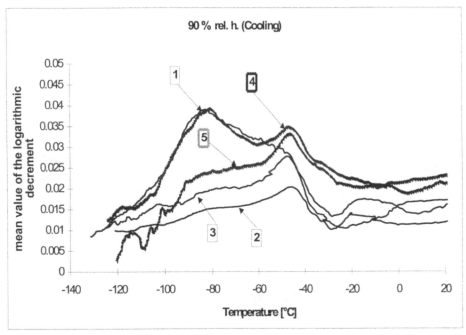

Fig. 3.   Plot of the mean value of the logarithmic decrement (damping) from cooling as a function of the temperature. The samples were stored at 90% rel.h. and the sample numbers denote from table 1.

In figure 3 the mean value of the logarithmic decrement as a measure of damping is plotted versus the temperature. The samples 1 and 4 without gypsum show two distinct damping maxima in the regions of about -45°C and -85°C. The latter value is in good agreement with data of hardened cement paste as described in detail by Xu and Setzer [5] or Zech and Setzer [10] [11]. For the samples 2,3 and 5 with additional gypsum this peak vanishes, whereas the damping maximum at -45°C is lowered but is still kept. The dynamic elastic modulus decreases with increasing amount of gypsum [6].

Regarding the water vapor adsorption isotherms in Fig. 4 the difference of autoclaved concrete with and without gypsum becomes more obvious. The adsorption branch is evaluated with standard BET [7] and with the new ESW (excess surface work) method [8], [9]. The results are listed in table 3.

Table 3. Specific surface areas of sample 1 (0% gypsum) and sample 2 (15% gypsum) from BET and ESW

Sample	Gypsum	BET	ESW
1	0%	103 m^2/g	45 m^2/g
2	15%	45 m^2/g	33 m^2/g

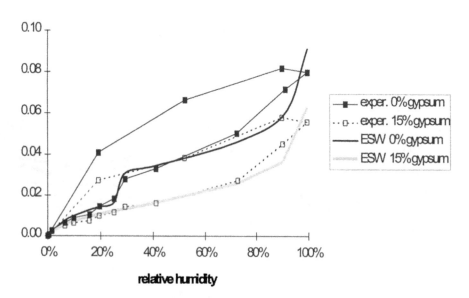

Fig. 4   Water vapor adsorption isotherms (20°C) from samples 1 and 2. The adsorption branch is evaluated with the ESW method.

Obviously the specific surface area of sample 1 calculated with BET is too large. The ESW values are much more realistic and correspond better to the low pressure branch of the isotherm in Fig. 4. Taking the ESW evaluation the specific surface area is lowered about 27% due to added gypsum. But the main difference between the two isotherms is to be found above 30% relative humidity. The step in the isotherm of sample 1 vanishes when gypsum is added (sample 2). Further the water uptake decreases to about half size, even at 90% relative humidity. Indeed the region of micro- and mesopores seems to be more influenced by added gypsum than the pure surface.

4 Summary

Taking into account all results we may resume that adding gypsum to autoclaved concrete leads to
- a shift of the mesopores from radii < 10 nm to radii of about 40 nm,
- increasing amount of mesopores with radii > 10 nm,
- increasing melting temperature,
- decreasing dynamic elastic modulus,
- decreasing water uptake and specific surface area,
- diminishing damping maximum at -85°C.

5 Acknowledgment

We thank the Sumitomo Metal Mining Co., LTD. for financial support and for supplying the specimens.

6 References

1. Xu, X. (1995) Tieftemperaturverhalten der Porenlösungen in hochporösen Werkstoffen, Ph.D. Thesis, University of Essen.
2. Adolphs, J. (1994) Thermodynamische Beschreibung der Adsorption, Ph.D. Thesis, University of Essen, *Forschungsbericht aus dem Fachbereich Bauwesen* 61, ISSN 0947-0921.
3. Stockhausen, N., Dorner, H., Zech, B., and Setzer, M. J. (1979) Untersuchung von Gefriervorgängen in Zementstein mit Hilfe der DTA.. In: *Cement and Concrete Res.* Vol 9. pp. 783-704.
4. Brun, M., Lallemand, A., Quinson, J.-F., and Eyraud, C. (1977) A New Method for the Simultaneous Determination of the Size and the Shape of Pores: The Thermoporometry. In: *Thermochimica Acta* Vol. 21. pp. 59-88.
5. Xu, X., Setzer, M. J.(1997) Damping Maximum of Hardened Cement Paste (hcp) in the region of -90°C - A Mechanical Relaxation Process *Adv. Cem. Based Mater.* Vol. 5, No 3/4. pp. 69-74.
6. Klosterköther, D.(1991) Dynamische E-Modul- und DSC-Messungen an ausgewählten Zementsteinproben, *Research Report*, Fachbereich Bauwesen-University of Essen.
7. Brunauer S., Emmet, P.H., Teller, E. (1938) Adsorption of Gases in Multimolecular layers. *J. Am. Chem. Soc.* Vol. 60. pp.309-319.
8. Adolphs, J., Setzer, M. J. (1996) A Model to Describe Adsorption Isotherms, *J. Colloid Interface Sci.*, Vol. 180. pp. 70-76.
9. Adolphs, J., Setzer, M. J. (1996) Energetic Classification of Adsorption Isotherms, *J. Colloid Interface Sci.*, Vol.184. pp. 443-448.
10. Zech, B., and Setzer, M. J. (1988) The Dynamic Elastic Modulus of Hardened Cement Paste. Part I: A New Statistical Model - Water and Ice Filled Pores. *Mat. Structures* Vol. 21. pp. 323-328.
11. Zech, B., and Setzer, M. J. (1989) The Dynamic Modulus of Hardened Cement Paste. Part 2: Ice Formation, Drying and Pore Size Distribution. *Mat. Structures* Vol. 22. pp. 125-132.

Damping measurements for nondestructive evaluation of concrete beams

E.A. VOKES
RSP-EQE Structural Engineering, Seattle, WA, USA
S.L. CLARKE
Archos Inc., Architectural Engineering and Management,
Olympia, WA, USA
D.J. JANSSEN
Department of Civil Engineering,
University of Washington, Seattle, WA, USA

Abstract
Change in relative dynamic modulus as determined by fundamental frequency measurements, is the most frequently used indicator for evaluating damage to concrete beams that are subjected to repeated cycles of freezing and thawing (ASTM C 666-92). While sinusoidal excitation has historically been the standard method for measuring fundamental frequency, impulse excitation has been approved as an alternate (ASTM C 215-91). An important advantage of the impulse method is that in addition to being rapid and quite reproducible, it also produces information on the damping characteristics of the vibrational modes with no additional testing. The work reported in this paper identifies linear changes in damping with early cycles of freezing and thawing before significant decreases in fundamental frequency can be identified. Comparisons of predicted and actual durability factors showed agreement within published testing errors for all of the mixtures tested. This work indicates that the durability factor (ASTM C 666-92) can be accurately predicted with damping measurements before the actual failure of the concrete beams due to of repeated cycles of freezing and thawing.
Keywords: Concrete, damping, durability factor, freezing and thawing, fundamental frequency, modal analysis, quality factor (Q)

1 Introduction

The frost resistance of concrete is most frequently determined by subjecting concrete beams to repeated cycles of freezing and thawing and periodically measuring the damage in the beams. Because the measurement of damage must be non-destructive, measurement of the concrete's vibration characteristics (modal analysis) has often been employed for damage evaluation. In 1938, Powers [1] presented a method comparing the musical tone of prisms after a hammer impact with a calibrated set of orchestra bells to determine the specimen's dynamic modulus of

Frost Resistance of Concrete, edited by M.J. Setzer and R. Auberg. Published in 1997 by E & FN Spon, 2–6 Boundary Row, London SE1 8HN, UK. ISBN: 0 419 22900 0.

elasticity (proportional to fundamental frequency). In his closure to written discussion of his paper, Powers suggests that monitoring the change in the dynamic modulus of elasticity may indicate deterioration of a sample subjected to repeated cycles of freezing and thawing. The following year Hornibrook [2] reported on the use of electronics to more accurately match frequencies. This report was followed by the discovery of the use of acoustic transducers to excite the specimens and scan for the frequency with maximum amplitude. Prominent early research included Thomson [3] in 1940, Obert and Duvall [4] in 1941, Long and Kurtz [5] in 1943, and Stanton [6] in 1944. In the next twenty years many investigations were performed to determine the results of this dynamic vibration on cement based materials [7-10]. The method most commonly used today to evaluate damage in concrete beams subjected to repeated cycles of freezing and thawing (ASTM C 666-92) still involves the determination of the fundamental frequency of the beam (ASTM C 215-91). This paper reports on work with damping measurements to measure damage in concrete beams and predict failure caused by repeated cycles of freezing and thawing in substantially fewer cycles than required by current procedures.

2 Background

2.1 Modal analysis

Modal analysis has frequently been employed for non-destructive damage evaluation and determination of the dynamic characteristics of a structure in terms of the structure's modes of vibration. A mode of vibration is defined as the deformation pattern of a structure at a natural frequency. Each mode of vibration is identified by a fundamental frequency, damping factor, and mode shape. A mode shape describes the deformed shape of a sample when subjected to a dynamic system. Fundamental frequency is the characteristic frequency at which a maximum response occurs for a given mode of vibration. The damping factor is a measure of the energy dissipated in each cycle of vibration. One method of modal testing uses sinusoidal excitation for the input signal. This method forces a structure to vibrate at a frequency while the response of the structure is monitored. The excitation frequency is varied until a maximum amplitude is observed, which gives the fundamental frequency for a particular mode of vibration. The sinusoidal excitation method is the most commonly used method for examining damage in concrete prisms (ASTM C 215-91) [11].

An alternative method of modal testing has recently been suggested as a possible way of determining the fundamental frequencies of concrete beams by investigators such as Malhotra and Carino [11] and Gaidis and Rosenburg [12]. The method uses impulse excitation; for instance, a hammer strike can be used to excite vibrations in a beam. A load cell on the hammer measures the impact force (input signal), and the response of the beam is recorded with an accelerometer (output signal). A Fast Fourier Transform (FFT) transforms the time domain data (amplitude as a function of time) into the frequency domain (amplitude as a function of frequency). The ratio of the Fourier transform of the output signal to the Fourier transform of the input signal is called the frequency response function, and is a complete mathematical description of the linear vibration characteristics of the prism over the range of frequencies determined by the hardware and software used.

2.2 Fundamental frequency, relative dynamic modulus, and durability factor

As mentioned above, a mode of vibration is identified by a fundamental frequency and a damping factor. While a beam can vibrate at a number of frequencies at the same time, the primary or fundamental frequency is the lowest frequency that has an amplitude substantially greater than both higher and lower frequencies. Figure 1 shows a portion of an idealized frequency response curve with the fundamental frequency labeled. Isolating a portion of the frequency response curve, as in Figure 1, so that only a single response "peak" is included is essentially treating the vibration response as a single degree of freedom system (SDOF). While vibrations outside of the limited range around the fundamental frequency contribute to the overall response of the beam, they are often ignored in modal analysis.

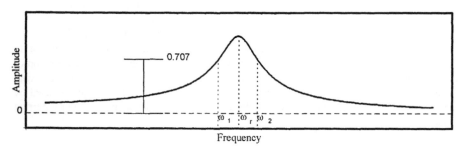

Fig. 1. Idealized frequency response curve.

The dynamic modulus of elasticity for a concrete beam can be calculated from the fundamental frequency by the equation given in ASTM C 215-91:

$$E_D = C * m * \omega_r^2 \tag{1}$$

where E_D is the dynamic modulus of elasticity, Pa,

$\quad C$ is a constant that takes into account the shape, dimensions, and Poisson's ratio of the beam, (details can be found in ASTM C 215-91)

$\quad m$ is the mass of the beam, kg, and

$\quad \omega_r$ is the fundamental frequency of the beam.

Both C and ω depend upon the mode of vibration tested, e.g., longitudinal or transverse. Most tests of concrete beams subjected to freezing and thawing use transverse vibration.

As a concrete beam deteriorates from repeated cycles of freezing and thawing, E_D and, therefore, ω_r decrease (ignoring mass and dimensional effects). The relative dynamic modulus is a measure of this deterioration and is defined by ASTM C 666-92 as follows:

$$P_n = (\omega_{r\,n}^2/\omega_{r0}^2) * 100 \tag{2}$$

where P_n is the relative dynamic modulus after n cycles of freezing and thawing,

$\quad \omega_{r\,n}$ is the fundamental frequency after n cycles of freezing and thawing, and

$\quad \omega_{r0}$ is the fundamental frequency at 0 cycles of freezing and thawing.

The effects of mass and dimension changes are ignored in the above equation but should be considered if there is significant change in mass.

A durability factor can be calculated for a concrete beam at the end of freezing and thawing testing as follows (ASTM C 666-92):

$$DF = P_n * N/M \tag{3}$$

where DF is the durability factor,

$\quad P_n$ is as defined above in Equation 2,

$\quad N$ is the number of cycles at which the testing is terminated (either because a minimum P_n is reached or because the specified number of cycles is reached), and

$\quad M$ is the specified number of cycles.

A typical value for M is 300 cycles and a typical minimum value for P_n is 60 %. ASTM C 666-92 also allows the termination of the freezing and thawing testing if the specimen exceeds a 0.10% length expansion.

2.3 Damping and quality factor

Malhotra and Carino [11] reviewed the damping properties of concrete and research involving damping in concrete. In the field of material science, research by Coppola and Bradt [13] has suggested that viscous damping is more sensitive to thermal damage than elastic modulus. For analysis purposes, a quality factor (Q) is frequently used instead of viscous damping. While viscous damping increases with deterioration, the quality factor decreases and is easier to quantitatively monitor. Q is related to viscous damping by the following:

$$Q = \omega_r / (2 * \sigma) \tag{4}$$

where   Q   is the quality factor,
  ω_r is the fundamental frequency, and
  σ is the damping coefficient.
Q is normally calculated as follows:

$$Q = \omega_r / (\omega_2 - \omega_1) \tag{5}$$

where ω_1, ω_2 are the frequencies on either side of ω_r at which the vibration amplitude of the
beam is 70.7 % of the amplitude of ω_r.
Figure 1 shows ω_r, ω_1, and ω_2.

As a concrete prism is subjected to cycles of freezing and thawing, microcracking occurs in the aggregate, in the matrix and at the aggregate-matrix interface. This microcracking creates damping [14] which causes the free vibrations to decrease in amplitude as a function of time. As microcracking occurs, damping increases, and the quality factor decreases. The damping is a result of pumping action as the cracks alternately open and close during the vibration cycle.

In 1969 Teodoru [15] suggested the use of damping along with fundamental frequency for the evaluation of damage in concrete subjected to freezing and thawing. He used the logarithmic decrement method of determining the damping, and the difficulty of precisely making these measurements may explain why little use has been made of his method. Damping information obtained from the sinusoidal method (as in ASTM C 215-91) has also not been used much. This may be attributed to the lack of precision in making the required measurements of the fundamental frequency and the corresponding baseline and maximum amplitudes necessary for determining the frequencies at 70.7 % of the fundamental frequency amplitude. While automated data acquisition processes could obtain the above information quickly, this process has a potential error associated with each measurement.

The impulse excitation method, however, involves no extra laboratory work to get viscous damping information. Once the modal information of the test has been processed by the FFT, the data are in the form of a series of points giving amplitude and frequency. By looking at a limited frequency range, researchers can treat the data as an SDOF system (no interference from other vibrational modes), and the frequency response function can be written in its real and imaginary parts [16, 17, 20].

In practice, the fundamental frequency and damping of the data from the FFT can be determined with a least squares "circle fitting" method in the Nyquist plane. The imaginary part of the frequency response function is plotted against the real part, and each mode shows up as a circle. The result is convenient for curve fitting to determine parameter values. "Each resonance arc is approximately tangent with, and lies below, the real axis" [18]. Fundamental frequency is determined from the point on the circle that is at a maximum distance from the origin in the Nyquist plane. Damping is related to the diameter of the circle.

A weighting function is used to increase the accuracy of the circle fitting. Each point in the Nyquist plot is multiplied by the square of the distance from the origin. Weighting is necessary for two reasons:

1. The data points nearest to the fundamental frequency are located in the half of the circle that is farthest away from the origin. This portion of the circle corresponds to the portion of the frequency response curve with higher amplitudes (Figure 1), and which is least affected by background noise in the measurements. There are fewer points in this region of the circle, and curve fitting without weighting would allow the circle fit to be more influenced by the data points closest to the origin (which are more influenced by background noise).
2. Though SDOF is assumed, modes that are outside the range of use can influence the examined mode. Data points near the fundamental frequency are affected less by other modes and are therefore weighted more.

Using the circle fitting method for determining Q is more accurate than measuring using the three points determined from a frequency response curve, as in Figure 1 and then applying Equation 5 [10]. All the data for a given mode are appropriately weighted and used in fitting a circle, while Equation 5 uses only three points to determine the fundamental frequency and quality factor. As the frequency response curve is rarely as symmetrical as shown in Figure 1, potential error exists in determining the baseline amplitude. This asymmetry or skewness is taken into account by the circle fit in the Nyquist plot. The Nyquist plot also has the advantage of having fewer points close to fundamental frequency, so that the plot is focused more on the resonance area [19]. Repeated tests of undamaged and damaged (DF≈60) concrete prisms with dimensions of 76x102x406mm tested in transverse vibrational mode showed repeatabilities (d2s limit expressed as a percentage of the mean) of 0.1 and 0.5 % for the fundamental frequencies of the undamaged and damaged prisms, and repeatabilities (d2s limit expressed as a percentage of the mean) of 12.3 and 25.5 % for the Q-values for the undamaged and damaged prisms, respectively [17, 20].

3 Testing Program

A testing program in which 26 marginally air-entrained concrete mixtures were subjected to repeated cycles of freezing and thawing in accordance with ASTM C 666-92 was conducted. The beams were periodically evaluated by an impact method of modal analysis. Details of the testing program are given below.

3.1 Equipment

A modally tuned hammer with a flat frequency response of up to 8 kHz was used to impact the beams. The hammer had a mass of 140 g. and a fundamental frequency of 31 kHz. A load cell in the hammer with a sensitivity of 12 mV/N was used to measure the magnitude of the hammer impact.

Vibrations in the beams were measured with an accelerometer with a sensitivity of 10 mV/g (acceleration). It had a mass of 1.9 g. and a fundamental frequency of 70 kHz. Output from the accelerometer was amplified by a factor of 10 before input into the analyzer.

A Fourier analyzer was used for data acquisition and initial analysis. Sampling rate, bandwidth, and resolution were variable and inter-related. For the testing described in the following section, a bandwidth of 400 Hz was used. This provided a sampling rate of 1,024 Hz and a resolution of slightly less than 0.4 Hz. Inputs were provided for both the load cell from the hammer and the amplified signal from the accelerometer. An exponential window was used in the analyzer, and three consecutive tests were averaged before the circle fitting described in Section 2.3 was performed in order to improve the precision of the testing.

3.2 Concrete Beams

Each of the 26 mixtures studied in this program contained five concrete beams per batch. The beams had dimensions of 76 mm x 102 mm x 406 mm. The mixtures were made with a Type-I cement and had water/cement (w/c) ratios of either 0.40 or 0.45. All of the mixtures contained one of three air-entraining admixtures; based on either neutralized vinsol resin, tall oil, or an organic acid salt. Some of the mixtures also contained either a water-reducer based on a salt of hydrocarboxilic acid, or a naphthalene- or a melamine-based high range water reducer. All but three of the mixtures contained a crushed limestone coarse aggregate with a maximum size of 25 mm. In one of the mixtures without the 25-mm crushed limestone, all of the aggregate larger than 12.5 mm had been sieved out of the fresh concrete mixture before specimen consolidation. The other two mixtures contained a glacial gravel with a maximum size of 22 mm in place of the crushed limestone. All mixtures were cured for one day in their molds at room temperature and then removed from their molds and placed in a saturated lime water bath at 23°C until they reached an age of 28 days.

3.3 Beam Support

Two options were considered for supporting the concrete prisms: fixing one end (the grounded method) or the unrestrained method (free boundary). The chosen method was the unrestrained method, which requires supporting the prism at the bending nodes. This allows the prism to comply with ASTM C 215-91's requirement of allowing the prism "to vibrate freely." The unrestrained method also produces more consistent results than the grounded method because of the difficulty in clamping one end of the beam [19].

The vibrational mode tested for was the first transverse mode, which has two nodes located 22.4 % of the length of the prism from each end. A testing support system was constructed with piano wire to support the prisms at each node, as suggested by Obert and Duvall [4]. The tension of the piano wires was adjusted to prevent the support's fundamental frequency from interfering with the fundamental frequencies of the prisms. Details of the support system used in this study can be found elsewhere [17, 20, 21].

3.4 Test Procedure

All concrete beams were subjected to freeze-thaw damage similar to ASTM C 666-92 Procedure A. The beams underwent six freeze-thaw cycles per day; a freeze-thaw cycle involved cooling the beam so that the center changed from 4°C to -18°C and then warming it back to 4°C, in four hours. At various cycle intervals, the beams were withdrawn from the testing chamber at the end of the thaw cycle and tested for changes in mass, transverse fundamental frequency, and quality factor. For the latter two variables, modal analysis was the testing procedure used. The beams were tested until the relative dynamic modulus (P_n, Equation 2) reached 50 % of its initial value, or until the beam had been subjected to at least 300 freeze-thaw cycles.

4 Analysis

Typical results including the average relative dynamic modulus and average relative Q (described in Equation 6, below) for a concrete mixture are plotted versus number of cycles of freezing and thawing in Figures 2 and 3.

$$\text{Rel } Q_n \quad = \quad 100 * (Q_n/Q_0) \tag{6}$$

where Rel Q_n is the relative quality factor after n cycles of freezing and thawing,
      Q_0 is the quality factor at zero cycles of freezing and thawing, and
      Q_n is the quality factor after n cycles of freezing and thawing.

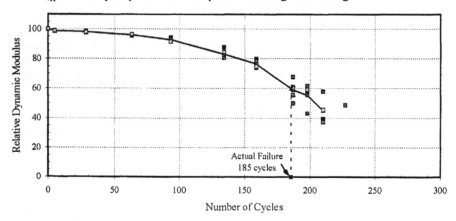

Fig. 2. Changes in relative dynamic modulus.

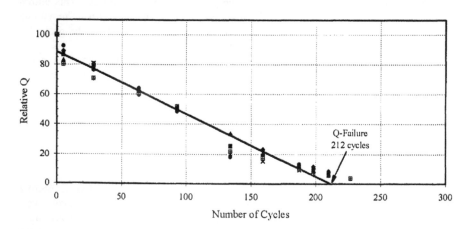

Fig. 3. Changes in relative Q.

In all of the beams, regardless of mixture characteristics or durability factors, Rel Q was found to drop about 15 to 20 % in the first few cycles. The reason for this drop in unclear, but in occurred in all mixtures. This effect is shown in Figure 3. After this drop, Rel Q decreased linearly for a number of cycles of freezing and thawing and eventually leveled out and/or decreased only slightly thereafter (Figure 3). It is important to note that the linear decrease occurs during early freezing and thawing cycles. The magnitudes of the slopes of the decreasing portions of the Rel Q plots appeared to be greater for mixtures that showed greater eventual damage caused by exposure to freezing and thawing.

4.1 Determining Q-Failure

The data was analyzed by least squares fitting the early linear portion of the five beam average Rel Q against the number of cycles data. This best fit line was then extrapolated to get the x-intercept of the Rel Q line. This value will be referred to as the Q-failure value (Figure 3). Q-failure is a first estimate of the failure cycle of the beam. For all but two of the mixtures tested, the best fit line used had a correlation coefficient of greater than 0.90.

4.2 Correlation with failure

Figure 4 shows the average Q-failure and the actual number of cycles necessary to reduce the relative dynamic modulus (RDM) to 60% for the five beams of each mixture tested. The linear regression equation for the best fit is as follows:

$$\text{N-fail} \quad = \quad 0.77 * (\text{Q-failure}) + 8.9 \tag{7}$$

where N-fail is the actual number of cycles of freezing and thawing needed to produce a relative dynamic modulus of 60 %.
The coefficient of determination (r^2) for the above regression was 0.96.

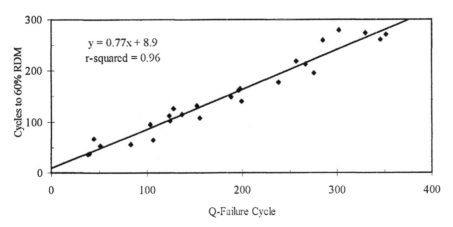

Fig. 4. Q-failure cycle versus cycles to 60 % RDM.

4.3 Predicting durability factor

The good correlation found with Equation 7 suggested that the Q values obtained from modal analysis could be used to predict cycles to a relative dynamic modulus of 60%. This could then be used to predict the durability factor, as in Equation 3. Figure 5 shows the predicted versus measured durability factors for the mixtures tested.

4.4 Expected precision

Laboratory durability factor precision are given in ASTM C 666-92 for the acceptable range between two sets of five beams tested by Procedure A. This acceptable range varies in relation to the durability factor value, with smaller ranges for very high and very low durability factors. In addition to the predicted and actual durability factors, Figure 5 shows the acceptable DF range for two sets of five beams. The predicted values were within the acceptable range of the actual values for all mixtures tested.

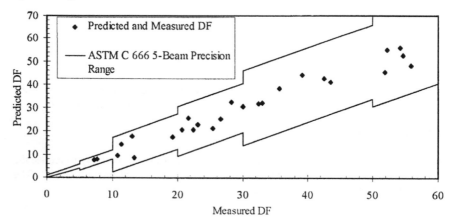

Fig. 5. Predicted and measured durability factors.

5 Summary and conclusions

The change in relative dynamic modulus is often measured to evaluate durability-related changes in concrete. The impulse excitation method of determining relative dynamic modulus is better than the sinusoidal method because it is easier, quicker, and also produces data about changes in damping in the beams.

The change in damping with increasing cycles of freezing and thawing includes a region with a linear decrease in Q for freezing and thawing cycles before the beams reach 60% RDM. If sufficient data are collected during this early portion of the freezing and thawing cycling, a least squares fit can be extrapolated to give estimated cycles to 60% RDM of the beam. This estimation has been experimentally shown to be closely correlated to the actual cycles to 60% RDM of the beam. In addition, the estimated failure cycle can used to predict the durability factor of a beam, and therefore could greatly reduce the amount of testing time needed to assess the extent of freeze-thaw damage in concrete beams.

This relationship between actual and predicted cycle failure is not unique to the chemical admixtures or coarse aggregate in the concrete. Additional testing is required to evaluate mixtures containing less durable aggregate, or other aggregates, both non-durable and durable. Also, mixtures with higher air contents and a wider variety of water/cement ratios should be investigated to determine if the damping results found in this study are consistent with other concrete mixtures.

6 References

1.   Powers, T.C. (1938) Measuring Young's modulus of elasticity by means of sonic vibrations. *Proc. ASTM*, 38, Part II. p. 460.
2.   Hornibrook, F.B. (1939) Application of sonic method to freezing and thawing studies of concrete. *ASTM Bulletin*, No. 101. p. 5.
3.   Thomson, W. T. (1940) Measuring changes in physical properties of concrete by the dynamic method. *Proc. ASTM*, 40. p. 1113.
4.   Obert, L., and Duvall, W.I. (1941) Discussion of dynamic methods of testing concrete with suggestions for standardization. *Proc.. ASTM*, Vol. 41. pp. 1053-1070.

5. Long, B.G., and Kurtz, H.J. (1943) Effect of curing method on the durability of concrete as measured by changes in the dynamic modulus of elasticity. *Proc., ASTM*, Vol. 43. pp. 1051-1065.

6. Stanton, T. E. (1944) Tests comparing the modulus of elasticity of portland cement concrete as determined by the dynamic (sonic) and compression (secant at 1000 psi) methods. *ASTM Bull.* No. 131. p. 17.

7. Axon, E.O., Willis, T.F., and Reagel, F.V. (1943) Effect of air-entrapping portland cement on the resistance to freezing and thawing of concrete containing inferior coarse aggregate. *Proc. ASTM*, Vol 43. pp. 981-1000.

8. Pickett, G. (1945) Equations for computing elastic constants from flexural and torsional resonant frequencies of vibration of prisms and cylinders. *Proc. ASTM*, Vol. 45. p. 846.

9. Batchelder, G.M. and Lewis, D.W. (1953) Comparison of dynamic methods of testing concretes subjected to freezing and thawing. *Proc. ASTM*, Vol. 53. pp. 1053-1068.

10. Kesler, C.E., and Higuchi, Y. (1953) Determination of compressive strength of concrete by using its sonic properties. *Proc. ASTM*, Vol. 53. pp. 1044-1051.

11. Malhotra, V.M. and Sivasundaram, V. (1991) Resonance frequency methods. CRC *Handbook on Nondestructive Testing of Concrete*.

12. Gaidis, J. M. and Rosenburg, M. (1986) New test for determining fundamental frequencies of concrete. *Cement and Concrete Aggregates*, CCAGDP, Vol. 8, No. 2. pp. 117-119.

13. Coppola, J.A., and Bradt, R.C. (1973) Thermal shock damage in SiC. *Journal of the American Ceramic Society*, Vol. 56, No. 4. pp. 214-218.

14. Swamy, N. and Rigby, G. (1971) Dynamic properties of hardened paste, mortar, and concrete. *Materials and Structures/Research and Testing* (Paris), 4(19). p. 13.

15. Teoduro, G. (1969) Assessment of concrete strength under freezing-thawing cycles by non-destructive methods. *RILEM Symposium Concrete Durability*, Prague, Vol. III. pp. 161-173.

16. Richardson, M. (1975) Modal analysis using digital systems. in *Seminar on Understanding Digital Control and Analysis in Vibration Test Systems*, The Shock and Vibration Information Center.

17. Clarke, S.L. (1991) *Improved Method for Non-Destructive Testing of Concrete Prisms*. MS Thesis, Department of Mechanical Engineering, University of Washington.

18. Halvorsen, W.G. and Brown, D.L. (1977) Impulse technique for structural frequency response testing. *Sound and Vibration*, Vol. 11 Number 11. pp. 7-21.

19. Ewins, D.J. (1985) *Modal Testing and Practice*.

20. Vokes, E.A. (1992) *Damping Measurements for Nondestructive Evaluation of Concrete Prisms*. MS Thesis, Department of Civil Engineering, University of Washington.

21. Janssen, D.J. and Snyder, M.B. (1994) *Resistance of Concrete to Freezing and Thawing*. SHRP-C-391, Strategic Highway Research Program, National Research Council, Washington, D.C.

7 Acknowledgments

This paper was prepared from a study sponsored by the Strategic Highway Research Program with some of the equipment provided by the National Science Foundation. The assistance of J.D. Chalupnik and D.W. Storti, Department of Mechanical Engineering and W.D. Scott, Department of Material Science and Engineering, in developing the initial impulse-excitation procedures is greatly appreciated.

Standard methods for freeze-thaw tests: a European research programme

E. SIEBEL and H. GRÄF
Research Center of the German Cement Industry,
Düsseldorf, Germany

Abstract
If structures and their components are exposed directly to the effects of weather they must have an adequate resistance to freezing and thawing with or without the presence of de-icing agents. Knowing the wide variety of basic materials for concrete and the great number of concrete compositions used in Europe, it is necessary to develop a common European standard for testing the freeze-thaw resistance of concrete. Several freeze-thaw test methods exist in several countries. The objective of this project was to test four of these methods and to clarify the parameters influencing the results of the test. The following methods were taken into account.
For testing scaling and internal deterioration:
1. The Slab method
2. the Cube method
3. the CDF and CF method
additionally only for internal deterioration:
4. the Beam test
The research work was carried out in three parts:
    In the first part of the programme the relevant conditions for the test procedures were investigated in greater detail in view of the scaling induced by freeze-thaw attack with or without de-icing agents. This involved the production, preliminary storage and preparation of the test specimens as well as the temperature cycle.
    The second part examined internal structural damage. For this purpose, it was necessary to improve the different freeze-thaw methods and the different measurements for the determination of the internal deterioration.
    The two parts were followed by an intercomparison of the four freeze-thaw tests between laboratories in 14 European countries in the third part.

Frost Resistance of Concrete, edited by M.J. Setzer and R. Auberg. Published in 1997 by E & FN Spon, 2–6 Boundary Row, London SE1 8HN, UK. ISBN: 0 419 22900 0.

Frost failure and rapid test method of concrete frost resistance

A.I. PANCHENKO
Rostov-on-Don State Building University, Rostov-on-Don, Russia

Abstract

Failure of concrete due to freezing occurs under the effect of stretching stresses, caused by the hydraulic pressure in its structure. Microcracks in the volume of the material accumulate with subsequent gradual merging and formation of larger cracks, which join to form through cracks dividing the concrete into separate parts, thus leading to its complete failure. Such a parameter of materials as crack resistance characterized by the value of the stress intensity factor K_c is the most sensitive to cyclic strains, which include intermittent freezing and thawing of concrete.

It is shown that the frost resistance of concrete develops following the fatigue mechanism and can be described with utilization of the basic provisions of fracture mechanics and discussed a rapid test method of concrete frost resistance.

Keywords: frost failure, frost resistance, tests, fracture mechanics, critical crack length, stress intensity factor.

Frost Resistance of Concrete, edited by M.J. Setzer and R. Auberg. Published in 1997 by E & FN Spon, 2–6 Boundary Row, London SE1 8HN, UK. ISBN: 0 419 22900 0.

Introduction

Assuming that the structure of any concrete contains a certain number of microcracks, the frost failure of concrete can be expediently regarded from the viewpoint of fracture mechanics based on analysis of the stressed state, the conditions and the nature of fracture in cracked bodies. Any crack in the concrete structure becomes a stress concentrator. The value of the stress concentration factor depends on the size of the initial cracks, the nature of the internal stress field formed during the hardening of the concrete, and on the intensity of the external impact. The cyclicity of impact leads to the emergence of a special failure mechanism manifested by a slow propagation of the initial crack(s) as the number of cycles grows and the crack reaches its critical length [1]. The crack's length becomes critical when it has reached the threshold, after which the crack's development becomes unstable, i.e., irreversible at a constant level of the existing load. The critical length of cracks is individual for each material and depends on the latter's properties [2]:

$$l_{cr} = \frac{K_c}{\pi\sigma^2} \tag{1}$$

where: l_{cr} - critical length of cracks; K_c - stress intensity factor; σ - internal stress.

As soon as the mean length of cracks reach the value critical for the cement stone, the fracture becomes unstable with the cracks growing at an accelerated rate even without any additional stresses in the structure, which significantly expedites the concrete failure process. Since the structure of both the original concrete and the concrete exposed to any external strain has a large number of chaotically oriented microcracks with different initial lengths, in this case what is meant by the initial and the critical length is the mean length of cracks in the structure.

Discussion

The stresses in the structure (σ_ϕ) emerging as a result of hydraulic pressure P in each capillary were obtained by using the Lame's solutions, applied to strength of materials. A bundle of capillaries was taken as the calculation scheme, which allowed to take account of the mutual interaction of pressures emerging in each of the adjacent capillaries:

$$\sigma_\phi = P \frac{(a-b)^2}{b^2 - a^2},\qquad(2)$$

where *a* and *b* are the geometric characteristics of the capillary bundle.

Resistance of concrete to cyclic atmospheric factors, in particular to freezing and thawing, is expressed, as a rule, by the number of cycles, after which one or more properties change to a certain standard value. Such properties, in accordance with different countries' standards, are reduction in strength, loss of weight, change in the elasticity modulus, residual deformations, etc. Failure of concrete due to freezing occurs under the effect of stretching stresses, caused by the hydraulic pressure in its structure. First, microcracks in the volume of the material accumulate with subsequent gradual merging and formation of larger cracks, which at a certain stage join to form through cracks dividing the concrete into separate parts, thus leading to its complete failure.

Analysis of freezing as a non-stationary temperature process has shown that the structure's ability to filtrate water in these conditions is dependent on the residual porosity Π_i, i.e., the number of capillaries with the calculation diameter d_i less than the diameter of capillaries in which water freezes at the given below-zero temperature. In accordance with Darcy's law and on the assumption that the flow of water filtrating through the capillaries can be determined, from the condition of equality to the ice volume gain per unit of time, which in its turn is proportional to the flow removing the ice melting heat. Such an approach allowed to obtain the equation to determine the positive hydraulic pressure in the form :

$$P = 32 \cdot n \cdot \eta \cdot \frac{\Delta W}{Q} \cdot \frac{dT}{dl} \cdot \frac{\lambda(1 + \Pi_i)}{\Pi_i^2} \cdot \frac{l}{d_i^2},\qquad(3)$$

where ΔW is the water volume gain after its transformation into ice; Q - the ice specific melting heat; dT/dl is the temperature gradient; λ is the thermal conductivity of the material; n - sinuosity factor of the capillaries ($n=1.4...1.5$); η - viscosity of water at the given temperature; l - the mean length of capillaries; Π_i - residual porosity; d_i - calculation diameter.

It is obvious that attainment of the limiting defectiveness of the concrete structure exposed to different types of physical factors, i.e., exhaustion of its resistance (D), occurs due to the growth of the aggregate or the reduced mean length of microcracks after each cycle (Δl_i), i.e., in the generalized form we can write:

$$D = \sum_{i=1}^{N} \Delta l_i , \qquad (4)$$

where N - cycles of freezing and thawing.

To describe the distribution in a body of stresses and displacements caused by the presence of a crack, fracture mechanics uses the stress intensity factor K_c as a quantitative characteristic of the stress fields in the near vicinity of the crack front. As is well-known, one of the most important domains of application of fracture mechanics, in particular the concept of stress intensity factor, is the mechanism of fatigue failure of materials [2], which can be used in its general form also to study the process of concrete failure under cyclic physical action of the environment. In (4) Δl_i is determined in the form of the ratio:

$$\Delta l_i = \frac{l_o - l_c}{l_o} . \qquad (5)$$

where l_o is the mean initial length of cracks in the structure prior to beginning of the action, while l_c is the same parameter after one cycle.

Then, considering that (1) the crack length can be determined by the value of the stress intensity factor, (4) can be written as :

$$D = f\left(\frac{K_c - K_{cc}}{K_c}\right), \qquad (6)$$

where K_c and K_{cc} are stress intensity factors of the concrete before and after the action, respectively.

Thus, the structure defectiveness criterion in the form of an increment to the reduced length of cracks as expressed by (4) contains parameters determined by experimental or calculation and experimental techniques, and can be used to predict or evaluate the frost resistance of concrete.

The summing up of the theoretical and experimental studies performed by the author, and by other researchers, became the basis for the development of the concrete fracture under a cyclic atmospheric action and in particular freezing and thawing [1]. The number of cycles, which characterize the resistance of concrete depends on the defectiveness level (the number of microcracks), the nature of internal stresses in the original concrete structure, the intensity of the external actions, and the capability of the structure to resist the development of cracks. The limit criteria for the concrete resistance is the reaching by the developing cracks of the length critical for the cement matrix, which leads to their unstable growth and an intensive fracture process in the concrete.

In view of the above, assessment of frost resistance can be reduced to the problem of determination of the reduced length of cracks after the first cycle of freezing. Direct determination of the length increment is currently a rather complicated task. The residual deformation of concrete after freezing and thawing can be used as an indirect parameter for assessment of the internal crack formation. However, the magnitude of the residual deformation cannot be taken as an indicator of the length increment in the structural cracks, because the change in the concrete volume in this case is also due to the plastic and creep flow of concrete. Besides, similar increments in the volume (or linear dimensions) may be caused in one case by a small number of bigger cracks, and in another case by a larger number of small cracks. Naturally, in the first case the reduction in the strength characteristics of concrete will be more apparent.

The development of existing or formation of new cracks will affect the concrete properties, above all, its crack resistance. After a single freezing down to minus $50^{\circ}C$ during comparative trials, the compression strength and the modulus of elasticity changed insignificantly, and in concretes with F300 and higher no such changes were observed at all. In certain cases, in concretes with cement consumptions more than 400 kg/m^3, there was even an improvement (by 3-6%) of the compression strength. As regards crack resistance, the outcome was quite different.

The criterion for fatigue failure of metals, plastics and certain composites is the change of the value of the stress intensity factor in each of the loading cycles. This presupposes that such a parameter of materials as crack resistance characterized by the value of the stress intensity factor K_c is the most sensitive to cyclic strains, which include intermittent freezing and thawing of concrete.

Results

Special experimental studies were conducted to assess the interrelations between the frost resistance of concretes and the modification of the value of the stress intensity factor K_c after a single freezing. Twelve different types of concrete with compression strength from 20 to 55 MPa and cement consumption 230-500 kg/m^3 were tested. The binding agents used were Portland cement (concretes 1-7 in Table 1), pozzolanic Portland cement (concretes 8 and 9), slag Portland cement (concretes 11 and 12) with an air-entraining admixture (concretes 3 and 7), and with an expanding admixture (concrete 4). The water to cement ratio varied from 0.72 to 0.38. The main properties of the concrete are shown in Table 1.

Table 1. Properties of the concretes and reduction in the value of the stress intensity factor

Concretes	R_c, MPa	R_t, MPa	F, cycles	K_c, MPa·m$^{1/2}$	K_{cc}, MPa·m$^{1/2}$
1	55.6	4.82	300	0.826	0.785
2	44.2	3.31	150	0.673	0.60
3	31.6	3.03	400	0.622	0.597
4	43.8	4.28	500	0.683	0.674
5	38.6	3.36	200	o.671	0.622
6	21.2	2.30	75	0.332	0.231
7	27.8	2.63	400	0.520	0.506
8	40.2	3.61	150	0.654	0.602
9	29.1	3.18	75	0.50	0.325
10	28.0	2.82	75	0.386	0.247
11	36.4	3.78	100	0.652	0.487
12	32.2	3.44	100	0.562	0.431

Note: R_c - compressive strength; R_t - tensile strength; F - frost resistance according to Russian standard; K_c and K_{cc} - stress intensity factors before and after freezing, respectively.

The tests were performed as follows. A series of six bars measuring 10 x 10 x 40 cm were divided, after saturation with water, into two parts. The first part was tested for crack resistance before freezing, and the K_c value was determined. The second part was tested after a single freezing to minus 50°C and thawing to determine the K_{cc} value. Testing for crack resistance both before and after freezing should be made on water-saturated samples, which provides for manifestation of the Rebinder effect known in physico-chemical mechanics. In this case the reduction in the value of the stress intensity factor (ΔK_c) is the effect of the internal stresses of the phase transition, which develop after a single freezing.

One specific aspect in the technique for preparation of the samples should be noted. The initial notches in the samples for crack resistance tests can be made either with a cutting tool before the test, or during the molding by means of putting of a special metal plate into the mold. Based on the preliminary results of comparative tests it was found that the second method was preferable, because in that case the internal variation coefficient did not exceed 5%, while when the initial notch was made by means of a diamond disk, the above coefficient was as high as 8-9%, and the K_c value in twin samples was reduced by 7-10%. The probable cause for that is the additional uncontrolled formation of microcracks in the process of cutting of the concrete.

The results of testing made it possible to establish the dependence of frost resistance on ΔK_c, shown in Fig.1. Mathematical processing of the experimental data allowed to obtain the mathematical expression describing this dependence:

$$F = 624.0 \cdot (\Delta K_c)^{-0.61} \qquad (7)$$

with the correlation ratio $R_s = 0{,}93$.

The obtained dependence (7) coincides with the formula of P.Paris: $dl/dN = C \cdot (\Delta K_c)^n$, which is fundamental in fracture mechanics for evaluation of fatigue longevity of materials (mostly metals). The closeness of the formula of P.Paris to the obtained dependence (7) is another confirmation of the assumption that the frost resistance of concrete develops following the fatigue mechanism and can be described with utilization of the basic provisions of fracture mechanics.

The proposed method for accelerated determination of frost resistance with the use of the dependence (7) is fast, since just one freezing cycle is enough, and not labor-intensive, because it requires only six prism samples to be double-tested for crack resistance before and after freezing.

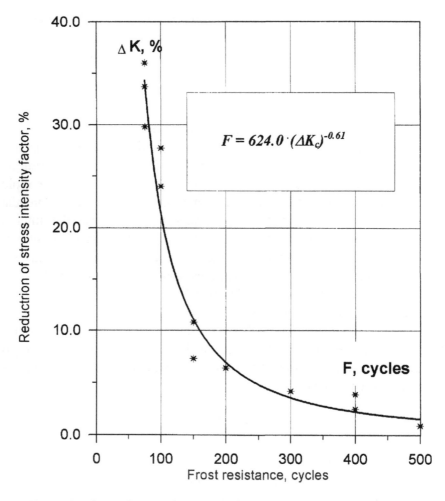

Fig.1 The dependence of concrete frost resistance on modification of the value of the stress intensity factor

References

1. Panchenko A.I. Some Aspects of the Concrete Frost Resistance. Proc. of Fouth CANMET/ACI Int. Conf. On Durability of Concrete. Sydney, Australia, August 17-22, 1997 (accepted to publish).
2. Zajtsev Yu.V. Modelling of Concrete Deformations and Strength by Fracture Mechanics Method. Moscow, 1982, 196 pp. (in Russian).

Concrete frost resistance test methods

N.K. ROZENTAHL
Research Institute of Concrete and Reinforced Concrete,
Moscow, Russia

Abstract

Frost resistance of concrete is an important problem for a large part of Russia with its severe climate. There are many cases of early failure of reinforced concrete structures under the action of water and deicing salts in cold weather.

The country adopted a system of codes and standards that specify requirements for concrete in terms of frost resistance. The system includes a parametric series of frost resistance grades from F10 to F1000, design specifications for reinforced concrete structures taking into account frost resistance requirements according to service conditions, requirements for frost resistance of concrete in standards for specific types of reinforced concrete structures, and methods used to determine the frost resistance grade of a particular concrete.

The development of methods for testing the frost resistance of concrete is an aspect of this problem. Many test methods based on different principles were developed. The majority of Russian experts believe the methods that involve repeated freezing and thawing of concrete specimens to be the most reliable.

Keywords: concrete, deicing salt, design specifications, frost resistance, test methods.

1 Introduction

The problem of frost resistance of concrete arose in Russia when large hydraulic structures were built. At the turn of the century, examinations of port facilities, such as moorings and breakwaters, found that some concrete components quickly deteriorated under the action of sea water, especially when concrete saturated with sea water was frozen. The first commercial hydraulic power plant was built in Russia in 1895 and 1896 under the leadership of the engineers V.N. Chikolev and R.E. Klassen. Major hy-

Frost Resistance of Concrete, edited by M.J. Setzer and R. Auberg. Published in 1997
by E & FN Spon, 2–6 Boundary Row, London SE1 8HN, UK. ISBN: 0 419 22900 0.

dropower projects, including Volkhov, Dnepr, and Lower Svir plants among others, and the Moscow-Volga Canal and the White Sea-Baltic Canal were completed before World War II. After the War a complex of hydroplants on the Volga, the Don and the Dnepr and large power plants in Siberia, including Irkutsk, Bratsk, Krasnoyarsk, Sayano-Shushenskoe, Zeya and dozens of others, were built. The construction of large unique facilities designed for operation for hundreds of years in regions with severe climatic conditions needed research into the resistance and durability of basic building materials, concrete in particular. Methods of testing concrete for frost resistance were called for. N.A. Belelyubsky was the first to suggest a method for determining frost resistance of concrete in Russia. The tests consisted in freezing water-saturated concrete in air and thawing it in water. The method was recognized to be basic at the International Conference on Testing Materials in 1886. The processes of frost destruction of concrete when acted upon by water and frost were originally studied. As the marine and road construction advanced and industrial salt processing facilities were built, it became necessary to develop methods for evaluating the resistance of concrete to solutions of salts and below-zero temperatures.

2 The system of codes and standards

The design and construction of reinforced concrete facilities affected by water and below-zero temperatures required a system of regulations reflecting specifications for frost resistance of concrete. This system has been developed for decades and now includes the following basic documents that regulate climatic effects and frost resistance of concrete:

- SNiP (Building Code) 2.01.01-82. Building climatology and geophysics.
- SNiP 2.01.07-85. Loads and actions.
- SNiP 2.03.01-84. Design specifications. Plain and reinforced concrete structures.
- SNiP 2.06.08-87. Plain and reinforced concrete structures of hydraulic facilities.
- SNiP 2.06.02-85. Motor roads.
- SNiP 2.05.03-84. Bridges and pipes.
- SNiP 2.04.02-84. Water supply. External networks and facilities.
- SNiP 2.11.02-87 Refrigerators.
- SNiP 2.05.08-85. Airfields.
- GOST 19804.0-78. Driven reinforced concrete piles.
- GOST 17608-91. Concrete pavement slabs.
- GOST 6665-91. Plain and reinforced concrete curb stones
- GOST 25912.0-91. Reinforced concrete slabs for airfield pavements.
- GOST 20425-75. Tetrapods for coast-protecting and other protecting structures.
- GOST 9561-76. Reinforced concrete panels for floors of buildings and structures.
- GOST 10060-96. Concretes. Methods for determining frost resistance.
- GOST 26134-84. Concretes. An ultrasonic method for determining frost resistance.

   Under this system, when reinforced concrete structures are designed, the service conditions of the facility are determined, including the design ambient air temperature in winter and conditions of wetting its reinforced concrete components, and then a frost resistance grade of the structural concrete is assigned according to the importance of the facility.

Appropriate chapters of the SNiPs contain more accurate data for particular types of facilities, including hydraulic structures, bridges, pipes, roads, refrigerators, and water supply facilities. Moreover, they specify requirements for frost resistance, including test methods, for some types of structures, such as piles, road and pavement slabs, and balcony slabs. It is pointed out, for example, that road-building structures should be tested for frost resistance in salt solutions.

3 Hypotheses of frost resistance

The development of tests for frost resistance of concrete is based on hypotheses for the mechanism of frost destruction of concrete. The well-known hypothesis of Powers that concrete is destructed by hydraulic pressure of water in freezing does not cover all known factors that affect the change of physico-mechanical properties of concrete under freezing and thawing. The pressure of ice crystals thus formed; nonuniform temperature strains in the concrete body when there is a temperature gradient; the differences between coefficients of thermal expansion of concrete components, including hydraulic binder, aggregate, and water; changes in the phase composition of crystal hydrates of hardened cement paste caused by the variation of temperature; chemical reactions with components of salt solutions that saturate the concrete, the porous structure of concrete (the presence of pores as elements of the structure which slow down microcracking), and other factors also play a part. Professor S.V. Shestoperov [1] who conducted extensive research on frost resistance of concrete listed 325 quality indices of initial materials, of the composition, and of the techniques, which affect frost resistance of concrete. He specified as the most important the quantity of water and the density of concrete, the presence of plasticizers and air-entraining admixtures, the quantity and type of pozzolanic additives in cement, and the quantity of aluminate phases in cement clinker. It was shown that high strength does not guarantee high frost resistance. The fact that air-entraining admixtures and cements of a certain quality significantly increase frost resistance favours the hydraulic pressure hypothesis and the hypothesis of the importance of the phase composition of hardened cement paste in concrete saturated with sweet water. When concrete is saturated with salt solutions and freezes, the mechanism of its deterioration changes significantly.

The difference between the mechanisms whereby internal stresses arise in freezing concrete saturated with fresh water and that saturated with salt solutions is as follows: When water-saturated concrete cools down to 0°C, the water in large pores turns into ice first and then water in micropores does as the concrete cools further. The freezing region gradually moves from the surface layer deeper into the concrete and the water-to-ice transition region moves down accordingly. A frozen outer layer impermeable to water is formed and the unfrozen water is expelled inward. The freezing of concrete saturated with a salt solution produces two freezing regions, including the internal region where the temperature is higher than that on the surface of the concrete and where water from the solution turns into ice while a concentrated salt solution remains in the liquid phase and the external region where the temperature reaches the eutectic point and the saturated salt solution passes into the solid phase. An internal layer and an external layer of low permeability are formed. The result is that the expulsion of water freezing between these layers is greatly hindered and high pressure arises accelerating the failure of the concrete. In addition, the freezing of concrete saturated with salt solutions creates conditions for an additional formation of crystal hydrates and for chemical transformations of hardened cement paste. The rate of concrete deterioration was found to be the greatest when the concentration of salt solutions was 5%.

4 The status of developing frost resistance test methods in the Russian Federation

Frost resistance of concrete is understood as its ability to maintain physico-mechanical properties when exposed to freezing and thawing. It is defined by frost resistance grades. Numerically, the frost resistance grade of concrete is equal to the maximum number of freezing and thawing cycles, under which the concrete maintains its initial strength. The test conditions are regulated. The dimensions of specimens, the conditions of saturation with water or a salt solution, and the temperature of freezing and thawing are set. These parameters have not always been the same. For example, in the past, the strength of concrete after cyclical freezing and thawing was compared with that of check specimens kept in humid conditions during the entire testing period. Testing of concrete of high frost resistance could continue for 1 year or more while the strength of the check specimens generally increased. By comparing the strength of test specimens and that of check specimens one could find how the frost action reduced the strength of concrete. But the ratio of this strength to the strength of concrete at the design age or to the design strength of the concrete was not indicated. In hydraulic construction, specimens were saturated with water of the same composition which was expected in the area where reinforced concrete structures were to be located.

Thus, an attempt was made to bring test conditions closer to the service environment. It allowed in effect each tester to conduct testing in his own way and quite often made it impossible to compare the results of tests conducted by different laboratories.

In addition to the need to make the test conditions similar to the service conditions, another difficulty is the practical necessity to accelerate tests as much as possible and to use them for on-line control of frost resistance of concrete, which would allow to adjust the composition of the concrete and the placing procedures during the fabrication of structures and during construction. It is especially important where the quality of cement and aggregate is inconsistent and the conditions of transportation, placing, and hardening of concrete vary.

There is an objective contradiction between the requirement to accelerate the tests and the need to take into account the slow processes of concrete destruction. The tests can be accelerated if they are conducted in more severe conditions, such as lower temperature, higher degree of water saturation of concrete, and use of salt solutions. When we make test conditions more severe, we must establish to what extent we accelerate the process of deterioration of concrete, i.e. we must introduce correction coefficients. We need a standard test method under constant specific conditions to serve as a reference for evaluation of accelerated test results. Testing concrete saturated with water at a moderately low below-zero temperature remains to be this method.

Many attempts are known to have been made to estimate frost resistance of concrete from small changes in the characteristics of the concrete during the first freezing and thawing cycle. Changes in the properties were assumed to be demonstrated by strains, variations in strength and in dynamic modulus of elasticity, by heat evolution data, and by acoustic characteristics during freezing. Considering the well-known connection of frost destruction with an increase in the volume of water as it turns into ice, some researchers suggest that the resistance of concrete be evaluated according to the quantity of water frozen in the concrete and the volume of the pores that are not filled with water. There is a technological approach to the problem, which consists in strict regulation of properties of initial materials and of the composition and mixing and placing conditions to obtain concrete of a specified frost resistance. Such regulation is useful but it cannot completely guarantee that concrete of a specific frost resistance can be produced in all cases in view of inevitable production errors. This method does not eliminate the necessity to test concrete for frost resistance.

5 Concrete test methods

GOST 10060, a Russian standard, stipulated tests of concrete using both the basic and accelerated methods. The basic method for ordinary concretes is saturation of concrete specimens with water and freezing in the air at -18°C. They thaw in water at 18°C. The specimens are cubes 100, 150, and 200 mm on edge. Specimens are tested at the designed age after they attained the strength that corresponds to their strength class. It was established that specimens passed a test if the average value of compressive strength did not drop lower than by 5% as compared with the strength of check specimens saturated with water and tested for compressive strength. The criterion was established under the general condition that freezing and thawing of concrete during a specified number of cycles should not reduce the strength of the concrete as compared with the original strength.

The basic test method for concrete of components to be used in construction of roads and airfields is the one where specimens are saturated with 5% solution of NaCl, freeze in the air at -18°C and thaw in the same solution. An additional requirement in the evaluation of test data is a limited reduction of the weight of the specimens, which should not exceed 3%. This method can be used as an accelerated technique for ordinary concretes. The tests of ordinary concrete thus run 2 to 6 times faster depending on the frost resistance grade of concrete.

The third method involves accelerated tests of ordinary and road concrete. With this method, specimens are saturated with 5% NaCl solution and freeze at -50°C in the same solution. And they thaw in the same solution. The evaluation criteria are the same as in the second method. The tests are accelerated 30 to 37 times for ordinary concrete depending on its frost resistance grade. The method accelerates testing 5 to 20 times for roadway concrete. The acceleration is faster when concretes of lower frost resistance are tested.

An accelerated frost resistance test method has been proposed recently on the basis of measuring dilatational strains in specimens as they freeze. The frost resistance grade of concrete is found from a table or chart that reflects the relationship of the grade versus the peak of dilatational strains at the moment when water freezes. The tests are conducted by means of a dilatometer. Still another method determines the frost resistance grade of concrete according to its capillary porosity and an increase in the compressive strength of the concrete in frozen condition. The method is based on the assumption that a large increase in strength of frozen concrete is evidence of a large volume of ice formed and thus shows a low frost resistance of the concrete. So, the above methods determine the state of concrete in the first freezing cycle and do not evaluate the processes that develop in the concrete later on. We believe that elements of these techniques can be used in research but have a limited value as methods for technical tests of specific concretes.

A method of fast freezing and thawing of specimens immersed with one edge in a solution of sodium chloride was developed in Russia to determine the resistance of concrete to scaling in road pavements.

Field tests of concretes for frost resistance in specific climatic conditions are widely used. S.V. Shestoperov and F.M. Ivanov [1] investigated concrete specimens of various compositions in the tailrace canal of the Shirokov power plant in the Urals. The plant operated under peak load conditions evacuating water several times a day. The specimens were subjected in winter to 3 or 4 freezing and thawing cycles per day. This method was used for comparison tests of the impact of different cements, aggregates and admixtures on frost resistance of concrete. Significant tests were conducted on marine stands in the Kola Bay and in the Kislaya Gulf of the Barents Sea and in the Sea of Japan [2]. The specimens were placed in the tidal zone and they froze and thawed two times a day in winter. The stand in the Kola Bay was used to test concretes of various compositions, including those with plasticizers and air-entraining admixtures, and inhibitors of corrosion of steel in concrete. Concrete prisms and reinforced concrete beams were investigated. Studies conducted by V.M. Moskvin and A.M. Podvalny [3] demonstrated the variation of the frost destruction rate of concrete according to the type and magnitude of stresses that were created by an outside load.

The test station of TsNIIS (Central Research Institute of Construction) in the Kola Bay studied the frost resistance of concrete specimens and of structural components in special pools located on the sea coast [4-6]. Sea water was pumped into the pools which were drained when gates were opened after the specimens thawed. Cold outside air was blown in by fans to accelerate the freezing of the specimens. The temperature in the concrete was controlled remotely with the help of thermocouples inserted in the specimens.

Concrete specimens have been tested on the stand in the Kislaya Gulf tidal power plant for more than 25 years now, including specimens of industrial concrete made when the plant was built. The tests have shown in particular that concrete in massive structures is less subject to damage than concrete in specimens of standard dimensions. The reason is largely the difference between stress states of concrete and temperature fields in massive structures and in specimens. Concretes with microsilica, superplasticizers and air-entraining admixtures are being tested.

Field tests of concrete for frost resistance are limited to particular climatic conditions of the region where the tests are carried out. The results of these tests cannot be transferred to conditions of another climatic region but they are quite valuable for the given construction site. In addition, the method allows to conduct comparison tests of concretes of different compositions and mixed in different ways.

Valuable information can be gleaned from the examination of the state of concrete in marine and river facilities. By comparing the actual condition of structures with design data and with data on how the construction progressed we can check with high reliability the resistance of concrete compositions thus used under specific service conditions and to reveal the effects of concomitant factors such as microclimatic characteristics and structural features on the frost resistance of the concrete.

Aktuganov and Ivanov [7, 8] developed a method of assigning a frost resistance grade to concrete according to the characteristics of a structure, conditions of wetting the concrete, and climate in the region of the construction project.

6 Conclusion

A system of normative documents was created in Russia to specify requirements for frost resistance of concrete at the structural design stage. Specifications for frost resistance of structural concrete in various types of facilities were established taking into account the peculiarities of climate in the region of construction. Methods of testing concrete for frost resistance were developed. However, the problem of concrete frost resistance tests remains. New fast test methods are required for the on-line control of frost resistance at the construction site.

7 References

1. Shestoperov, S.V. (1966) *Durability of concrete in structures for transportation*, Transport, Moscow.
2. Sviridov, V.N. (1979) Estimation of aggressiveness of weathering factors on Sakhalin Island. *Proceedings: VNII Transportnogo Stroitel'stva*, Moscow.
3. Moskvin V.M., Podvalny A.M. (1960) Frost resistance of concrete in a stress state. *Beton i Zhelezobeton*, No. 2.
4. Baklanov, A.S. (1966) Accelerated field tests of concrete for frost resistance. *Transportnoe Stroitel'stvo*, No. 9.
5. Ivanov, F.M., Baklanov, A.S. (1967) *Experience in construction of reinforced concrete sea structures in severe climate*, Orgtransstroy, Moscow.
6. Gladkov, V.S., Ivanov, F.M. (1969) Estimation of climate harshness when assigning concrete frost resistance. *Proceedings: VNII Transportnogo Stroitel'stva*, Vol. 70, Moscow.
7. Aktuganov, V.Z. (1983) Methods of evaluation of climate temperature and humidity effects on durability of concrete. *Izvestia VUZov, Stroitel'stvo i Architecture*, No. 4.
8. Ivanov, F.M. (1995) Testing methods and frost resistance standardization of concrete, in *Concrete under severe conditions*, E & FN Spon, London, pp. 107-13.

Temperature shock test for the determination of the freeze-thaw resistance of concrete

M. MAULTZSCH and K. GÜNTHER
Department of "Safety of Structures",
Federal Institute for Materials Research and Testing (BAM),
Berlin, Germany

Abstract
A large number of freeze-thaw-tests is known. Many of them claim to simulate natural conditions of freezing and thawing. The temperature shock test which is now established in several guidelines and draft European standards for concrete repairs is far away from real temperature changes in winter. But its results seem to be similar to those of well-known long-term tests. It uses a saturated sodium chloride solution at -15 °C and a water bath at +20 °C. The samples are alternately submerged in both liquids for 2 hours each with very quick changes. This simple principle needs special techniques. An automatically working equipment is described. Investigations and results are presented. The comparison with results achieved with the cube method according to CEN TC 51 drafts indicate sufficient reliability of the shock method for judging the freeze-thaw resistance of concrete with a considerable acceleration of the test.
Keywords: Freeze-thaw resistance, test method, accelerated test

1 Introduction

Numerous methods for the assessment of the freeze-thaw resistance of concrete with or without impact of deicing agents are well-known. The tests consist mostly of temperature cycles within the range of +20 °C and -15...20 °C, each lasting 24 hours with respect to usual working hours. It is often claimed that these cycles simulate natural conditions. But long term measurements in structures have demonstrated that the real temperature range is much more limited and the rate of cooling or warming is much lower on site than it is in the tests [1]. So might be concluded that not the impact but the effect of freezing and thawing in the presence of salt or salt solutions is simulated. On that condition, even a temperature shock impact can be applied if the detrimental effect on concrete is similar to the one observed on site. One of these methods originally based on a RILEM proposal for testing the

Frost Resistance of Concrete, edited by M.J. Setzer and R. Auberg. Published in 1997 by E & FN Spon, 2–6 Boundary Row, London SE1 8HN, UK. ISBN: 0 419 22900 0.

thermal compatibility of polymer bearing products with concrete substrates will be described, with its principles, the technique and previous experiences. Considering the test results, the advantages and disadvantages will be discussed with regard to the suitability for testing conventional concrete.

2 Principle of the test method and requirements

The original RILEM proposal [2] provided the alternating submersion of the specimens in water of +20 °C and salt-solution of -18 °C for one hour each, i.e. one cycle lasts 2 hours. Establishing the first German guidelines for testing concrete repair products [3, 4] and gathering experiences in respective test and research work [5, 6], the method was modified. The lower temperature was fixed to -15 °C, and each period was extended to 2 hours to guarantee sufficient periods of freezing and thawing even in the center of concrete specimens. The requested course of temperatures follows to a rectangular curve. The mentioned guidelines [4] completed the prescription by a range of temperature tolerance: the deviation of the bath temperature from the target line shall not exceed 3 K. The changes from one bath to the other one shall be made as quick as possible. Meanwhile the method has been incorporated in the CEN test programme for concrete repairs as a draft standard [7]. The temperature tolerance has been reduced to $\pm$ 2 K.

According to the guidelines and draft standards, the assessment criteria are the cracking, the scaling and above all the adherence of the tested material to concrete substrates.

3 Development of special equipment

As the duration of one cycle is 4 hours with the necessity to handle the specimens each 2 hours, only 1 - 2 cycles per working day are possible with interruption for at least 16 hours overnight and more than 60 hours on weekends. The BRITE/EURAM research project [6] made it possible for the BAM to develop improved test facilities and to reduce the manpower input decisively, resulting in an enormous acceleration of the test procedure. Furthermore the improvement aimed at the reliable maintenance of the required temperatures in the liquids even when large specimens with high heat capacity are submerged.

The design for the test facilities is shown in Fig. 1. The idea is that the specimens, positioned on a grid, can alternately be dipped into the water or salt solution by the help of an automatically controlled motor-driven lever arm. On the left side the draft shows the temperature-controlled water tank with 600 dm^3 capacity and the samples on the stainless steel grid, mounted at the end of the lever arm. The motor and gearing is provided for a maximum load by specimens of 200 kg. Therefore a respective counter-balance is necessary. On the other side the insulated container with the cooled salt solution is positioned. It has the same capacity as the water tank. Both the water and the saturated sodium chloride solution are circulated with external pumps. Hereby the salt solution passes a heat exchanger to be cooled to - 15 °C. The organic cooling liquid - about 100 dm^3 - needs only to be cooled to about -17 °C outflow temperature in a 12 kW cooling machine to guarantee the requested constant temperature in the open container.

It proved to be absolutely necessary to keep the sodium chloride concentration close to the eutectic composition, i.e. about 23,3 percent by mass NaCl in the solution instead of 26,7 percent which is the content of the saturated solution at room temperature. Otherwise the risk occurs that ice or solid $NaCl.2H_2O$ is precipitated within the pipes of the heat exchanger. Therefore the condensation of moisture on

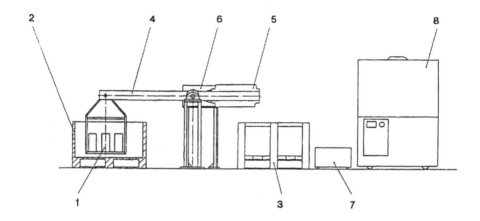

Fig. 1.   Temperature shock test facilities developed in the BAM. 1: specimens; 2: water tank +20 °C; 3: salt solution tank -15 °C;   4: lever-arm; 5: counterbalance; 6: motor and gearing; 7: heat exchanger; 8: cooling machine.

the surface of the cooled solution had to be prevented. The problem was solved by small plastic balls swimming on the surfaces of the liquids and covering them, but not hindering the specimens to dip in. A high corrosion resistance must be provided for the pumps and valves.

The complete test facilities needing an area of more than 2 x 6 m was constructed and mounted in the BAM. Also the electronic control unit was designed and made in the institute. It allows to steer the movement of the lever arm very exactly for each position with programmable periods. In the reported cases the cycle was fixed to the requested immersion periods of exactly 2 hours in each tank. The time needed to change from one tank to the other is about 20 seconds. By this the cycling can be carried out without any interruption and the total test time for e.g. 100 cycles is reduced to only 17 days.

4 Test results

The method and new facilities have been developed and proved at first for composites of concrete substrates and repairs. Fig. 2 demonstrates the course of temperatures in two usual cycles when the machine was loaded with 6 concrete/repair mortar composites with the dimensions 120 x 300 x 300 mm^3, i.e. a mass of the specimens of about 150 kg. The temperature was measured in the water or salt solution near to the specimens and within one specimen in 0, 1, 2, and 7 cm depth. The example makes obvious that in any moment the required temperature is kept within the tolerance range. The cooling effect when the frozen specimens are submerged in the tempered water does not exceed 1.5 K, and vice versa the warming effect in the moment of submersion in the cooled salt solution. Each part of the specimens passes the 0 °C line within a period of less than 30 minutes. After 1 hour the temperature in the specimens exceeds 17 °C in the warming period and falls below 11 °C in the freezing period, and at the end of each period the specimens have totally the same temperature as the ambient liquids. So the technical preconditions are given.

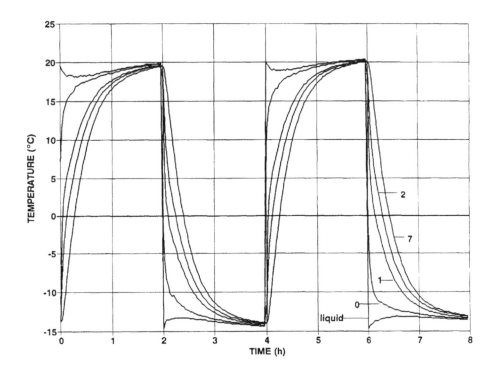

Fig. 2.   Course of temperatures during two cycles, measured in the liquids and in large concrete slabs in different distances (in cm) to the surface.

The method has also been applied to plain concrete specimens. For this purpose 100 mm cubes were used. These are the same as are requested for the method which is recommended by the DAfStb [8] now introduced in CEN TC 51 drafts [9]. The assessment criteria are in both cases besides cracking the loss of mass by scaling. Fig. 3 shows a typical result of mass loss achieved with the DAfStb method using a 3 % NaCl solution. Fig. 4 gives the results for specimens which are made of the same concrete batch but put to the temperature shock test. The scaling effect appears - assessed as well visually as by weight loss curves - very similar. There is a higher number of cycles necessary for a comparable degree of deterioration. But due to the shorter term of each cycle, the total time is much lower.

The shock method and the DAfStb method have been applied in comparison on different concrete samples including a highly resistant concrete with air-entrainment. Up to now, in all cases the scaling process occurred in a very similar way. The ranking of the materials proved to be the same, if a certain criterion is chosen. Fig. 5 demonstrates the time effort to reach the assessment criterion "95 mass-%" (related to the initial mass) in both methods for same materials. The needed time is

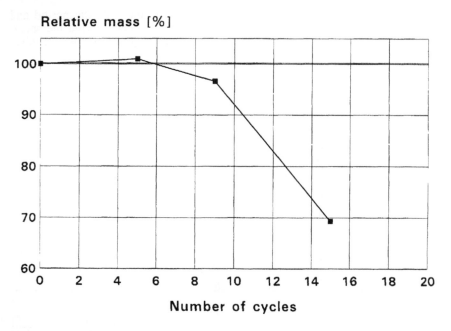

Fig. 3.   Average mass loss of 4 concrete cubes tested according to the DAfStb-Recommendation

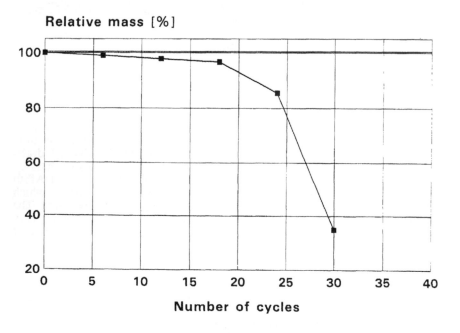

Fig. 4.   Average mass loss of 4 concrete cubes submitted to the shock test.

calculated only with the plain cycling time, i.e. automatized devices are provided in both cases. The dashed line marks the equivalence values. It is evident that the shock method means a considerable acceleration of the test. The comparability seems to be given, though further provements by more test series remain desirable.

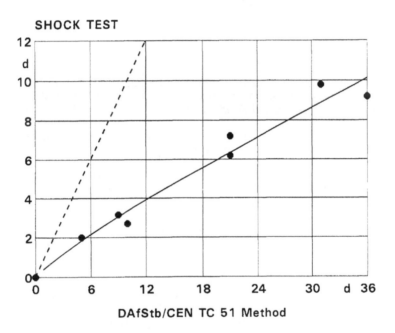

SHOCK TEST

DAfStb/CEN TC 51 Method

Fig. 5.   Comparison of the time needed for the same criterion in the shock test and in the DAfStb test

5 Conclusions

From the academic point of view, the physical and chemical affects of slow temperature changes and salt solutions with low concentration should be quite different from those caused by temperature shock and nearly saturated sodium chloride solution. The reported tests and results, however, demonstrate that the detrimental effect is comparable. So it might be concluded that the shock test with 4-hours-cycles is as well suitable as conventional tests with 24-hours-cycles to assess the resistance of concrete against freezing and thawing in presence of deicing agents. The limited effect of the saturated sodium chloride solution should not be neglected. At least in those cases of original application of the shock test on concrete repairs, the chloride ingress and corrosion of embedded rebars are used as additional performance criteria [3, 4, 6].

The advantage of the shock test is at first the considerable accelaration and moreover the simple procedure - if facilities of such capacity and automatization are provided as they were just described. Of course the disadvantages have to be mentioned as well. The method cannot simulate the uniaxial frost attack on concrete surfaces, and the influence of different deicing agents, of sea water etc. cannot be determined. It is restricted to the attestation of a certain resistance against freeze-thaw attack including some additional effects by chloride.

6 Acknowledgements

The design, development and construction of the test facilities was performed by M. Manthey and K. Oppat in the BAM division Building Materials. Financial support was given by the European Commission within the framework of a BRITE/EURAM project and by the BAM.

7 References

1. Schimmelwitz, P., Hoffmann, D. and Maultzsch, M. (1982) Untersuchungen zur Einwirkung von Tausalzen auf Brückenbauwerke aus Stahlbeton. *Forschung Straßenbau und Straßenverkehrstechnik*, No. 370, pp. 23 - 60.

2. RILEM TC 52-RAC Resin Adherence to Concrete. (1986) *Draft Test Recommendations, Thermal Compatibility Test I.*

3. Bundesminister für Verkehr. (1987) *Zusätzliche Technische Vorschriften und Richtlinien für Schutz und Instandsetzung von Betonbauteilen ZTV-SIB 87.* Verkehrsblattverlag, Dortmund.

4. Deutscher Ausschuß für Stahlbeton DAfStb. (1992) *Richtlinie für Schutz und Instandsetzung von Betonbauteilen, Teil 4: Qualitätssicherung der Bauprodukte.* Beuth-Verlag, Berlin.

5. Rößler, G., and Sasse, H.R. (1988) Durability of composite systems made of normal concrete and repair mortars, in *Durability of non-metallic inorganic building materials,* (ed. J. Eibl and H.K. Hilsdorf), Schriftenreihe des Instituts für Massivbau und Baustofftechnologie, Heft 6, Karlsruhe, pp. 149 - 160.

6. Maultzsch, M. (1995) Freeze/thaw cycling tests on concrete repairs. BRITE/EURAM II Project 3291 The development of standardized performance tests and criteria for concrete repair, Final Technical Report on Sub-Task 4.4 Freeze/thaw cycling with salt impact, 25 pp.

7. CEN TC 104 / SC 8 / WG 1 and 2. (1996) Draft European standard prEN 104-840-1 Products and systems for the protection and repair of concrete structures - Test methods - Determination of thermal compatibility - Part 1: Freeze-thaw cycling with deicing salt immersion.

8. Deutscher Ausschuß für Stahlbeton. (1991) *Prüfung von Beton. Empfehlungen und Hinweise als Ergänzung zu DIN 1048.* (ed. N. Bunke), DAfStb Heft 422, Beuth-Verlag, Berlin, pp. 12 - 15.

9. CEN TC 51 N 475 (1994) Draft European prestandard: Test method for the freeze-thaw resistance of concrete - Tests with water or sodium chloride solution - Part 1: Scaling - Reference method B (cube test).

Internal frost attack — state of the art

G. FAGERLUND
Division of Building Materials,
Lund Institute of Technology, Lund, Sweden

Abstract
Different internal frost damage mechanisms are described briefly. For each mechanism, some of the most relevant applications and consequences are described, together with a discussion of research required for acquiring a deeper knowledge of the mechanism, and its relevance for concrete destruction. The importance of understanding moisture uptake and moisture redistribution processes for clarifying the frost resistance problem is emphasized.
Keywords: Concrete, frost resistance, moisture, durability, service life

1 Introduction. Internal frost attack versus surface scaling

There are two main types of frost attack; (i) internal frost attack caused by freezing of moisture inside the material; (ii) surface scaling, normally caused by freezing of weak salt solutions at the surface. Probably, the two types of attack depend on the same basic mechanism, namely that too much moisture is present in the material, either in its interior, or at the surface. The salt-frost attack is, however, normally treated as a separate problem, because it only occurs during unsealed freezing where the concrete stays in contact with a salt solution. Therefore, it is to a very high degree coupled to the manner by which freeze/thaw is performed. Internal frost attack, on the other hand, can take place also when there is no moisture exchange, or moisture contact, between the material and its surroundings. In this paper, only internal frost attack is considered.

2 Destruction mechanisms

2.1 Introduction
In the following, an isolated representative unit cell of the concrete -or cement paste- is considered. By "representative unit cell" is meant a material volume that is big enough to be representative of the material in bulk, but not much bigger than that; i.e. it shall be big enough to contain the same porosity and the same pore-size distribution

Frost Resistance of Concrete, edited by M.J. Setzer and R. Auberg. Published in 1997
by E & FN Spon, 2–6 Boundary Row, London SE1 8HN, UK. ISBN: 0 419 22900 0.

as the material in bulk. For cement paste, this means a volume that might be 500 mm^3 (sphere with radius about 5 mm). For a coarse aggregate particle with its surrounding mortar, "belonging" to the particle, it might be 50 000 mm^3 (sphere with radius about 25 mm). This means that moisture movements to, or from, an adjacent unit cell is not considered. Besides, the cell is so small that the effect of moisture gradients can be neglected.

The three mechanisms, that are briefly described below, might very well be active simultaneously within the same concrete. So for example, porous saturated coarse aggregate particles, or porous interface zones, might freeze as "closed containers", while the cement paste within the same concrete might, when it freezes, cause both hydraulic pressure and ice-lens growth. Thus, the effect of freezing on the behaviour of the concrete is the sum of of more than one destruction type.

2.2 The closed container

2.2.1 Theory

The simplest material model is the "closed container" according to which the concrete (or rather, the cement paste) is supposed to be built up by "unit cells" consisting of hole-spheres with impermeable walls and with the external radius R, and the internal radius r. All evaporable water (freezable and non-freezable) is supposed to be located to the centrical hole. When water freezes, a pressure p [Pa] is built up in the ice/water phase. The magnitude of the pressure depends on the freezing temperature, θ_f [°C]:

$$p \approx -10^6 \cdot \theta_f \tag{1}$$

This pressure is transferred to the wall where tensile stresses appear. The magnitude of these depends on the degree of saturation of the container.

(a) Completely saturated container:
If no consideration is taken to the compressibility and ductility of the three phases solid wall, ice, non-freezable water, the following average tensile stress σ [Pa] is built up in the solid wall, provided the wall is completely plasticized:

$$\sigma \approx -10^6 \cdot \theta_f \cdot \pi \cdot r^2 / \pi (R^2 - r^2) = -10^6 \cdot \theta_f \cdot P^{2/3} / (1 - P^{2/3}) \tag{2}$$

where θ_f is the temperature of the unit cell, and P is the porosity [m^3/m^3]. Let us assume that the temperature θ_f is -10°C and the tensile strength of the wall is 8 MPa, which is a high value for cement paste. Then, the maximum allowable porosity is only 2% if the container shall not burst. *This calculation shows, that it is not possible for a concrete to survive freezing when it is completely saturated.*

(b) Unsaturated container:
The volume expansion of water is about 9% when it is transformed into ice. Consequently, no tensile stresses will occur when the *effective* degreee of saturation S_f is below a critical value, that is:

$$S_{f,cr} = w_f / (w_f + 0.09 \cdot w_f) = 0.917 \tag{3}$$

where w_f is the amount of freezable water $[m^3/m^3]$. The *total* degree of saturation S is defined:

$$S = w_e/P \tag{4}$$

where w_e is the total evaporable water $[m^3/m^3]$. Let us assume that the fraction of non-freezable water, w_{nf}, at the temperature θ is k_θ $(k_\theta=w_{nf}/w_e=(w_e-w_f)/w_e=1-w_f/w_e)$. Then, the critical total degree of saturation is:

$$S_{cr} = S_{f,cr}/\{1-k_\theta(1-S_{f,cr})\}=0.917/\{1-k_\theta \cdot 0.083\} \tag{5}$$

If 50% lf the evaporable water is unfreezable, the maximum allowable degree of saturation is 0.957, etc.

Equations (3) and (5) can be used for a calculation of the minimum amount of air-filled space required for frost resistance. Let us assume that the total evaporable water in a cement paste is 300 l/m^3 and the non-freezable water is 100 l/m^3. Then, the minimum allowable amount of air-filled pore volume is 18 l/m^3 or about 2%.

Note: The closed container model is a special case of the hydraulic pressure model described below. The only difference is that water in the closed container model has to freeze in-situ, because the pore walls are supposed to be impermeable.

2.2.2 Applications and consequences

The closed container model might be used for the following cases:

1. *Porous coarse aggregate grains.* The freezable water must be contained inside the grain. Therefore, this must have an effective degree of saturation below 0.917. If the grains are small, or if the porosity of the grains is very low, they can be allowed to be completely saturated. In such cases, the tensile stresses in the cement paste, imposed by the aggregate grains that freeze, are too small to cause damage. The criterions for maximum aggregate size and porosity are theoretically treated in [1].
2. *Freezing of water in porous interfaces between aggregate and cement paste.* The total volume of such porous interfaces might be rather high. A calculation performed in [2], based on direct observations of the size of the interface zone, indicates that the total volume might be as high as 4 litres/m^3 if only the coarse aggregate is considered, but as much as 45 litres/m^3 if all aggregate is considered. This means that considerable pressure can be exerted on the cement paste. Probably the pressure will be sufficiently high to severely harm the concrete in cases where the air content is low. The best way to avoid damage is to dry the concrete once. Then, water will leave the interface zones irreversibly, so that in the future, the degree of saturation of the interfaces will always be below 0.917.
3. *High performance concrete with extremely low water/cement ratio.* Such concrete might be so dense that water cannot flow from sites where it freezes to air-filled spaces. Thus, the water expansion at freezing must be taken care of locally. This means, that the effective degree of saturation of the cement paste, air pores excluded, must be below 0.917.
4. *Freezing of the green concrete.* In this case, the tensile strength of the cement paste is so low, that water expansion due to freezing must be taken care of locally; air-entrained pores have no function at this stage. Therefore, a self-desiccation by hydration, corresponding to at least 9 % of the freezable water volume, must take place before the concrete can be allowed to freeze. The positive effect of self desiccation was shown for the first time by Powers [3], and has been further inve-

stigated in [4]. In [4], it was shown theoretically, that the required self-desiccation corresponds to a degree of hydration α_{req} of:

$$\alpha_{req} \approx 0.5 \cdot w/c \tag{6}$$

This value corresponds well with experiments, [5].

5. *Freezing of water in deep cracks.* The expansion can normally be taken care of by air-pores in the cement paste. If the cracks are too wide, and the air content is too low, freezing will occur as in a closed container. Thus, it will be difficult to avoid damage since the cracks will probably be saturated during moist conditions.

6. *Freezing of water absorbed in air-pores.* Under certain circumstances, air-pores can become water-filled before, or during, freeze/thaw. The absorption mechanism can be a dissolution-of-air and replacement-by-water process of the type described in [6]. This process will, however, be very slow when the pores are big, and, therefore, complete water-filling of numerous air-pores is unlikely unless the concrete is stored in water for a very long period. Another, and more rapid pore-filling mechanism, is pure capillary suction due to a collapse of the air-pore system before the concrete was set. By "collapse" is meant that the air pore system of the fresh concrete is unstable causing continuous channels linking the individual air-pores to the outside water source. When water in air-pores freezes, there will normally be no place for the expelled water. Therefore, freezing of each individual "air-pore" will take place more or less as in a closed container, that is completely saturated. This is a dangerous situation since air-entrainment of this inferior type will worsen, the situation. High performance concretes ought to be especially vulnerable, due to their low permeability.

2.2.3 Needed research

The closed container model ought to be studied by systematic investigations of the different cases described in 2.2.2. The tests are not simple, because one has to make one variation at a time. The following example indicates the difficulties; if one wants to investigate the application of the model to coarse aggregate, one must be able to vary the degree of saturation of the aggregate without changing the degree of saturation of the cement mortar. Similarly, if one wants to test the model on high performance concrete by determining the critical effective degree of saturation, one has to know the freezable water inside the cement paste and to be able to distinguish freezable water in the cement paste from freezable water in air-pores, cracks, interfaces etc. If one wants to investigate the effect of freezable water in interfaces, one has to know that the water content in the other phases -aggregate and cement paste- is the same in all tests.

2.3 The hydraulic pressure

2.3.1 Theory

The closed container model predicts low required air contents when concrete as a whole is seen as a container. The real air requirement is considerably higher. One plausible reason is that excess water caused by freezing has to flow from saturated areas to air-filled spaces. This water flow creates a hydraulic pressure, [7], which in turn produces tensile stresses in the material. A simple model is shown in Fig 1. A piece of the saturated cement paste is freezing. The pressure is at maximum at point C at the centre of the piece. The hydraulic pressure p [Pa] inside the material sector considered is; [8]:

$$p = (0.09/K)\cdot(dw_f/d\theta)/(d\theta/dt) \int_o^X [v(x)/a(x)]\cdot dx \qquad (7)$$

where K is the effective permeability [m²/(Pa·s)], $dw_f/d\theta$ is the change in freezable water at a change in temperature [m³/(m³·degree)], $d\theta/dt$ is the rate of temperature lowering (the freeezing rate) [degree/s], $v(x)$ is the volume of the sector considered between x=x and x=X, where X is the distance from the surface of the piece to point C, $a(x)$ is the cross section of flow at x=x.

The material fractures when the hydraulic pressure p equals the tensile strength of the material, f_t [Pa]. This means that there will be a maximum allowable distance X, or a maximum allowable size 2·X.

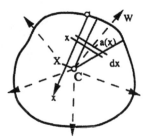

Fig 1: A piece of a saturated material. Moisture flow caused by freezing.

The maximum size depends on the geometry of the piece. Two geometries are considered in more detail, and one geometry in more principally.

(a) The critical thickness: For a slice of thickness D, the critical thickness D_{cr} [m] is [a(x)=a; v(x)= a(D_{cr}/2-x)]:

$$D_{cr} = \{8\cdot K\cdot f_t / [0.09\cdot(dw_f/d\theta)/(d\theta/dt)]\}^{1/2} \qquad (8)$$

(b) The critical wall thickness, or the critical spacing factor: For a hole-sphere with a wall thickness L, an impermeable outer periphery and a centrical hole with the specific area α (α=3/r where r is the radius of the hole), the critical shell thickness L_{cr} is obtained by; [a(x)= =4·π·(r+x); v(x)=(4/3)·π·{(r+L)³-(r+x)³}]:

$$L_{cr}^2\cdot\{(L_{cr}\cdot\alpha)/9+1/2\} = f_t\cdot K/[0.09\cdot(dw_f/d\theta)/(d\theta/dt)] \qquad (9)$$

L_{cr} is the critical spacing factor, derived originally by Powers [7]. L_{cr} is half the maximum distance water has to be transported from a place where it freeezes in the cement paste to the nearest air void; see Fig 2(b).

(c) The critical flow distance
A real material is neither composed of slices, as in (a), nor of hole-spheres, as in (b). It consists of a water saturated matrix enclosing a complicated array of more or less spherical air-filled pores. For each material, there will be numerous flow paths for expelled water; Fig 2(a). Each flow path will give rise to a hydraulic pressure. Theoretically, there will be a certain distance L′ causing the biggest pressure, thus de-

termining if the material shall be frost resistant or not. This distance is a measure of the biggest distance water can flow from a point where freezing occurs to the periphery of the nearsest air-filled pore. When L´ transgresses a critical value L'_{cr} fracture oc-curs. This value L'_{cr} cannot be calculated theoretically due to the big geometrical complexity of the matrix and air-pore system. It must be determined experimentally from the critical degree of saturation; see below. For the simple geometry in Fig 2(b), L'_{cr} is equal to the critical wall thickness L_{cr} defined by eq (9). Then all air-filled pores are supposed to be of equal size (specific area α) and placed in a cubic array. L_{cr} is, as mentioned above, the critical Powers spacing factor.

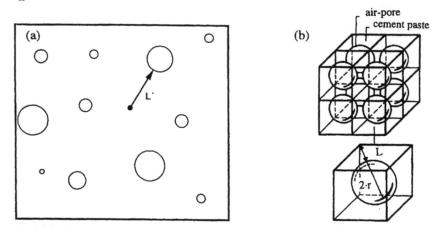

Fig 2: (a) The real material; the "flow distance", L´. (b) The Powers model, [7].

2.3.2 Applications and consequences

As mentioned above, the hydraulic pressure model is a special case of the closed con-tainer model. The difference is that the container walls are supposed to be permeable in the hydraulic pressure model. This model is probably applicable to most concretes, and especially to concretes with high water/cement ratio, where the amount of non-freezable water is low. According to the model, the concrete shall stop to expand when the temperature lowering, $d\theta/dt$, is zero, because then no more excess water is then generated. Such a behaviour has also been observed; see Fig 3(a), [9]. When the temperature is kept constant at about -8°C, the cement paste stops to expand. When cooling is resumed, expansion starts again. According to the theory, the specimen should even contract a bit when temperature lowering is stopped, since the hydraulic pressure ceases. It is reasonable to assume, however, that so much ice has been for-med before cooling is stopped, that contraction is physically hindered, since the ice keeps the specimen in an expanded condition.

The hydraulic pressure model leads to the existence of a critical degree of saturation of the cement paste, or a critical content of air-filled pores. The critical degree of satu-ration is reached when half the distance between pores that are still air-filled equals the critical flow distance. Thus, in order to calculate the critical degree of saturation one must know how water is distributed in the air-pore system.

For the general model in Fig 2(a) the following relation between the air content of the cement paste, a $[m^3/m^3]$, the specific area of the air-pores, α, $[m^2/m^3=m^{-1}]$, and the flow distance L´ [m], can be used; [10]. (*Note:* only the cement paste is considered. Therefore, the values a, α and L´ in the equation are valid for the cement paste phase):

$$a\{1+L´\cdot\alpha+2\cdot L´^2\cdot\alpha\cdot[u]_1/[u]_2+1.33\cdot L´^3\cdot\alpha\cdot[u]_0/[u]_2\} = C \qquad (10)$$

where $[u]_i$ is the i:th statistical moment of the size distribution of air-filled pores. C is a constant determining the volume fraction of the matrix within which all points have a distance shorter than L´ to the nearest air-pore. Thus, C describes the "protected volume fraction" of the paste. C=1 for the probability 63% (it can be proven, [10], that L´corresponds to the so called Philleo spacing factor, [11], when C=1). C= 2.3 for the probability 90%. Thus, the value of C increases rapidly with increasing protected volume fraction.

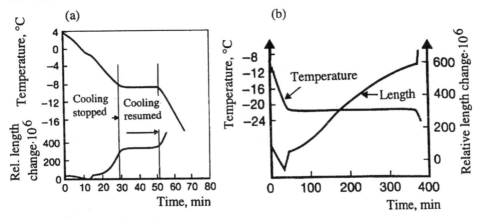

Fig 3: Length-temperature-time curves at freezing of cement paste; [9].
(a) w/c-ratio 0,60. (b) w/c-ratio 0.45.

For the model in Fig 2(b) the following relation between a, α and L is valid; [7]. (Like eq (10,) only the cement paste phase is considered in the equation):

$$L=(3/\alpha)\{1.4(1/a)^{1/3}-1\} \qquad (11a)$$
or:
$$a=1/\{0.364\cdot(L\cdot\alpha/3+1)^3\} \qquad (11b)$$

Increase of the water content of the cement paste leads to a decrease of the air content. Besides, the specific area of the air-pore system is changed. It is reasonable to assume that the smallest air-pores are water-filled at first; [6]. Therefore, a gradual water absorption leads to a reduction of α. Consequently, according to eq (10) or (11), the flow distance, or the Powers spacing factor, increases. At a certain water-filling, corresponding to the critical degree of saturation, or the critical air content, the critical flow distance $L´_{cr}$, or the critical Powers spacing factor L_{cr} is reached. Theoretically, the critical degree of saturation can be calculated if the size distribution of coarse pores, or, in other words, air-pores, (radius >1 µm) is known. A hypothetical example is shown in Fig 4. The smallest pore-class is 40 µm and the biggest 480 µm. The total volume of all coarse pores is 7%. The total porosity is 35% (including coarse pores). The critical Powers spacing factor L_α is 0.6 mm. The water absorption is supposed to

start with the smallest pores. The result of the calculation is shown in Fig 4(b). The critical distance is reached when all pores smaller than 140 μm are water-filled. This corresponds to a residual air content of 5%, which is the critical air content, and a residual specific area of 14 000 m^{-1}. The critical degree of saturation is;
$$S_{cr} = (0.35-0.05)/0.35 = 0.86.$$

In the normal case, the critical degree of saturation of a cement paste is considerably lower than 0.917 which was valid for the closed container model. The reason for this is that expelled water is not taken care of locally, where freezing occurs, but on a certain distance from the freezing site. The degree of saturation will, therefore, be a function not only of the critical distance, but also of the size and shape of the recipient pores (the air-pores).

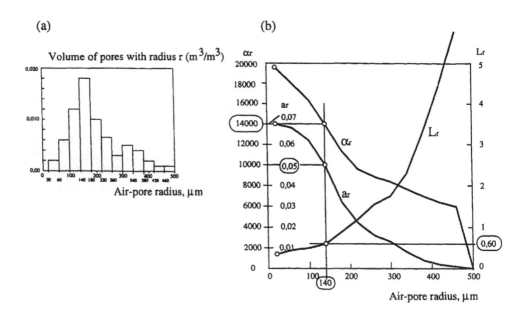

Fig 4: Example of a calculation of the critical air content and the critical degree of saturation. (a) Assumed air-pore size distribution. (b) Effect of a gradual water absorption on the air content, the specific area of air-filled pores, and the Powers spacing factor.

2.3.3 Needed research

The hydraulic pressure model ought to be studied by tests where the ice formation, the specimen dilation and the permeability are studied at the same time, using many different temperature curves for freezing and thawing. A combination of dilatometry and calorimetry was suggested by Verbeck and Klieger [12]. It was applied by the author in [13]. An example is shown in Fig 5. The material is a cement paste with a high degree of saturation.

From the ice formation curve, the parameter $(dw_f/d\theta)/(d\theta/dt)=dw_f/dt$, which is the freezing rate, can be evaluated. The expansion curve is a measure of the internal pressure caused by ice formation. A comparison between the measured expansion curve and a calculated expansion curve based on the hydraulic pressure model will, therefore, tell whether this is a reasonable destruction mechanism or not. The tests should be made on sealed specimens, pre-conditioned to different moisture contents. In an unsealed experiment, the spacing factor will gradually decrease during the test due to water absorption, and therefore, the experiment will be difficult to interprete.

It is remarkable, that the biggest expansions in Fig 5 occur when the rate of ice formation is at its lowest. One possible explanation is that ice already formed at higher temperatures block the pores so that the parameter K in eq (8) or (9) is reduced. The net effect might be that the biggest pressure occurs at the lowest temperature despite the fact that the rate of ice formation then is the lowest . In order to verify the hydraulic pressure model it is, therefore, necessary to estimate the effect of temperature on permeability. This can be done theoretically or experimentally.

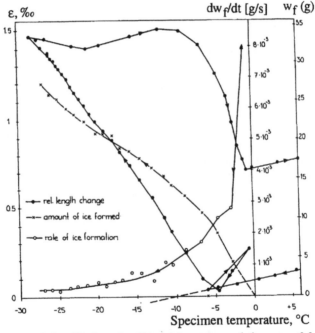

Fig 5: Measurements of the dilation, the freezable water and the rate of freezing of a cement paste with w/c-ratio 0,83; [13]. S=0,90, $S_{cr}\approx0,86$.

A simple theoretical calculation of the effect of ice formation on the hydraulic pressure can be made by assuming that all pores are straight parallell cylinders assembled in a bunch with pore radii varying from r_{min} to $r_{max,\theta}$ where r_{min} is the smallest radius of pores containing unfrozen water, that is "movable". $r_{max,\theta}$ is the biggest pore containing unfrozen water. Before freezing, r_{max} is the biggest of all pores in the material.

The hydraulic prsssure p [Pa] is:

$$p=k\cdot L\cdot(dw/dt) /\Sigma\, r_i^4\cdot n_i \qquad (12)$$

where k is a constant including the viscosity of the flowing substance (the ice-water mixture), L is the distance of flow, dw/dt is the total rate of flow [m^3/s], r$_i$ is the radius of the pore, and n$_i$ is the number of pores with radius r$_i$. n$_i$=f(r)·dr [number/m^3] where f(r) is the frequency function of pore radii. The water flow can be replaced by the rate of formation of excess water, dw/dt = 0.09·dw$_f$/dt = 0.09·(dw$_f$/dθ)/(dθ/dt). Thus eq (12) can be written:

$$p=k'·L·(dw_f/dθ)/(dθ/dt) \Big/ \int_{r_{min}}^{r_{max,θ}} r^4·f(r)·dr \qquad (13)$$

r$_{max,θ}$ is a function of the freezing temperature; the lower the temperature, the smaller the value of r$_{max,θ}$. A purely theoretical equation for r$_{max,q}$, assuming equilibrium conditions (no local supercooling) is:

$$r_{max,θ} = -σ·v_s/[ΔH·ln(T/T_o)]+t \qquad (14)$$

Where σ is the surface tension between air and water [N/m], v$_s$ is the molar volume of water [m^3/mole], ΔH is the molar heat of fusion (a function of temperature) [Nm/mole], T is the freezing temperature [K] and T$_o$ the normal freezing temperature [273 K]. t is the thickness of the adsorbed layer [m].

r$_{max,θ}$ can also be obtained directly from the measurement of the amount of frozen water, combined with a determination of the total pore size distribution.

Another important research task is to determine the critical distance experimentally. This can be done by experimental determinations of the critical degree of saturation, in combination with a determination of the size distribution of coarse pores (bigger than about 1 μm). The method is the inverse of calculating the critical degree of saturation; see Fig 4. Some determinations have been made on basis of freeze experiments and pore size analyses, using image analysis of polished surfaces; [14]. Examples are shown in Fig 6.

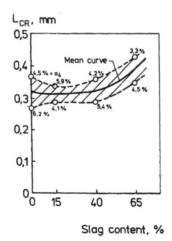

Slag content, %

Fig 6: Experimentally/theoretically determined critical Powers spacing factors for concrete made with different types of cement and with different air contents; [14].

The materials tested were concretes made with cements of different type. The water/cement ratio is 0.45, and different air contents were used. It is interesting to note that the calculated Powers spacing factor is considerably higher than values normally given in the literature. The reason is, that such values are based on measurements of the spacing factor of the entire air-pore system; also pores that were water-filled during the freeze test on basis of which the determination of the critical spacing factor was made. There are many indications that the true critical Powers spacing factor for concrete is closer to 0.35 mm than to 0.25 mm, a value often maintained in the literature.

2.4 The microscopic ice lens growth mechanism

2.4.1 Theory

A cement paste will always contain both unfrozen water and ice; the lower the water/cement ratio, the higher the fraction of unfrozen water. Differences in free energy between water and ice will draw water to ice that was formed in the coarser capillaries and to ice formed in air-pores by freezing of water, that was expelled during the initial freezing phase. Consequently, the ice-bodies in filled capillaries will grow and exert a pressure on the pore walls. Thereby, the free energy of the ice increases. Ice in air-pores, that are only partly filled, can grow without obstruction from the pore walls; i.e. they grow in a stress-less condition. Water is primarily drawn from very small capillaries and from gel pores. This will cause a drying which will decrease the residual free energy of unfrozen water. Consequently, the energy differences between ice in filled pores and unfrozen water will finally disappear, whereby ice growth stops. Before that happens, considerable damage might occur. So far, it has not been possible to derive an expression for the stress in the material. Some estimates are made in [15] showing that the pressure can be very high in a concrete with free access to moisture from outside.

Ice in partly filled air voids can grow without hinder from the pore walls. Therefore, water transport to air-pore ice reduces the possibility of water transport to capillary ice. If the air pores are numerous, capillary ice growth will not take place at all since the free energy of air-pore ice is much lower promoting their growth. This means, that also the microscopic ice lens growth mechanism predicts the existence of critical distances between air-filled spaces; [9]. In contrast to the hydraulic pressure model, it is not possible at present to derive a theoretical expression for the critical distance. It has to be determined experimentally through a determination of the critical degree of saturation combined with a determination of the size distribution of the coarse pore structure (the "air-pores").

Examples of measurements of length changes of cement paste during freezing is shown in Fig 7; [9]. The Powers spacing factor was different in different specimens. It is interesting to note the big contractions when the spacing factor is low. This indicates that shrinkage due to the drying effect caused by withdrawal of water from the finer pores dominates over the expansion due to the pressure caused by ice lens growth. The most plausible explanation is that ice growth takes place primarily (or only) in air-pores, that are only partly filled. When the spacing factor is high, the specimen expands considerably, indicating that ice lens growth takes place in completely filled capillaries.

2.4.2 Applications and consequences

Like the hydraulic pressure model, the ice lens growth model is probably applicable to most concretes, and especially to concrete with low water/cement ratio, in which the amount of non-freezable water is high. According to the model, the concrete shall not stop to expand when temperature lowering is zero, because stresses are not promoted by a high rate of temperature lowering, but by a long duration of low temperature. It has also been observed, for cement paste with low w/c-ratio, that damage continues to increase despite the temperature is kept constant; see Fig 3(b). When the temperature is kept constant at about -22°C, the cement paste continues to expand. This behaviour is contradictory to the behaviour shown in Fig 3(a) for the cement paste with the higher water/cement ratio.

The ice lens growth model also leads to the existence of a critical degree of saturation of the cement paste, or a critical content of air-filled pores. This is a consequence of the fact that critical distances pobably exist.

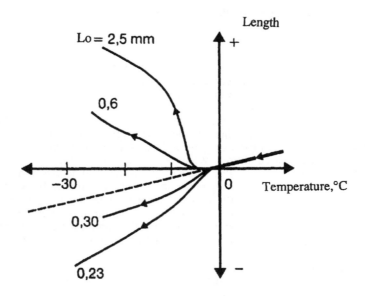

Fig 7: Length-temperature curves for air-entrained cement pastes with different spacing factor; [9].

2.4.3 Needed research

The ice lens segregation model could be studied by the same type of tests as was suggested for the hydraulic pressure model; i.e. tests where ice formation, specimen dilation and permeability are studied simultaneously, using many different temperature curves for freezing and thawing. A combined dilatometer and calorimeter is the most suitable instrument. By keeping the temperature constant for longer or shorter times and noting the ice formation and the expansion (or contraction), the significance of the destruction mechanism can be estimated. By knowledge of the elastic constants of the material, the pressures exerted by the growing ice can be estimated.

The tests should be made on specimens conditioned to different moisture contents and frozen in sealed condition. Ice lens growth can also occur when the specimen is placed in contact with unfrozen water. This is, however, a much more complex (and much more dangerous) situation, and it is not so well suited for mechanism studies. By a gradual water uptake, the spacing factor will gradually decrease during the experiment which makes the interpretation very difficult.

3 Effect of external parameters on the critical moisture content

3.1 Effect of repeated freeze/thaw cycles
Repeated freeze/thaw cycles will have a marginal effect on the value of the critical degree of saturation. This has been shown experimentally in [16], [17], and [18]. On the other hand, for water contents above the critical, there seems to be a fatigue effect. The following relation seems to be valid; [19]:

$$S<S_{cr}: \quad D=0 \text{ for all N} \tag{15a}$$

$$S>S_{cr}: \quad D=K_N(S-S_{cr})=K_N \cdot \Delta S \tag{15b}$$

Where D is damage expressed as loss in dynamic E-modulus, K_N is a "coefficient of fatigue". Experiments indicate that there exists a fatigue limit that is reached after a rather small number of freezings; [19]:

$$K_N \approx A \cdot N/(B+N) \tag{16}$$

Where the "fatigue limit" A is of the order 1 to 10, and B is about 4. This means that very small additional damage occurs after about 20 freeze/thaw cycles. For a typical concrete with A=8 and ΔS=0,05 the loss in E-modulus is 29% after 10 cycles, 33% after 20 cycles, 39% after 100 cycles and 40% after 1000 cycles.

The damage equation (15) can possibly be used for an estimation of the future deterioration of frost damaged concrete; see [20].

3.2 Effect of the freezing rate
The freezing rate has small influence on the critical degree of saturation; [21]. This can be explained by the hydraulic pressure model. A doubling of the freezing rate will according to eq (9) decrease the critical Powers spacing factor by a factor of about 0.2. This will not influence the critical degree of saturation so much. For the hypothetical example in Fig 4, the critical spacing factor will be reduced from 0.60 mm to 0.48 mm. This will increase the critical air content from 5% to 6.1% and decrease the critical degree of saturation from 0.86 to 0.83. The effect on the service life might, however, be significant because it takes longer time to fill a concrete by water to a degree of saturation 0.86 than to 0.83.

3.3 Conclusions. Needed research
The critical degree of saturation can be looked upon as a material constant that is not so much influenced by external climatic conditions. This is important because it opens the possibility of using the critical moisture content for service life predictions; [22].

There is however more experiments to be done in order to verify the results so far obtained, and described above. One cannot exclude that prolonged freezing temperatures increase the ice lens growth, which in turn might reduce the critical distance more than what was calculated in paragraph 3.2 on basis of the hydraulic pressure mechanism. Consequently, the critical degree of saturation will be reduced more than calculated above.

4   Freezing involving moisture transfer over longer distances

4.1   Introduction.
In sections 2 and 3, unit cells without moisture exchange with other unit cells were considered. There are freezing conditions, however, where water is taken up by, or transferred to, other unit cells, increasing their degree of saturation before they freeze, or when they freeze. These cases must be treated separately. Probably, it is the same basic mechanisms that determine the inner stresses in these cases as the mechanisms described in section 2.

4.2   Moving ice front. Theory and needed research
Let us assume, that the interior of the material is not critically saturated when freezing starts. Ice is at first formed in the surface layer (or surface unit cells) because freezing temperature is reached there at first. Therefore, if the surface is supposed to be impermeable due to formation of ice, excess water will be forced into the unfrozen layer that is closest to the frozen layer. As the ice front moves, more and more water is forced inwards. At a certain instant, the critical degree of saturation is reached in the unfrozen layer close to the ice front. Then, damage will occur, and possibly the frozen layer will scale off.

One can calculate a "critical depth of unsaturation", $D_{cr,us}$, which is the distance from the ice front to the depth where there must be air space enough to take care of the expelled water. It is a complicated task to calculate this value. It will also depend on the amount of expelled water. Therefore, $D_{cr,us}$ will depend on the depth x from the surface to the ice front. Application of Darcys law to the flow inwards give; cf. eq (8):

$$p \approx 0.09 \cdot [(dw_f/d\theta)/(d\theta/dt)] \cdot dx \cdot (x_{us}-x)/K \qquad (17)$$

Where dx is the layer within which freezing occurs, x is the distance from the surface to the ice front, and x_{us} is the depth from the surface to the place where expelled water can be accomodated. It has been assumed that the amount of freezable water is constant for all layers dx, which cannot be the real case since expelled water is taken up in each layer from previous freezings of outer layers. Thus, the freezable water increases with increasing depth x. It has also been assumed, that the pressure gradient is linear from the ice front to the depth of unsaturation. This is a simplification since the water content varies in this zone. Fracture occurs when $p=f_t$ which is the tensile strength. Introducing eq (8) gives:

$$(x_{us}-x)_{cr} = D_{cr,us} = (Dcr^2/8) \cdot (1/dx) \qquad (18)$$

This equation is not easy to apply. Warris [23] suggested, on basis of an analysis of the heat balance equation for the surface, that the parameter 1/dx could be replaced by the surface/volume ratio, S/V [m^{-1}] for the specimen.

$$D_{cr,us} \approx (D_{cr}^2/8) \cdot (S/V) \qquad (19)$$

This equation can only be applied when the moisture profile is known. This will change during the freezing process. Scaling will be avoided when the criterion (18) is fulfilled over the entire frozen part of the specimen.

It is doubtful whether the moving ice front mechanism, as described above, is valid for concrete. There is always unfreezable water, that makes it possible for expelled water to flow outwards. It seems clear, however, by experiments on brick, [24], that the mechanism can be valid for dense materials with low amount of freezable water.

It should be interesting to study the suggested mechanism by letting long specimens conditioned to different initial moisture contents freeze unidirectionally, and measure the damage occurring at different distance from the cold surface.

4.3 Frost heave. Theory and needed research

When unfrozen parts of a concrete has almost unlimited access to free water, a destruction mechanism of the the same type as frost heave in roads might occur. The prerequisites are; [4]:

1: A zero-degree isotherm, separating a frozen surface part from an unfrozen interior part, lies still for a long time. This means that there must be a lasting heat balance between heat leaving the ice front and heat arriving at this. Heat arriving is caused by flow of water to the ice front from the interior of the concrete, or from an outer water source. Heat is brought both by convection and by freezing. The driving mechanism for water transport is the differences in free energy between ice and water that are described in paragraph 2.4.1.
2: The permeability must be so high, that the moisture flow, required for an immobile ice front, can be fulfilled.
3: The tensile strength of the concrete must be so low, that a macroscopic ice lens is able grow at the ice front
4: The concrete must have access to large amounts of unfrozen water

One cannot exclude, that this destruction mechanism is valid for some types of concrete with low quality, such as concrete dams or reservoir walls, foundations founded below the ground water level, etc. It is quite clear, that the mechanism occurs in green concrete that freezes before the concrete is set; [4], [15].

The mechanism ought to be studied by experiments on concrete specimens simulating real cases, such as a dam wall exposed to frost at one face and to free water at the other. Different concrete qualities should be tested.

5 Moisture mechanics and frost damage

5.1 Introduction

Damage of a unit cell will only occur when the water content in the cell exceeds the critical moisture content. Only a few freeze/thaw cycles are needed in order to more or less completely destroy the cell when this is more than critically saturated. Macroscopic frost damage will occur when many unit cells are destroyed simultaneously. *The frost resistance problem is, therefore, to a great extent a moisture mechanics problem, and it can only be solved when the relevant water uptake processes are understood.*

There are many possibilities for water absorption causing frost damage:

1: Moisture is taken up before the concrete is exposed to freezing. This can occur due to isothermal water uptake, or due to water uptake during natural variations of the outer climate.
2: Moisture is taken up from outside during the freeze/thaw process. This might happen in nature, but is more significant during traditional unsealed freeze/thaw tests.
3: Moisture is redistributed internally during the freeze/thaw process. This might occur during natural freezing, but more often during freeze/thaw tests.

All these cases must be considered. So for example, by slight changes in the pre-treatment before freezing, or in the freeze/thaw process in a labtest, the water absorption changes. This might cause very big differences in the test results. The same concrete might be judged both frost resistant and non-frost resistant depending on how the test was performed. The critical moisture content is not changed by variations in the test procedure, only the real moisture content. This is enough to change the test results.

It is a difficult task to predict moisture conditions that are high enough to cause frost damage. This depends on the fact that frost damage almost always occur in, what can be called, the "over-capillary range", where we have no calculation or measurement methods, neither for moisture transport, nor for moisture fixation. One can define three "moisture ranges" for a porous material:

1: The hygroscopic range (below about 98% RH). In this range, moisture fixation is described by the sorption isotherms, and moisture transport is described by normal diffusion equations containing transport coefficients, that can be easily determined by many different test methods.
2: The capillary range (from about 98% RH to "capillary saturation"). Moisture fixation is described by "suction curves" describing the relation between capillary pressure and moisture content. Moisture transport is described by the same type of equations as for the hygroscopic range. The transport coefficients can be evaluated from a series of capillary absorption experiments; [25].
3: The over-capillary range (moisture content higher than that corresponding to "capillary saturation"). Moisture transport in this range is governed by quite other mechanisms than the mechanisms acting in the other two ranges. There are no capillary suction gradients or vapour pressure gradients causing water transport. Water is instead forced by differences in external pressure (e.g. "ice pressure" as in the hydraulic pressure mechanism, or "ice suction" as in the ice lens growth mechanism), or by dissolution and diffusion of entrapped air; [6]. The latter mechanism is described in the next paragraph.

5.2   Water uptake before freeze/thaw

Concrete placed for a long time in water will absorb more water than that corresponding to capillary saturation, S_{cap} (S_{cap} is indicated as a nick-point on a water absorption square-root of time curve for a thin specimen). The most plausible mechanism is a dissolution of air due to the over-pressure in enclosed air-bubbles contained in air-pores. The smaller the bubble, the bigger the over-pressure, and the more rapid the dissolution. Consequently, there will be a concentration gradient in dissolved air between small bubbles and coarse. The smallest bubbles will therefore "disappear" at first, and air from them migrates to coarser bubbles. From there, air migrates to bubbles that are even bigger, and finally to the surface. Air, that is dissolved, is replaced by water. It can be shown theoretically, that this inter-pore diffusion process causes an internal "suction" over a large portion of the concrete. Therefore, the water absorption process can be assumed to occur simultaneously across a rather big volume adjacent to the water source. The process is described in detail in [6].

The most simple case is isothermal long-term water absorption, where the specimen absorbs water unidirectionally, or when it is completely immersed. It can be shown, both theoretically and experimentally, that the water absorption process can be described by an equation of the following type:

$$S(t) = S_{cap} + A \cdot t^B \tag{20}$$

Where $S(t)$ is the total degree of saturation after a period t of water absorption, S_{cap} is the capillary saturation (the nick-point in the absorption curve; see above), A is a coefficient determined mainly by the diffusivity of dissolved air through the saturated matrix, B is a coefficient determined mainly by the size distribution of the air-bubbles; $B < 0,50$.

This expression can be used for calculating a sort of potential service life, which is the time the concrete can be allowed to absorb water. The potential service life is:

$$t_{pot} = \{(S_{cr}-S_{cap})/A\}^{1/B} \tag{21}$$

The potential service life is of course not the same as the real service life in the real environment. It only giver en expression for the risk of frost damage in a "standard environment" which is very moist.

5.3 Needed research
Since moisture transport processes are so vital for understanding the frost destruction process, and for estimating the frost resistance, it is important that all kinds of moisture uptake and moisture movements, before and during freeze/thaw, are investigated. This requires many types of experiments, most of which are very difficult to perform, since it is moisture movements on the microscale that determines whether the concrete shall be frost resistant or not.

For composite materials, composed of two or more frost sensitive phases, it is also necessary to investigate the moisture conditions at interfaces, and moisture flow across interfaces. Concrete is such a material, since it always contains both porous interface zones between cement paste and aggregate, and in many cases also porous aggregate. Surface treated concrete or repaired concrete are other composite materials where water at the interface between the phases is to a great extent determining the frost resistance of the composite. The same might be true for carbonated concrete.

In high performance concrete with very low water/cement ratio, self desiccation could make the concrete frost resistant for a certain period; e.g the first years after production. One cannot exclude, however, that this effect is only temporary; c.f. paragraph 2.2.2. Therefore, the effect of self desiccation on the long-term moisture condition and the effect on the frost resistance ought to be studied.

6 References

1. Fagerlund, G. (1973) *Frost Resistance of Concrete with Porous Aggregate*, Swedish Cement and Concrete Research Institute, Research Fo 2:78, Stockholm.
2. Fagerlund, G. (1995) *Freeze-Thaw Resistance of Concrete*, Div. of Building Materials, Lund Insitute of Technology, Report TVBM-3060, Lund
3. Powers, T.C. (1962) Prevention of Concrete to Frost Damage to Green Concrete. *Bulletin RILEM*, No 14.
4. Fagerlund, G. (1980) *Influence of Slag Cement on the Frost Resistance of Concrete- a Theoretical Analysis*, Swedish Cement and Concrete Research Institute, Research Fo 1:80, Stockholm
5. Möller, G. (1962) Early Freezing of Concrete. Swedish Cement and Concrete Research Institute, *Applied Studies* No 5, Stockholm.
6. Fagerlund, G (1996) Predicting the Service Life of Concrete Exposed to Frost Action through a Modelling of the Water Absorption Process in the Air-Pore System, In*The modelling of Microstructure and its Potential for Studying Transport Properties and Durability*, (eds. H. Jennings et al.), Kluwer Academic Publishers.
7. Powers, T.C. (1949) The Air Requirement of of Frost Resistant Concrete, *Highway Res. Board*, Proceedings No 29.
8. Fagerlund, G. (1986) *The critical Size in Connection with Freezing of Porous Materials*, Cementa Report CMT 86039, Danderyd, Sweden.
9. Powers, T.C. and Helmuth, R.A. (1953) Theory of Volume Changes in Hardened Portland Cement Paste, *Highway Res. Board*, Proceedings, No 32.

10. Fagerlund, G. (1977) *Equations for Calculationg the Mean Free Distance between Aggregate Particles or Air-Pores in Concrete*, Swedish Cement and Concrete Research Institute, Research Fo 8:77, Stockholm

11. Philleo, R.E. (1955) *A Method for Analyzing Void Distribution in Air-Entrained Concrete*, Portland Cement Association, Tentative Paper, Chicago.

12. Verbeck, G. and Klieger, P. (1958) Calorimeter-Strain Apparatus for Study of Freezing and Thawing of Concrete, Res. and Devel. Labs. of PCA, Bull. 95.

13. Fagerlund, G. (1973) The Significance of Critical Degrees of Saturation at Freezing of Porous and Brittle Materials, In *Durability of Concrete*, ACI, Special Publication, SP 47.

14. Fagerlund, G. (1982) *The Influence of Slag Cement on the Frost Resistance of the Hardened Concrete*, Swedish Cement and Concrete Research Institute, Research Fo 1:82, Stockholm.

15. Powers, T.C. (1956) Resistance of Concrete to Frost at Early Ages, In *Winter Concreting*, Proc. of a RILEM Symposium, Copenhagen.

16. Rombén, L. (1973) Test results presented in *Report of Activities 1972/73*, Swedish Cement and Concrete Research Institute, Stockholm.

17. Klamrowski, G. and Neustupny, P. (1984) *Untersuchungen zur Prüfung von Beton auf Frostwiderstand*, Bundesanstalt für Materialprüfung (BAM), Forschungsbericht 100, Berlin.

18. Fagerlund, G. (1972) *Critical Degrees of Saturation at Freezing of Porous and Brittle Materials*, Lund Institute of Technology, Div. of Building Technology, Report 34, Lund. (In Swedish).

19. Fagerlund, G. (1994) *Influence of Environmental Factors on the Frost resistance of Concrete*, Div. of Building Materials, Lund Institute of Technology, Report TVBM-3059, Lund.

20. Fagerlund, G. (1995) *Frost Resistance of Concrete. Estimation of the Future Deterioration*, Div. of Building Materials, Lund Institute of Technology, Report TVBM-3067, Lund.

21. Fagerlund, G. (1992) *Effect of the Freezing Rate on the Frost Resistance of Concrete*, Nordic Concrete Research, Publ. No 11, Oslo

22   Fagerlund, G. (1996) Assessment of the Service Life of Materials Exposed to Frost, In Durability of Building Materials and Components (ed. C. Sjöström), Proc. of the 7:th Int. Conf. Stockholm, May 19-23, E & F Spon, London.

23. Warris, B. (1964) *The Influence of Air-Entrainment on the Frost-Reistance of Concrete*, Swedish Cement and Concrete Research Institute, Proceedings Nr 36, Stockholm.

24. Sandin, K. (1989) Moisture Conditions and Frost Effects in Cavity Brick Walls, *Bygg & Teknik,* Nr 4, Stockholm. (In Swedish)

25. Janz. M. (1997) .*Methods of Measuring the Moisture Diffusivity at High Moisture Levels*, Div. of Building Materials, Lund Institute of Technology, Report TVBM-3076, 1997.

APPENDIX

RILEM Technical Committee 117-FDC

Recommendation

First published in *Materials and Structures*, Vol. 29 (1996) pp 523–528

RILEM TC 117-FDC:
FREEZE-THAW AND DEICING RESISTANCE OF CONCRETE

CDF TEST - TEST METHOD FOR THE FREEZE-THAW RESISTANCE OF CONCRETE - TESTS WITH SODIUM CHLORIDE SOLUTION (CDF)

RECOMMENDATION

Prepared by M.J. Setzer, G. Fagerlund and D.J. Janssen

TC MEMBERSHIP: **Chairman**: M J. Setzer, Germany; **Secretary**: R. Auberg, Germany; **Members:** C. Dubois, France; G. Fagerlund, Sweden; V. Hartmann, Germany; S. Jacobsen, Norway; D. J. Janssen, USA; H. Kukko, Finland; J. Marchand, Canada; T. Miura, Japan; P.-E. Petersson, Sweden; M. Pigeon, Canada; J. Prost, France; T. F. Rônning, Norway; E. Sellevold, Norway; E. Siebel, Germany; J. Stark, Germany; W. Studer, Switzerland;

FOREWORD
During the meetings of RILEM TC 117-FDC; the drafts of the three procedures recommended by RILEM TC 117 and published in MATERIALS & STRUCTURES (Vol. 28, No. 177) were discussed extensively. As already stated in the draft edition, TC 117-FDC will not issue a final recommendation concerning any procedure for which the precision following ISO 5725 is not given. Before the meeting in Sapporo (July 31st and August 1st, 1995) precision data were published only for the CDF test;the situation has not changed since then. Therefore, RILEM TC-117 FDC unanimously passed the following text proposed by Prof. Janssen:

„TC 117 directs RILEM to recommend the CDF test for the determination of the scaling resistance of concrete. However, the committee 117 cannot decide whether the draft procedures for the CF, Slab or Cube tests should be recommended until the precision data for these tests have been made available and the committee has an opportunity to examine these data.“

A RILEM recommendation should be restricted to the description and evaluation of the test procedure itself. The regulations should have a substantial scientific basis. Since an acceptance criterion can be modified by reasons which are valid only under certain conditions and economic considerations, a RILEM recommendation could contain at most guideline values.

Due to these considerations the drafts for the CDF test had to be modified (without CF test). However, the necessary modifications were not so important, that they required approval by the complete committee. Therefore, TC 117 established an editorial group responsible for these modifications, which consisted of Prof. Fagerlund, Prof. Janssen and Prof. Setzer, the Chairman of the Committee.

This work by the editorial group has now been completed.

As Chairman of RILEM Technical Committee TC 117-FDC, it is an honour for me to thank all the members, and especially the editorial committee, for their very serious and engaged work, which led to a recommendation with sufficient precision for the freeze-thaw and deicing salt resistance of concrete.

M:J. Setzer - July 1996

1 Introduction

Adequate resistance against freeze-thaw attack with deicing chemicals should be tested.[1] This testing requires both a test procedure with demonstrated precision, i.e., repeatability and reproducibility[2], as well as an acceptable correlation of test results to the performance of concrete in practice.

As in other fields such as strength testing, a test procedure for freeze-thaw and deicing salt resistance cannot cover all the variations possible under exposure conditions in practice. This would increase the expenditure. At the same time the scatter would increase and the results would become non-interpretable. The `CDF´ test has been developed to attain a high degree of precision with minimum expenditure on equipment and labour. CDF means **C**apillary suction of **D**e-icing solution and **F**reeze thaw test.[3]

2 Scope

This procedure allows us to measure the amount of scaling per unit surface area due to a number of well defined freezing and thawing cycles in the presence of deicing salt - as a rule sodium chloride solution - (CDF), and leads to an estimate of the freeze-thaw and deicing salt resistance of the concrete tested.

It can be used to test an exposed surface of concrete structures or of precast concrete products, and to test the constituents of concrete as well as the concrete mix.

3 Standards

prEN-ISO 2736/2-1993 Testing hardened concrete-test specimens. Part 2: Making and curing of test specimens.

ISO 5725/1 to 6-1990: Accuracy (trueness and precision) of measurement methods and results.

---

[1]   Freeze thaw and deicing salt resistance has been achieved to date by rules, based on long term practical experience, that prescribe the water/cement ratio, the types of constituent and the amount of artificially entrained air voids. However, new constituents or untried compositions are required for certain practical demands. There are concrete applications where the practical design rules cannot be adopted, such as in dry concrete production of blocks, kerbs and flags. In these cases a reliable test procedure of the freeze-thaw and of the-freeze thaw and deicing salt resistance is necessary.

[2]   Any test procedure should be made as precise and reliable as possible by a sufficiently exact definition of the relevant parameters. Its results should be related to practical experience, such as the above mentioned rules of mix design. The precision should be assessed following ISO 5725. By this methodology the freeze-thaw or the freeze-thaw and de-icing salt resistance, even under unfavourable practical conditions, should be predictable with sufficient reliability.

[3]   Several different procedures have been developed. However, there is still a lack of precision i.e., reproducibility and repeatability. The CDF test has been developed by analysing and optimising the existing procedures and by adopting basic research work on adequate restriction of test parameters.

4 Definitions

1. The **freeze-thaw and deicing agent resistance is the** resistance against alternating freezing and thawing in the presence of deicing agent solution as test solution as test liquid.
2. The **test solution** is the solution of 3 mass% sodium chloride and 97 mass% demineralized water.
3. **Scaling** is the loss of material at the surface of concrete due to freeze-thaw or freeze-thaw and deicing attack.
4. The **reference point** is the physical measuring point at which the temperature cycle is controlled.
5. The **reference temperature** is the temperature measured at the reference point.
6. The **test surface** is the surface of the test specimens on which the freeze-thaw and deicing agent resistance is established in this test specification.

5 Equipment

1. **Climate chamber** with a temperature of 20±2°C and a relative humidity of 65±5 %.
2. Evaporation is measured using an **evaporation bowl** with a depth of approximately 40 mm and a cross-sectional area of 225±25 cm².
3. **Lateral sealing** is by epoxy resin or aluminium foil with butyl rubber. Both must be durable at temperatures of -20°C and resistant against the attack of the deicing solution. They cannot be brittle at the minimum temperature reached.
4. **Test liquid,** consisting of deicing agent solution e.g., 97% by weight of demineralized or distilled water and 3% by weight of NaCl.
5. **Test containers (Figs 1 and 2).** The specimens are stored in stainless steel containers during the freeze-thaw cycles. The size of the test container should be selected in such a way that the thickness of the air layer between the vertical side of the specimen and the test container is restricted to 20±10 mm.[4,5] On the container bottom a 10±0.1 mm high spacer is arranged to support the specimen and to guarantee a defined thickness of the liquid layer between the test surface and the container.
   The same test containers can be used for capillary suction. Other containers can to be used if they assure an equivalent arrangement for capillary suction. During the capillary suction the test container must be closed with a cover. If condensation is anticipated, the cover should have an incline to prevent any possible condensation water from dripping onto the specimens.

---

[4]   The air layer between the vertical side of the specimen and the test container acts as thermal insulation.

[5]   The stainless steel containers are matched in several modular sizes so that the same boundary conditions can be met for each specimen form.

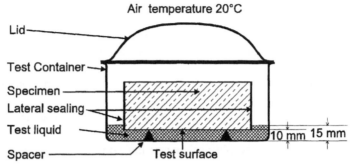

Fig. 1 Capillary suction.

6. **Temperature controlled chest (Fig.3).** For temperature control during the freeze-thaw cycle and to guarantee uniaxial heat flux, a chest with a liquid cooling bath is used. The temperature of the cooling bath is controlled by an appropriate device. The heating and cooling capacity and the control unit must be capable of maintaining the temperature regime at the reference point within ±0.5 K with full loading by the test containers with the test specimens. The temperature in the bath must be uniform within a limit of ±0.5 K at least at the minimum temperature and of ±1 K at other temperatures. A constant time shift between the test containers is acceptable.

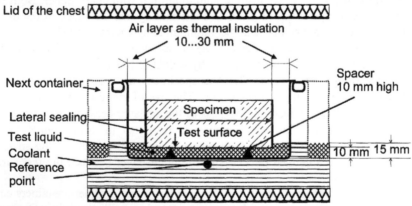

Fig. 2 Test container with specimen in the liquid cooling bath.

The chest must be equipped with supports for the test containers above the cooling bath to ensure an immersion depth of the bottom of the test containers of approximately 20 mm. During the freeze-thaw cycles the upper space of the chest containing the specimens must be separated from the cooling bath either by the test containers or by other lids.[6]

---

[6]   When conducting the test in a cryogenic bath, no cover of the test containers is required, since the chest lid will provide a sufficient evaporation barrier, while the container wall will serve as a cold trap. The geometry of liquid coolant bath ensures a uniaxial thermal attack. No lateral thermal isolation is required, because the thermal conductivity of the ambient air is sufficiently low and because the liquid coolant is capable of adding and removing larger amounts of heat.

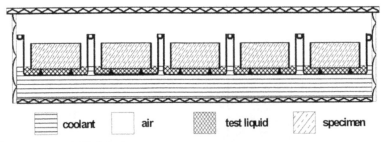

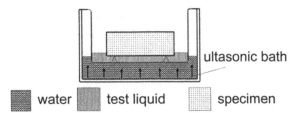

Fig. 3 Temperature controlled chest.

The reference temperature is measured in the cooling bath liquid below the bottom of the test container. The gauge is fixed to the container in such a way that a precise distance of 1 mm between the bottom and the measuring point is ensured. A standardised platinum resistance temperature gauge with an electrical resistance of 100 Ω at 0°C (PT100) is recommended for measurement. For calibration the minimum temperature of -20°C should be used. For monitoring and controlling the reference temperature a test container in the centre of the bath is used.[7]

7. **Ultrasonic bath (Fig. 4).** The size of the ultrasonic bath must be sufficiently large. The test containers have to fit in the ultrasonic bath without mechanical contact. Additionally, a minimum distance between the test container and the bottom of the bath of 15 mm must be ensured. The bath should provide the following power data: ERS power 250 W; HF peak power 450 W under double half-wave operation; frequency 35 kHz.

Fig. 4 Ultrasonic bath.

8. **Paper filter** for collecting scaled material.
9. **Unit for adjusting liquid level,** i.e., by a suction device (Fig. 5). The suction device may consist of a capillary tube with a spacer of 15±1 mm that is connected with a water jet pump to suck up the excessive liquid in the test containers.

---

[7] Basically, the freeze-thaw test can also be performed in an air cooled chest. However, a uniaxial heat flux must be ensured as well as a temperature cycle with sufficient precision to establish the same scaling at the test surface. The temperature regime cannot be controlled as precisely as in the liquid cooling bath and must be adjusted.

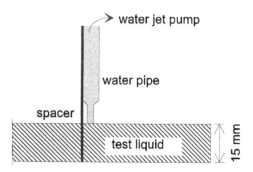

Fig. 5 Suction device to remove the liquid exceeding the level of 15 mm in the test containers.

10. **Drying cabinet** for a temperature of $(110\pm5)°C$.
11. **Balance** with an accuracy within ±0.01 g.
12. **Vernier callipers**, with an accuracy within ±0.1 mm.

6 Test specimens

6.1 Required total test surface area and number of specimens

For one series a number of ≥5 specimens[8] with a total test surface area of ≥0.08 m²
is recommended.

6.2 Making test specimens for testing concrete mixes or concrete constituents in a mix (reference specimens)

For testing concrete mixes or concrete constituents in a mix, the test specimens are
cast and compacted on a vibrating table in 150 mm cube moulds according to prEN-
ISO 2736/2. Centred in the mould is a vertical teflon disk which separates the mould
into two halves. The teflon plate must not be treated with any demoulding agent. The
concrete surface at the teflon plate is the test surface. For a larger aggregate size the
teflon disk can be arranged at one side.

After 24 ± 2 hours of curing the specimens are removed from the mould and stored
for 6 days (until the age of 7 days) in tap water at 20 ± 2 °C. (If strength development of
the specimens is low the curing in the mould can be increased. The storage in tap water
is then decreased by the same amount.) This treatment precedes the dry storage.

6.3 Test specimens for testing the surface of concrete structures

The test surface should correspond to the surface of the real structure exposed to
weathering. The test surface should be plane and can be of any kind - cast, screeded,
sawed or of different texture.[9]

---

[8]   A minimum number of 5 specimens is recommended for statistical evaluation and for finding
possible stragglers.

[9]   Test specimen not covered by the curing procedure as described in § 6.2 should be equivalently
treated.

The test surface should be free from any demoulding agent. If a demoulding agent cannot be avoided this can affect the scaling during the first freeze-thaw cycles and must be taken into account in the assessment.

The height of the test specimens should be comparable to the height of the specimens described in section 6.2. Specimen heights between 50 mm and 150 mm are acceptable.

6.4 Test specimens for testing precast concrete elements

Small precast concrete elements, such as concrete blocks and flags, can be tested directly, independent from the outer shape. If the dimension exceeds 200 mm, the element should be cut. The test surface is the weathered surface of the element and should be plane.

The test surface should be free from any demoulding agent. If a demoulding agent cannot be avoided, this can affect the scaling during the first freeze-thaw cycles and must be taken into account in the assessment.

The height of the test specimens should be comparable to the height of the specimens described in section 6.2. Specimen heights between 50 mm and 150 mm are acceptable.

7 Test procedure

The test procedure consists of three steps: the dry storage, the presaturation by capillary suction and the freeze-thaw cycles. The test procedure starts immediately after the curing period. For test specimens made following Section 6.2 this is at the age of 7 days.

7.1 Dry storage

The concrete specimens are stored in the climate chamber (20°C 65% RH) for surface drying for 21 days. Monitoring of the weight is recommended.

In the climate chamber the evaporation from a free water surface shall be 45±15 g/m²h. The evaporation is measured by weight loss of a water filled evaporation bowl with a depth of approximately 40 mm and a cross-sectional area of 225±25 cm². The bowl shall be filled up to 10±1 mm from the brim. Monitoring of the weight is recommended.[10]

7.2 Preparation of specimens

Between 7 and 2 days before presaturation the specimens are sealed on their lateral surfaces either with aluminium foil with butyl rubber or with a solvent free epoxy resin.[11] The specimens must be clean and dry. Before sealing the lateral surfaces, they are recommended to be treated with an adequate primer.

---

10   After the drying period, all freezable water, at least near the test surface, should have evaporated by reaching a water content equivalent to an equilibrium below 70% RH.

11   When conducting frost/deicing salt tests, this prevents falsified results due to possible lateral scaling.

7.2.1 Sealing with aluminium foil with butyl rubber[11]

A piece of aluminium foil with butyl rubber is glued tightly on the lateral surfaces with an overlap of 20 mm.

7.2.2 Sealing with epoxy resin

A solvent free epoxy resin is laid on the lateral surfaces, whereas the bottom of the specimens and the test surface must be kept free.

7.3 Presaturation of test liquid by capillary suction

Following dry storage the specimens are placed in the test containers on the 10 mm high spacers with the test surface underneath. Subsequently, the test liquid is filled into the container to a height of 15±1 mm without wetting the specimen's top. (This can be achieved by filling to approx. 17 mm and removing the surplus solution by a capillary tube combined with a spacer of 15 mm and a connected to water jet pump.)

During the capillary suction the test container must be closed with a cover that should have an incline to prevent any possible condensation water from dripping onto the specimen's top surface.

The capillary suction period is 7 days at a temperature of 20±2°C. During capillary suction the liquid level should be checked and adjusted as described above at regular intervals, depending on the suction capacity of the material.

The weight gain of the specimens should be measured.

7.4 Cleaning of test surface before starting the freeze-thaw cycles

Before starting the freeze-thaw cycles, loosely adhering particles and dirt should be removed from the test surfaces of the specimens by treatment in the ultrasonic bath as described in section 7.6. The material removed is discarded.

7.5 Freeze-thaw testing

7.5.1 Temperature cycle

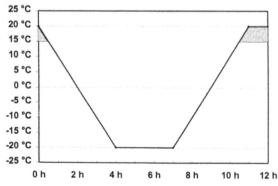

Fig. 6 Control temperature cycle.

A 12 h freeze-thaw cycle is applied (Fig. 6). Starting at +20 °C the temperature is lowered in 4 h with a constant cooling rate of 10 K/h. It is kept constant for 3 h at -20 °C and increased in 4 h with a constant heating rate of 10 K/h. It is kept constant for 1 h at +20 °C. [12] The temperature cycle is monitored at the reference point. The deviation of temperature measured at the reference point should not be more than ±0.5 K at least at the minimum temperature and more than ±1 K at other temperatures. A constant time shift between the test containers is acceptable.

7.6 Determination of surface scaling

The surface scaling can be measured while the temperature is above 15 °C (shaded area in Fig. 6).

To remove loosely adhering scaled material from the test surface, the test container is dipped into the contact liquid of an ultrasonic bath and subjected to ultrasonic cleaning for 3 minutes.

The solution comprising the scaled material is filtered. The paper filter is subsequently dried at 110±5 °C for 24 h and cooled for 1 h (±5 min) at 20±2 °C and 60±5 RH. The mass of the filter containing the dried scaled material μ_b is weighed to 0.01 g precision. The mass of the empty filter μ_f is determined before with the same accuracy.

The mass of the scaled material μ_s is then: $\mu_s = \mu_b - \mu_f$.

The amount of scaling should be determined after 14 and always after 28 freeze-thaw cycles. Additional measurements, e.g., after 4 or 6 cycles, are recommended.

8 Expression of test results

The following is to be calculated for each measuring occasion and each specimen: the total amount of scaled material related to the test surface after the nth cycle m_n:

$$m_n = \frac{\sum \mu_s}{A} *10^6 \, g/m^2 \qquad (1)$$

μ_s is the mass of scaled material of the measurement after n cycles (g) with an accuracy of 0.01 g. The sum is taken over all measurements until the nth cycle.

A   is the area of the test surface. It is calculated on the basis of the linear dimensions. They are taken as the average of at least two measurements determined to the nearest 0.5 mm.

The mean value and the standard deviation of the scaled material should be evaluated. The result should be checked for stragglers.

---

[12]   The cycle corresponds principally to a recommendation agreed upon at the meeting of RILEM TC 117 Freeze-thaw and deicing resistance of concrete in May 1990. The duration of minimum temperature is maintained as well as the maximum and minimum temperatures and the constancy of cooling and heating rates. However, the rate is increased and the duration at the maximum temperature decreased to reach a 12 h cycle. It is proved that the difference in scaling is small (Setzer, M.J., and Hartmann, V., `Verbesserung der Frost-Tausalz-Widerstands-Prüfung - Improved frost/deicing salt resistance testing', Betonwerk und Fertigteiltechnik, Vol. **57**, Heft 9, (1991) 73-82).

The mean value and the individual values for each specimen after 28 cycles are used for evaluating the scaling resistance.

9 Assessment of the CDF test

The freeze-thaw and deicing salt resistance of the CDF test is assessed after 28 freeze-thaw cycles.

The precision for freeze-thaw and deicing salt resistance (3 % sodium chloride solution) can be given in accordance to ISO 5725.[13]

9.1 Precision of the CDF test for plastic concrete mixes

Three types of precision are distinguished, repeatability, reproducibility and between laboratory scattering. The precision of the CDF test procedure with a 3% sodium chloride salt solution was evaluated according to ISO 5725 for plastic concrete mixes as described section 6.2. The coefficient of variation v depends on the mean scaling m related to the resistance level $m_0 = 1500$ g/m²:

$$v = v_0 \cdot \left(\frac{m}{m_0} \right)^d \tag{2}$$

The parameters in Equation 2 for reproducibility, repeatability and between laboratory variation as an exponential relationship of mean scaling m are as tabulated here

	Repeatability	Between laboratory	Reproducibility
d	-0.33	-0.26	-0.29
v_0	10.4%	14.0%	17.5%

---

[13] Further details about correlation with behaviour in practice, the CDF resistance limit and the precision data are in

Setzer, M. J., and Auberg, R., `Freeze Thaw and Deicing Salt Resistance of Concrete Testing by CDF Method; CDF Resistance Limit and Evaluation of Precision', Mater. Struct. 28 (1995) 16-31.

The practical behaviour is dealt with in

HARTMANN, V., `Optimierung und Kalibrierung der Frost-Tausalz-Prüfung von Beton - CDF Test', PhD Thesis, University of Essen, 1993

The statistical analysis is described in more detail in

AUBERG, R., PhD Thesis, University of Essen 1996.

The basic ideas were described first in

SETZER, M.J., `Prüfung des Frost-Tausalz-Widerstands von Betonwaren'. Essen, Universität-Gesamthochschule-Essen, Forschungsberichte aus dem Fachbereich Bauwesen Heft 49 (1990)

10 Report

The test report shall contain at least the following information:

a) A reference to this description.

b) Size, origin and marking of the specimens.

c) In the case of the testing of concrete mixes or constituents, the composition of the concrete.

d) The composition of the test liquid

e) The amount of scaled material for each specimen as well as the mean value and the standard deviation in g/m^2 rounded to the nearest 1 g/m^2, at least after 14 and 28 cycles.

f) The mass of solution sucked up during the capillary suction period for each specimen as well as the mean value and the standard deviation.

g) Visual assessment (cracks, scaling from aggregate particles) before the start and at least after 14 and 28 cycles.

h) Any deviations from the standard test procedure.

AUTHOR INDEX

KEYWORD INDEX